The Characteristics of Parallel Algorithms

MIT Press Series in Scientific Computation
Dennis Gannon, editor

The Characteristics of Parallel Algorithms

edited by
Leah H. Jamieson
Dennis Gannon
Robert J. Douglass

The MIT Press
Cambridge, Massachusetts
London, England

PUBLISHER'S NOTE

This format is intended to reduce the cost of publishing certain works in book form and to shorten the gap between editorial preparation and final publication. Detailed editing and composition have been avoided by photographing the text of this book directly from the author's prepared copy.

This book was printed and bound
in the United States of America.

Library of Congress Cataloging-in-Publication Data

The characteristics of parallel algorithms.

(MIT Press series in scientific computation)
Includes bibliographies and index.
1. Parallel programming (Computer science)
2. Algorithms. I. Jamieson, Leah H. II. Gannon,
Dennis B., 1947-   III. Douglas, Robert J.
IV. Series.
QA76.6.C4298 1987   005.1′2   87-2964
ISBN 0-262-10036-3

Contents

Preface. The Characteristics of Parallel Algorithms

There has been a tremendous growth in interest in parallel architectures and parallel processing in recent years, resulting in dozens of new machine designs, prototypes, and programming languages for parallel and distributed computing. This basic research in architecture has led to over a dozen commercially available parallel systems. However, comparatively little research has attacked the problem of how to characterize the parallelism in programs and algorithms.

This book is an outgrowth of a workshop that we helped organize in Santa Fe on the Taxonomy of Parallel Algorithms. That workshop indicated to us the need for a better understanding of where parallelism can be found and how it can be expressed in programs and algorithms (either explicitly by the programmer or automatically by a compiler or run-time system). While the Santa Fe workshop showed that the creation of a taxonomy of parallel algorithms might be too ambitious an undertaking at this point in time, it did bring out several different approaches to characterizing parallelism. Several rough taxonomies of parallel architectures and parallel languages are beginning to emerge, but there remains a lack of understanding of how to classify parallel algorithms and applications and how a class of parallel algorithms relates to a given architecture or programming language, or even to another class of parallel algorithms. Designers of parallel architectures and languages usually demonstrate the effectiveness of their systems by showing how a few algorithms could be implemented and speeded up on their machines or with their language, but, lacking some way to characterize classes of parallel programs, it is difficult to determine what broad class of algorithms or applications is most suitable for a given parallel system. We hope that this book will be a starting place for just such a characterization.

To address these issues, we requested contributions from a range of researchers who are known for their work on parallel algorithms and programs in numerical analysis, artificial intelligence, theory of computation, operating systems, programming languages, and image and speech processing. As a focal point for the book, we posed the following questions to the authors:

1. What attributes of algorithms are the most important in dictating the structure of a parallel algorithm and how can they be characterized or formalized?

2. What are the salient characteristics of parallel algorithms in your problem domain?

3. How much commonality in algorithm structure is there across problem domains?

4. How can algorithms be characterized for the purpose of matching algorithms to languages and to architectures?

5. On what basis should a taxonomy of parallel algorithms be formed?

6. To what extent should the design of new architectures and languages be based on models of classes of algorithms?

7. How should parallelism be represented? Is it best explicitly specified by the programmer or derived automatically from implicit data dependencies?

8. What are the characteristics of parallel algorithms that influence the design of the tools we use for programming?

This book is organized into three main sections. In the first section the authors consider general models of parallelism and the influence these models have on algorithm design. These chapters share a common theme of considering frameworks aimed at achieving the dual goals of unifying our knowledge about algorithms and providing the ability to discriminate between different kinds of algorithms. In the second section the characteristics of concurrency are addressed from the perspective of the applications specialists. Problems from the areas of speech recognition, image processing, partial differential equations, and numerical linear algebra are presented. The last section describes how the organization of parallelism in concurrent computations can result in new ways to organize software engineering tools to help the applications programmer.

Most of the contributors to this volume were at the Santa Fe workshop that started this project. All have been involved in parallel computation research for many years. The editors are delighted that they have taken the time to address the issues raised in our questions and share their experience and insight. We extend thanks to Bill Buzbee for providing the impetus and support for the Santa Fe workshop and Pat Kerkhoff for a tremendous amount of work in handling the organizational details for a book where the authors spanned distances of over 6000 miles and, at times, the editors of over 2000.

Leah Jamieson
West Lafayette, Indiana

Dennis Gannon
Bloomington, Indiana

Robert J. Douglass
Denver, Colorado

January 1987

The Characteristics of Parallel Algorithms

Section I. General Characteristics of Parallel Computation Models

This section contains five papers each of which addresses, in general terms, some aspect of the problem of characterizing parallel algorithms. The specific foci of the papers are quite different; however, the papers share a common goal of searching for means of improving our ability to think about parallel algorithms in an orderly way.

In the first chapter of this section, Nelson and Snyder examine programming paradigms, paying particular attention to communications patterns. Just as we are adept at programming serial computers because of the (large) repertoire of programming methods that we have come to understand well, so the identification of paradigms for programming parallel computers should assist us in the design of new parallel algorithms. Because of the critical importance of communications in parallel computers, they are concerned with paradigms for non-shared memory computers that encapsulate information about useful communications patterns. They consider three such paradigms: compute-aggregate-broadcast algorithms, composed of a compute phase, a combining phase, and a broadcast phase in which the aggregate information is returned to the processes; pipelined and systolic processes; and divide-and-conquer strategies. Through examples, they illustrate the principal characteristics of each. They present the paradigms as a means of structuring our understanding of known algorithms and of facilitating the development of new algorithms.

In the second chapter, Raphael Finkel examines a particular type of parallelism: large-grain parallelism in which processes collaborate to execute an algorithm, with communications performed by infrequent message passing. He uses three applications to explore different ways in which algorithms can be structured for distributed execution and to provide some insight into why perfect speedup is difficult to attain. He too makes use of the idea of programming paradigms: generate-and-solve problems, in which a problem is divided into independently solvable subproblems; iterative relaxation techniques, in which the data space is divided into communicating regions; passive data pool algorithms, where a data space serves many processes; and systolic processes. The three applications provide a rich framework for comparing different approaches.

In the third chapter of this section, Leah Jamieson adopts the premise that paradigms in parallel programming lead to patterns that can be observed in various aspects of parallel algorithms, and that these patterns can be exploited in mapping algorithms onto architectures. She identifies a set of thirteen algorithm characteristics that act as dimensions for describing a computation, and relates these characteristics to the types of parallel architectures that best support the different algorithm characteristics. She presents this characteristics-based approach as a means of improving our general understanding of the relationship between parallel algorithms and architectures, with specific application to the problem of selecting machine configurations in a reconfigurable parallel architecture.

In the fourth chapter, Levitan's emphasis is on communications: on characterizing algorithms in terms of their communications and on evaluating communications-related metrics as predictors of architecture/algorithm performance. He reports the results of experiments using four different metrics applied to six kernel algorithms running on eight different architectures. The objective of the study is to compare the performance predicted by the metrics with the performance observed in the actual execution of algorithms. He concludes that, while the metrics considered do not seem to capture the performance characteristics of the algorithms, "the performance of some parallel algorithms running on parallel machines correlates very well with the performance of other parallel algorithms on those same machines." This can be thought of as a specific instance of the point made by Nelson and Snyder: paradigms do exist, and they can be put to good use once we figure out what they are. Based on the study, Levitan proposes a set of kernel tasks to be used as a performance suite for evaluating communications structures.

The last chapter, by Manthy, addresses the concept of hierarchy as a means of understanding and explaining systems, and contrasts what constitutes useful hierarchies for sequential and concurrent systems. He hypotheses that, in a computational system, we are interested in the identity of processes and the causal relationships within and between processes. A hierarchy commonly used with respect to sequential systems is based on functional dependency between processes, and for such systems, functional dependency hierarchies give us an accurate view of a computation at varying levels of detail. However, he argues that such hierarchies fail to represent important aspects of message-passing concurrent systems. For concurrent systems, he proposes hierarchies based on shared conserved resources as better capturing the key notions of request/reply cycles, interprocess communication, and non-determinism.

Programming Paradigms for Nonshared Memory Parallel Computers

Philip A. Nelson and Lawrence Snyder

Computer Science Department
University of Washington
Seattle, Washington 98195

1. Introduction

In serial computation there are several recognized programming paradigms, such as the divide-and-conquer, the greedy and the dynamic programming techniques [2]. These computational methods are not algorithms *per se*, but rather they are problem solving strategies that are frequently used in structuring efficient algorithms. Thus, paradigms are the high level methodologies we recognize as common to many of our effective algorithms.

The benefit in identifying paradigms, besides the pleasant and perhaps scientifically important insight that apparently dissimilar problems can be solved by similar means, is the fact that the set of recognized programming paradigms becomes the programmers "tool kit." It is a set of known-to-be-useful strategies that he can use to structure the development of new or improved algorithms. Thus, the programmer not only has a place to begin when developing a new algorithm, but more importantly, experience and knowledge from earlier problem solutions can be directly transferred to a problem by means of the paradigm. Both benefits are important in the parallel computation arena, since to date programmers have generally had little direct experience with parallelism. But for parallel computation there is an even more compelling reason to concentrate on programming paradigms.

Programming paradigms generally encapsulate information about useful patterns of data reference, which in the case of parallel computation simply says that paradigms generally encapsulate information about useful *communication patterns*. Since providing efficient and effective communication *is* the problem in parallel computer architecture, the paradigms serve as a specification of which communication patterns are most useful and hence should be supported well. Thus, for example, the "to do list" paradigm -- keep a queue data structure of task descriptors and have server processes at the completion of a task select a new task from the head of the list -- requires all processors to fetch from the same memory cell (list head) and thus seems to favor a shared memory implementation with a combining network [9].

The paradigms presented here are directed towards nonshared memory implementations in the sense that we usually specify how the data is distributed among a set of processes and how it must be moved around to accomplish the computation. This formulation thus describes what kind of communication topology the paradigms *require*. Unlike an approach that presumes a particular given topology and endeavors to accommodate it, our results can provide useful data to computer architects on what communication topologies are most important.

2. The Paradigms

Three different parallel programming paradigms will be treated: the compute-aggregate-broadcast (CAB) paradigm, the pipelined and systolic paradigm, and divide-and-conquer. Though only the first is new in the sense of not having been mentioned previously in the published literature, there is much to be said about each one. We will describe the principle characteristics of each paradigm, identify several algorithms in which they are used, and describe variants of the paradigms that permit one to accommodate various situations. Along the way we will have the opportunity to introduce new algorithms.

There are two aspects of paradigms that will not be treated here. First, we will not be exhaustive; there are other paradigms that we do not treat [22]. Second, we will not touch on the analysis of paradigms, e.g. contraction analysis, that enables one to solve a problem for the paradigm and have the solution be applicable to all the algorithms that use it [21, 22].

Finally, there is a cautionary remark to be made on precision: it is difficult to define paradigms precisely. We know of no adequate and convenient formulation that can be used. Paradigms are not algorithms, so one does not present them in a programming language, nor are there algorithmic schemata that could be presented in some metalanguage. They can be illustrated by example (which we will do), but this is an inadequate definitional means since it is rarely clear which properties are special and which are general. Still, paradigms are a phenomenon that programmers understand, and in sequential computation, we have learned them by some means. Our approach will be to describe them verbally and augment the description with examples. Finding a precise and perspicuous medium to express paradigms in is a challenging and interesting research problem.

3. Compute, Aggregate, and Broadcast

Compute-aggregate-broadcast (CAB) algorithms are composed of three basic phases bearing those names: a compute phase which performs some basic computation, an aggregate phase which combines local data into one or a few global values, and a broadcast phase which returns global information back to each process. The compute phase varies widely from algorithm to algorithm. It may be as much as a complete algorithm, like matrix multiplication, or it may be as little as a single primitive operation performed in each process. The aggregate phase is usually a tree-based computation that combines data from the processes, producing a single global value, such as minimum or maximum trees. The broadcast phase returns the global information or a directive based upon it to all processes. This may be a "keep going" message for iterative algorithms or a value necessary for the following compute phase. Many

CAB algorithms iterate on the three phases. Although not all CAB algorithms start with the compute phase, they generally have the phases in the same order, aggregate following compute, broadcast following aggregate, and compute following broadcast as the names tend to indicate. The time for these algorithms is $O(k(c+a+b))$, where k is then number of iterations and c, a, and b are the times for the phases. With p processors, the a and b times are often $O(\log p)$ due to the tree interconnection structure used by the aggregate and broadcast phases.

A simple example of a CAB algorithm is a parallel implementation of the classical Jacobi iterative method for solving Laplace's equation on a square. An instance of this kind of problem, the electric field problem, is shown in Figure 1. The square is represented by n discrete values, V_{ij}, the voltage at the point. The boundary and the voltage sources are kept at a constant value. An initial guess is computed for each point. Iteratively, a new value, the average of a point's 4 neighbors, is computed for each of the n points. The iteration is terminated when the difference between the new value and the old value at every point is less than some tolerance.

For the CAB implementation using n processes, a process does the computation for a single point. The compute phase connects the processes as a $\sqrt{n} \times \sqrt{n}$ 4-neighbor mesh. Each process sends its value to its neighbors. The new value at a process is a weighted average of the values received from its neighbors. The aggregation phase connects the processes in a binary tree and computes the global maximum of the differences between old and new values using a tournament algorithm in which each process sends the maximum of its difference and the maxima received from its children. The broadcast phase, also using the binary tree connection, sends a signal that is relayed down the tree informing each process to continue processing, if the global maximum is larger than the tolerance, or to terminate. The times for the phases are $O(1)$ for compute and $O(\log n)$ for aggregate and broadcast, yielding a time of $O(k \log n)$ for this algorithm. The iterative nature of the algorithm phases might be

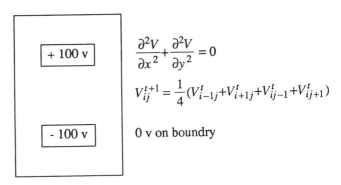

$$\frac{\partial^2 V}{\partial x^2} + \frac{\partial^2 V}{\partial y^2} = 0$$

$$V_{ij}^{t+1} = \frac{1}{4}(V_{i-1j}^t + V_{i+1j}^t + V_{ij-1}^t + V_{ij+1}^t)$$

0 v on boundry

Figure 1: Electric Field Problem

described as $(CAB)^k$.

Notice two features of this algorithm. First, the aggregate and broadcast (AB) phases are long when compared to the compute phase. Also, the compute phase does not depend on a value computed by the previous AB phases. For CAB algorithms with these two features, we can deviate from the standard CAB paradigm and do several compute phases before we do an AB phase, described as $(C^s AB)^t$. The AB phases use the difference value computed in the last compute phase. This will add at most $s-1$ extra compute phases, assuming that after all differences are below the tolerance, any more compute phases will only reduce the differences monotonically. We then save several AB steps at the cost of possibly doing a few extra compute phases. By a correct choice of s, the total running time may be reduced.

Not all CAB algorithms can use the removal of some AB phases to get improved running times. Consider the parallel conjugate gradient iteration algorithm [1, 8]. The conjugate gradient iteration is a method used to solve any linear system

$$Ax = y$$

where A is symmetric, positive definite. The main loop of the algorithm uses a CAB approach. The compute phase is a matrix product. The aggregate phase computes an inner product. The broadcast sends the value of the inner product back to all the processors. Because the next iteration requires the value of the inner product, we cannot remove any AB steps and still compute the correct value. Consequently, Van Rosendale [26] has investigated minimizing the inner product data dependencies.

The CAB paradigm is also useful when the AB phases do not perform a global aggregate and broadcast, but rather something less, perhaps a "regional" aggregate and broadcast. To illustrate this, we give a new algorithm for topological sort. We are given a directed, acyclic graph G, and we want to put an ordering on the nodes such that if node i is before node j in the ordering, there is no path from node j to node i.

Our solution is achieved in two steps. (See Figure 2.) The first step is to assign a

Algorithm Topological Sort:

$G_{ij} \leftarrow$ initial graph, $1 \le i,j \le n$
$Level_{ij} = 0, \ 1 \le i,j \le n$
UpdateLevel(0);
For indx = 1 to log n −1 do
    BooleanMatrixMultiplication; (compute G^2)
    UpdateLevel(indx);
end;
Sort($Level_{ij}$ is the key);

Figure 2: Topological Sort Algorithm

level number to every node, the next is to sort on level numbers. To assign level numbers, all nodes with no incoming edges are assigned a level of 0. All other nodes are assigned a level number that is the length of the longest path from a level 0 node. The last step sorts the nodes into an increasing level number order. It is the first step, the computing of level numbers, that can be solved using a CAB algorithm. The sorting step can be done by a suitable sorter.

We assume there are n^2 processes P, and P_{ij} has the corresponding entry in the adjacency matrix G. G_{ij} has the value 1 if there is a directed edge from node i to node j in the graph to be sorted and a 0 otherwise. The initial state has all nodes at level 0. The first step makes sure that the level numbers reflect the initial graph. The main loop consists of squaring the graph followed by updating the level numbers to reflect the current graph. The boolean matrix multiplication is the compute phase and the UpdateLevel step contains the AB phases.

The UpdateLevel step, using the current level number of the nodes and the current graph, G^k, calculates a new level for each node. G^k is the graph of paths of length k in the original graph. If node i has an edge to node j in G^k, then the level of j must be k greater than the level of i. The UpdateLevel algorithm is shown in Figure 3. It uses the n^2 processes connected in a variant of the mesh of trees. Each row and each column is connected into a tree with the root located at the process on the diagonal. At the start of UpdateLevel, $Level_{ij}$ is the level of node i. If G_{ij}^k is a 1, there is a path of length k from node i to node j. Therefore, the new level of node j is at least $Level_{ij} + k$. By taking a maximum across column j, we now have the new level for node j. This is done for all columns at the same time. The trees in the columns are used for this aggregation step. To set up for the next UpdateLevel, we

UpdateLevel(indx)

iv $= 2^{indx}$

Aggregate on columns: (on column trees)
    Leaf: If $G_{ij} = 1$, send iv $+ Level_{ij}$ to parent
        Otherwise, send 0 to parent
    Inner: if $G_{ij} = 1$, send max(iv $+Level_{ij}$,LeftChild,RightChild) to parent
        Otherwise, send max(LeftChild,RightChild) to parent
    Root: Set $Level_{ii}$ to max(LeftChild,RightChild)

Broadcast on rows: (on row trees)
    All processors set $Level_{ij}$ to $Level_{ii}$.
    (Get value from parent, and send to children.)

Figure 3: UpdateLevel Algorithm

must make sure that each $Level_{ij}$ is updated to the current level of node i. Because the roots of the trees are on the diagonal, the correct value for the level of node i is in process P_{ii}. A simple broadcast over the row provides the correct value of $Level_{ij}$ to all processes in row i. This is done using the trees in the rows.

The topological algorithm takes $O(n \log n)$ time with n^2 processors. This is because the boolean matrix multiply takes $O(n)$ time with n^2 processors and there are $\log n$ iterations. Due to the trees in the UpdateLevel, the AB phases run in $O(\log n)$ time and this is not the dominant factor. If n^3 processors are used, the matrix multiply time is reduced to $O(\log n)$ [20], yielding $O(\log^2 n + s)$ time for the algorithm, where s is the time for sort algorithm.

We make a few observations about this algorithm. First, as we said before, the aggregation and broadcast are not global, but localized to the rows and columns. Also, notice that the aggregate and broadcast phases are done first, before any computing. If we include the sort as the final compute phase, this algorithm is really an ABC (aggregate-broadcast-compute) algorithm. Finally, the compute phase is a complex phase that may be solved using a different paradigm, e.g. pipelining or divide and conquer.

The last CAB algorithm we look at is a numerical integration algorithm. We are given the function $f(x)$ and we would like to integrate between a and b using n processes. An easy solution would be to divide up the integration into n separate problems and let each process integrate a small section of the curve. The problem with this technique is that some sections of the curve may need more processing than other sections to get a satisfactory approximation. This provides an unequal load for the processes.

A better solution gives one process control of the work the other processes are performing. Consider the following algorithm.

Integrate $f(x)$ between a and b.

While "not done" do
    Control process: decide which section to integrate.
    Broadcast: Broadcast the section
    Compute: Each process integrates its $1/n$ section of the curve.
    Aggregate: Sum the n areas. Return area of section to control process.
end while

The broadcast uses a binary tree. An internal node receives the end points that its subtree is to integrate. It then divides the interval into 3 pieces, a section for the left subtree, a section for the right sub tree, and a section for itself, making sure that each process will have the same sized section. The compute uses no communication. Each node integrates its piece of the total section and decides if its value is a sufficiently accurate approximation. The integration can be done by various methods, such as the trapezoid method or Simpson's rule. The aggregation uses the tree to sum the areas computed in the compute phase. If the computed value for a subsection is not sufficiently accurate, the area is not added into the sum and the control process is sent a message identifying the subsection.

The control process is responsible for choosing the sections to integrate and collecting the final value for the area. The initial step is to broadcast the entire interval to be integrated. The results are a sum of the accurate subsection areas and a list of the subsections needing more processing. This list is assumed to be small. The control process then recursively integrates all of the subsections. This keeps all the processes busy and never recomputes any subsection that has been integrated with sufficient accuracy.

We make two observations about this algorithm. First, this algorithm is a member of the CAB paradigm, but the first phase executed is the broadcast, i.e. it is BCA. Also, the compute phase did not need any communication and therefore the algorithm needs only a single communication interconnection structure. This variant of the CAB paradigm has been observed in other algorithms, e.g. Dekel, Nassimi and Sahni's $O(\log n)$ matrix multiplication algorithm for the binary n-cube [6].

4. Pipelining and Systolic Computation

Perhaps the most widely used paradigm for nonshared memory parallel computation is the systolic approach [15]. Although definitions of systolic computing have been provided by the inventors [14, 16] and others [17], the concept remains imprecise. We will broaden the class presented here to include the general concept of pipelining, and we give examples of both systolic algorithms and pipelined algorithms that do not seem to meet all of the criteria of the systolic definition.

The key paradigmatic concept in pipelining and systolic algorithms seems to be the decomposition of the problem into subcomputations that are assigned to dedicated processes with the data "flowing" through the processes, visiting all or an appropriate subset of processes to complete the computation for that input. In addition, systolic algorithms are expected to exhibit additional properties: locality of communication, a regular communication structure, and have only a few different types of simple processes. The cost, l, of starting the algorithm, called *latency*, and the number, t, of input values generally determine the running time of $O(l+t)$.

One of the first systolic algorithms developed was the band matrix multiplication of Kung and Leiserson [12] shown in Figure 4. The algorithm multiplies two $n \times n$ matrices, A and B, with band widths l and m respectively, producing C. The processes are organized as an $l \times m$ hex-connected array. These matrices flow in different directions through the processes. To produce the correct result, the placement of the input data is very important. Since all data items move at the same time, there must be two "empty" spaces, or "bubbles," between the items. This insures that the correct data items meet at a process, e.g. a_{11}, b_{11}, and c_{11} in the figure. The C values are initialized to zero as they "enter" the array and therefore no input is required. Each data path can be considered as a pipeline that interacts with the other pipelines. Notice that the results do not exit from the end of the input pipelines, but from output

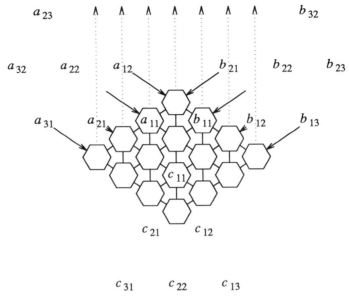

Figure 4: Band Matrix Multiply

pipelines that cross and interact with the input pipelines. The time required is $O(n)$.

The property of flow not only gives these types of algorithms their names, but it seems to be the property that distinguishes them from other parallel algorithms with similar properties. Specifically, in the last section we described the Jacobi iterative algorithm which exhibited many of the properties of the pipelined algorithms: It was decomposed into many small (approximation) subcomputations, each process was specialized (to computing the charge at a point), the communication was regularly structured and local, and there were a small number of processes (boundary, source and interior). But such algorithms are not normally designated as pipelined or systolic. Assuming that this observation is a reflection of the style of computing and not simply a statement about the historical order of the development of the Jacobi iteration and systolic computation, then what seems to be lacking in the Jacobi is the concept of flow.

The claim that the communication of the Jacobi algorithm does not exhibit flow and that, say, the Kung and Leiserson systolic matrix multiplication algorithm does is obviously a matter to which not all researchers would subscribe, but consider the following "flow test" which systolic algorithms generally pass and the Jacobi fails:

Flow Test: Select an arbitrary communication edge and "radioactively tag" a single transmission across that edge. (The concept of radioactive tags is due to Cuny and her students, where it is used in debugging parallel programs [4].) As the computation progresses from the time of tagging, let all values computed with one or more radioactive values become radioactive. Then define the "contaminated region" as the set of processes and communication channels touched by some radioactive value. An algorithm passes the flow test if the edge over which the initial value was transmitted is on the boundary of the region; otherwise it fails the test.

Clearly, the Flow Test captures the concept of flow in that the tagged value "flows out" to define the contaminated region. The reader can verify that the Jacobi algorithm fails the Flow Test while the Kung and Leiserson algorithm passes it. The Flow Test is not a perfect test, since some algorithms generally called systolic will fail, e.g. certain linear arrays, but it does seem to identify the presence of flow.

Consider the problem of computing the sums of n element vectors. (See Figure 5.) A single vector is summed using a tree of processes. The vector is input at the leaves and the sum of the vector comes out from the root. Successive vectors are input at the leaves when the previous vector has moved up the tree by one level. We can identify each level in the tree as a stage in the pipeline, with the data flowing from stage to stage. The latency is $O(\log n)$, yielding an execution time of $O(d+\log n)$ where d is the number of n element vectors to be summed. This algorithm passes the flow test and it appears to have the features of systolic algorithms. But trees cannot be laid out in the plane with edges of constant length independent of the size of the tree [23]. Thus, if communication time is proportional to distance, as is the case in the VLSI models normally associated with systolic arrays, then the time between stages increases as larger and larger trees are considered. One wonders, therefore, whether tree based pipeline algorithms meet the locality requirement of systolic computation.

Some algorithms do not produce results that flow out of the pipelines. Consider the matrix multiply algorithm of S. Y. Kung et al. [13] for the wavefront array processor. (See Figure 6.) The data arrives at every clock cycle and flows in vertical and horizontal pipelines. The results are accumulated at the processes so that when the

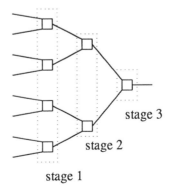

stage 3

stage 2

stage 1

Figure 5: Vector Sum Pipeline.

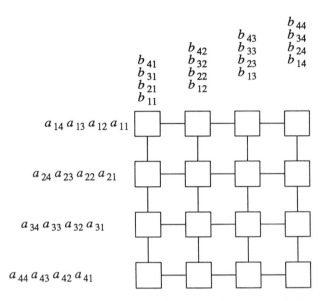

Figure 6: Data Staging for the WAP Matrix Multiply

input data has flowed through the input pipelines, the results are contained in the processes. The advantage to this algorithm is that no "bubbles" are required in the input pipelines. Although the results do not flow out of the pipelines, we nevertheless consider this to be a systolic algorithm.

Another interesting systolic algorithm is the dynamic programming algorithm of Guibas, Kung and Thompson [10]. The processes are arranged in a triangle with pipelines in the horizontal and vertical directions. (See Figure 7.) Data moves from the diagonal toward the top and right. At time $2t$, results are ready at every process that is t distance away from the diagonal. Processes (12), (23), etc. are defined to be distance 1 from the diagonal. For the next t time units these results move up and to the right at the rate of one process per time unit. After that, they move one process in the same direction every two time units. Depending on the problem to be solved, the processes may have preloaded data. The computation performed at each process will also vary depending on the problem to be solved.

An interesting part of the Guibas, Kung and Thompson algorithm is the use of the pipelines for both fast moving data elements and slow moving data elements. This feature provides an order reversal on the data, and it can be implemented either by a pair of connections between the processes or by a single connection doing double duty.

Another pipeline algorithm is the Hough Transform algorithm designed by Cypher and Sanz [5]. The Hough Transform can be used to detect lines and curves in picture data [7]. Given a $n \times n$ image of pixels, the Hough transform does

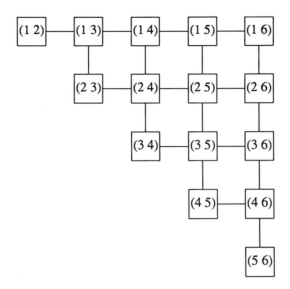

Figure 7: Dynamic Programming Process Structure

computations on bands, usually one pixel wide. (See Figure 8.) For a given θ, the complete picture will be processed by computing all bands. The bands are identified by the pair (θ,ρ). Most applications using the Hough transform require results for multiple values of θ.

The algorithm uses n^2 processes connected by the $n \times n$ torus. (A $n \times n$ mesh with end around connections.) The picture is distributed one pixel per process. A value of θ

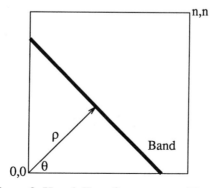

Figure 8: Hough Transform on $n \times n$ Picture

is input to the first column. (Actual implementations may use $\sin \theta$ and $\cos \theta$ as input.) The input, along with the values being computed will move to the right across the processes. All bands for one θ will be calculated by a single sweep. After the first value of θ has moved to the second column, another value of θ may be input.

For $\theta=90°$, the bands will be contained in the rows of processes. For other values of θ, the bands will not be horizontal, requiring the accumulated band data to follow the bands by moving up or down in the column. To guarantee that the entire column of data moves by at most one process, θ is limited to $45°\leq \theta \leq135°$. This keeps the bands more horizontal than vertical. For the other values of θ, the algorithm can be rotated by 90 degrees. The vertical movement requires the band data to include the band number, a value of ρ, and the accumulated value for the band to the current point in the computation.

Depending on θ, there is between n and $\sqrt{2}n$ bands in a picture. To accommodate this, each process contains data for two bands, the active band involved in the local computation and the band that does not pass through the current process. When the data is shifted off the top or bottom, one band is complete and a new band is started.

The time used by this algorithm depends on two variables, n, the picture size and p, the number of values of θ. One value of θ is calculated by a single sweep of the data across the $n \times n$ processes and requires $O(n)$ time. Since successive values of θ can be input at regular intervals, the last value of θ will be input at time $O(p)$. Therefore, the total time is $O(n+p)$.

There are two features of special interest. First, the vertical movement of the data is very input data dependent, unlike the other algorithms where the flow pattern is fixed. Even with this vertical movement, the algorithm passes the flow test. Also, computation involves only one of two data values being moved, swapping the active and passive data when moved off the top or bottom. The final results may have one or two computed values, and this is also input data dependent. Even with these features, it is easy to see that this is a pipelined algorithm. It may be too irregular in other respects to be classified as a systolic algorithm.

Even though these algorithms differ in many ways, all of them have similar features. The data flows in a regular fashion. The processing times at each process is approximately the same and the processing to be done is one of a small number of tasks. This is not true for all pipeline algorithms. There are algorithms where each stage has radically different work to do, different amounts of input data, and different amounts of processing time for that data. A good example of this kind of a pipeline is found in the funneled pipelines [11]. The funneled pipelines are used to implement several graph theoretic algorithms. The early stages take in large amounts of data and do quick processing, eliminating some of the data as unneeded by the next stage. Each successive stage takes in less data and spends more time processing its data than the previous stage. This is balanced so that no stage is given too much data or too little data to be processed in any given time interval.

5. Divide and Conquer

Divide and conquer is a well known paradigm in sequential algorithm development [2]. A problem is divided up into two or more smaller problems. Each of these problems is solved independently and their results are combined to give the final result. Often, the smaller problems are just smaller instances of the original problem, giving rise to a recursive solution. In the sequential world, the smaller problems are solved serially, but in the parallel world, these problems can be solved at the same time, given sufficient parallelism. Some processing is usually required either to split the original problem into subproblems or to combine the results of the subproblems for the final solution. This extra processing is also performed using parallelism.

As the first example of this paradigm we look at Batcher's bitonic merge sort [3]. The problem is that of sorting n data items using n processes. We assume that $n = 2^k$ for some k, and that each process initially contains one data item. The sort should leave the smallest data item in the process labelled 1, the next smallest data item in the process labelled 2, and so forth.

The algorithm divides the sequence up into two subsequences, performing an increasing sort on the first half and a decreasing sort on the second half. The result is a bitonic sequence. This bitonic sequence is transformed into a monotonic sequence using another application of the divide-and-conquer strategy. Figure 9 shows the data exchanges required by the algorithm. The horizontal lines represent a single process and the arrows point to the process which keeps the larger data element. Notice that only a single step compares items across the division point and that these comparisons are of corresponding items in the subsequences, e.g. in the figure process 1 exchanges data with process 9. The execution time $O(\log^2 n)$ assuming a unit cost transmission

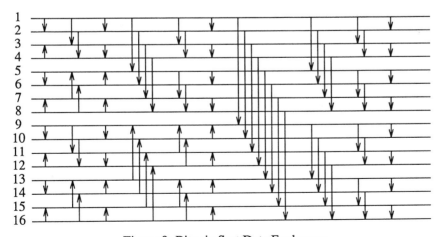

Figure 9: Bitonic Sort Data Exchanges

time for each data exchange.

The communication structure for this algorithm is the binary n-cube. This can be easily seen by projecting the horizontal lines in Figure 9 to a point and retaining the vertical edges. The binary n-cube is a minimum connection graph for any divide-and-conquer algorithm that divides the problem into halves and needs communication between corresponding processes of each half. The fast Fourier transform (FFT) is an algorithm with this property [18]. The sequential divide and conquer algorithm for the FFT can be easily parallelized. Since it combines corresponding elements of each half, it also uses the n-cube communication structure.

Our next divide-and-conquer algorithm is the multiplication of two dense $n \times n$ matrices,

$$AB = C$$

using n^2 processes, and assuming $n = 2^k$ for some constant k [20]. The processes are viewed as an $n \times n$ array where the processes are labelled PE_{ij} for $1 \le i,j \le n$. The matrices A and B are initially distributed in the n^2 processes such that a_{ij} and b_{ij} are contained in PE_{ij}. After the product we have c_{ij} contained in PE_{ij}.

To begin with, consider the 2×2 case. PE_{11} contains a_{11} and b_{11}. To compute c_{11} the values a_{12} and b_{21} are needed. Similarly, all other processes need only 2 elements not already stored at that process. To provide for direct communication, a grid interconnection structure is used. The processes then send their a_{ij} value to the other process in the same row, and their b_{ij} value to the other process in the same column (see Figure 10). After this communication, each process, PE_{ij}, has all the data

$$\begin{bmatrix} a_{11} & a_{12} \\ a_{21} & a_{22} \end{bmatrix} \begin{bmatrix} b_{11} & b_{12} \\ b_{21} & b_{22} \end{bmatrix} = \begin{bmatrix} c_{11} & c_{12} \\ c_{21} & c_{22} \end{bmatrix} = \begin{bmatrix} a_{11}b_{11}+a_{12}b_{21} & a_{11}b_{12}+a_{12}b_{22} \\ a_{21}b_{11}+a_{22}b_{21} & a_{21}b_{12}+a_{22}b_{22} \end{bmatrix}$$

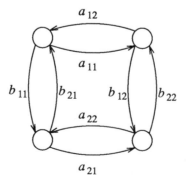

Figure 10: 2×2 Product and Communication Structure

required to compute c_{ij}.

Now consider the $n \times n$ case. The algorithm uses Strassen's [25] matrix decomposition where two $n \times n$ matrices can be viewed as two 2×2 matrices of $\frac{n}{2} \times \frac{n}{2}$ matrices. The 2×2 matrices are then multiplied using matrix product and matrix addition on $\frac{n}{2} \times \frac{n}{2}$ matrices.

Let A_{11} be the upper left $\frac{n}{2} \times \frac{n}{2}$ submatrix of A. Similarly define the other 3 submatrices, A_{12}, A_{21}, and A_{22}, the submatrices of B, the submatrices of C, and the subarrays of the processes, P. Then A_{ij} and B_{ij} are contained in P_{ij}, $1 \leq i,j \leq 2$. As in the 2×2 case, A_{12} and B_{21} are required to compute C_{11}. If the corresponding processes in P_{11} and P_{12} are directly connected (see Figure 11), A_{12} can be sent to P_{11} in parallel with one communication step. B_{21} can be sent to P_{11} using a similar connection scheme in one communication step. The full connection structure connects PE_{ij} with both $PE_{i \pm \frac{n}{2} j}$ and $PE_{ij \pm \frac{n}{2}}$. With $A_{11}, B_{11}, A_{12},$ and B_{12} in P_{11}, C_{11}

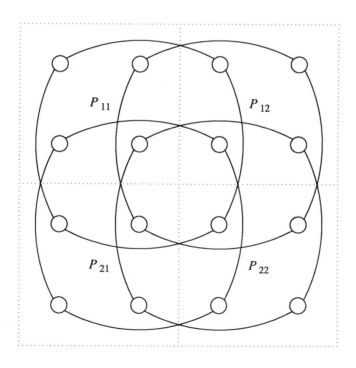

Figure 11: 4×4 Connections

can be computed by doing two $\frac{n}{2} \times \frac{n}{2}$ matrix products and one matrix addition. Analogous products can be done using this same algorithm on the other $\frac{n}{2} \times \frac{n}{2}$ matrices. The recursion will stop after $k-1$ levels when a 2×2 matrix product is done. The matrix addition is performed element by element.

Each recursion level requires it's own interconnection structure. The complete interconnection structure, supplying an edge for every communication in the algorithm on every level of recursion, is the n-cube.

The time required for this algorithm is $O(n)$. Nelson [20] also gives an $O(\log n)$ algorithm using $O(n^3)$ processes. They both employ the same divide and conquer paradigm. One matrix multiply is divided up into 8 matrix multiplies and 4 matrix additions.

Although the matrix multiply algorithm uses the binary n-cube, it is a result of dividing the problem into quarters. The utilization of edges for communication is quite different than the previous algorithms that divided the problem into halves. Given these differences, it is interesting that they all use the binary n-cube.

Not all divide-and-conquer algorithms require the binary n-cube interconnection structure. Stout discusses several divide-and-conquer algorithms for image processing [24]. One of these is the algorithm for connected ones (image data connected components) of Nassimi and Sahni [19] using the mesh interconnection structure. Given an $n \times n$ black and white pixel image, each connected set of black (or white) pixels are to be assigned a unique number. The solution takes the image and divides it into 4 quadrants, solving the connected ones problem for each quadrant independently. Once they are solved, only the boundaries with other quadrants need to be examined to find components that cross the boundary.

6. Summary and Conclusions

We have presented three paradigms for nonshared memory parallel algorithms: compute-aggregate-broadcast, pipelining, and divide-and-conquer. We included the systolic algorithms in the pipelining paradigm. For each paradigm, we described the principle characteristics and gave example algorithms. These algorithms illustrated the features and variations on the paradigms.

These paradigms, though not a complete set for nonshared memory algorithms, provide the algorithm designer with some tools to aid in the development of new algorithms. More importantly they unify the underlying structure of known algorithms. As research continues, we expect more paradigms to be discovered.

Acknowledgments

We thank Robert Cypher for the time he spent explaining his algorithm for the Hough transform. This work was supported in part by National Science Foundation Grant DCR-8416878 and by Office of Naval Research Contract No. N00014-86-K-0264.

References

1. L. M. Adams, *Iterative Algorithms for Large Sparse Linear Systems on Parallel Computers*, Ph.D. Thesis, University of Virginia, Charlottesville, 1982.

2. A. V. Aho, J. E. Hopcroft and J. D. Ullman, *The Design and Analysis of Computer Algorithms*, Addison-Wesley, Reading, Massachussets, 1974.

3. K. E. Batcher, "Sorting Networks and their Applications", *Proceedings of the AFIPS Spring Joint Computer Conference Vol 32* (1968), 307-314.

4. J. E. Cuny, Private Communication, Oct. 1986.

5. R. Cypher, Private Communication, Jan. 1987.

6. E. Dekel, D. Nassimi and S. Sahni, "Parallel Matrix and Graph Algorithms", *SIAM Journal on Computing 10(4)* (Nov. 1981), 657-675.

7. R. O. Duda and P. E. Hart, "Use of the Hough Transformation To Detect Lines and Curves in Pictures", *Comm. of the ACM 15* (Jan. 1972), 11-15.

8. D. Gannon, L. Snyder and J. Van Rosendale, "Programming Substructure Computations for Elliptic Problems on the CHiP System", in *Impact of New Computing Systems of Computational Mechanics*, A. K. Noor (editor), The American Society of Mechanical Engineers, 1983, 65-80.

9. A. Gottlieb, B. D. Lubacbevsky and L. Rudolph, "Basic Techniques for the Efficient Coordination of Very Large Numbers of Cooperating Sequential Processors", *Trans. Prog. Lang and Systems 5(2)* (Apr. 1983), 164-189.

10. L. J. Guibas, H. T. Kung and C. D. Thompson, "Direct VLSI Implementation of Combinatorial Algorithms", *Proceedings of the Conference on Very Large Scale Integration: Architecture, Design, Fabrication*, Cal Tech, Jan. 1979, 255-264.

11. P. H. Hochschild, E. W. Mayr and A. R. Siegel, "Techniques for Solving Graph Problems in Parallel Environments", *Proceedings of the 1983 International Conference on Parallel Processing*, 1983, 351-359.

12. H. T. Kung and C. E. Leiserson, "Systolic Arrays (for VLSI)", in *Sparse Matrix Proceedings 1978*, I. S. Duff and G. W. Stewart (editor), Society for Industrial and Applied Mathematics, 1979, 256-282.

13. S. Y. Kung, K. S. Arun, R. J. Gal-Ezer and D. B. B. Rao, "Wavefront Array Processor: Language, Architecture, and Applications", *IEEE Transactions on Computers C-31, No. 11* (Nov. 1982), 1054-1065.

14. H. T. Kung, "Why Systolic Architectures?", *Computer 15(1)* (Jan. 1982), 37-46.

15. H. T. Kung, *A Listing Of Systolic Papers*, Department of Computer Science, Carnegie-Mellon University, Pittsburgh, Pennsylvania, Sept 1986.

16. C. E. Leiserson, *Area-Efficient VLSI Computation*, MIT Press, Cambridge, Massachusetts, 1983.

17. H. F. Li and R. Jayakumar, "Systolic Structures: A Notion and Characterization", *Journal of Parallel and Distributed Computing 3* (1986), 373-397, Academic Press.

18. J. D. Lipson, *Elements of Algebra and Algebraic Computing*, Addison-Wesley, Reading, Massachussets, 1981.

19. D. Nassimi and S. Sahni, "Finding Connected Components and Connected Ones on a Mesh-Connected Parallel Computer", *SIAM Journal on Computing 9* (Nov. 1980), 744-757.

20. P. A. Nelson, *A Non-systolic Matrix Product Algorithm*, University of Washington, Department of Computer Science, Technical Report No. 85-11-02, Nov. 1985.

21. P. A. Nelson and L. Snyder, "Programming Solutions to the Algorithm Contraction Problem", *Proceedings of the 1986 International Conference on Parallel Processing*, Aug. 1986, 258-261.

22. P. A. Nelson, *Parallel Programming Paradigms*, Ph.D. Thesis, University of Washington, Seattle, WA, 1987.

23. M. S. Paterson, W. L. Ruzzo and L. Snyder, "Bounds on the Minimax Edge Length for Complete Binary Trees", *Proceedings of the Thirteenth Annual ACM Symp. on Theory of Computing*, 1981, 293-229.

24. Q. F. Stout, "Properties of Divide-And-Conquer Algorithms for Image Processing", *Proc. of the 1985 IEEE Computer Society Workshop on Computer Architecture for Pattern Analysis and Image Database Management*, Nov. 1985, 203-209.

25. V. Strassen, "Gaussian elimination is not optimal", *Numerische Mathematik 13* (1969), 354-356.

26. J. Van Rosendale, "Minimizing Inner Product Data Dependencies in Conjugate Gradient Iteration", *Proceedings of the 1983 International Conference on Parallel Processing*, 1983, 44-46.

Large-grain parallelism — Three case studies

Raphael A. Finkel

Computer Sciences Department
University of Wisconsin–Madison
1210 W. Dayton Street
Madison, WI 53706

This chapter deals specifically with large-grain parallelism, in which (1) a group of processes collaborate in an algorithm, (2) processes do not share memory but rather communicate by sending each other messages, and (3) messages are relatively infrequent.

We lay the groundwork for studying algorithms exhibiting large-grain parallelism by defining performance measures, identifying standard algorithmic decomposition techniques, and categorizing reasons why perfect speedup is not achieved. We then investigate three applications in detail, describing increasingly effective distributed algorithms aimed to solve those problems. The applications are a palindrome-generating puzzle, a general package for recursive tree search, and the simplex method for linear optimization. These examples demonstrate the decomposition techniques and the causes of imperfect speedup. They also highlight trade-offs involved in choosing distribution strategies and practical issues that arise when algorithms are implemented.

1. Introduction

Many different approaches to distributed computation are under active investigation on a variety of machines. In this chapter, we will deal specifically with **large-grain parallelism**, which we define to have the following components.

- A group of processes collaborate in an algorithm. When we discuss timings, we will assume that each process resides in a separate machine, so they may all execute at the same time. In fact, if a process is idle, we will assume that the processing power of its machine is wasted.

- Processes do not share memory; instead, they communicate by sending each other messages. We will not deal with the semantics of message passing; remote-procedure call is one of several adequate alternatives. However, we will assume that messages are relatively expensive, requiring times on the order of tens of milliseconds for delivery.[1]

- Because messages are so expensive, our algorithms will strive to limit the frequency of message traffic. As a rule of thumb, we will often have fewer than ten messages generated by any process per second.

Algorithms exhibiting large-grain parallelism are important because they can be implemented both on multiprocessors, where memory is shared, and multicomputers, where memory is not shared. As the size of a problem increases, such algorithms usually can take advantage of more and more processors and still achieve reasonable speedup.

Designing distributed algorithms is not easy. This chapter describes general techniques for distributing algorithms and then gives detailed examples of those techniques in action. The results are not always as efficient as one might hope. We will categorize obstacles to perfect efficiency and see how they apply to the algorithms we develop.

This chapter begins with a section on background information, including definitions of performance measures, a collection of algorithm styles, and a list of sources of performance degradation. We then turn to the Palindrome problem, which, although a toy problem, demonstrates many fundamental principles of large-grain parallelism. The DIB implementation of tree search is covered next, introducing new concepts and enlarging on the ones already presented. We close with at attempt to distribute the simplex method, which shows the difficulties of predicting speedup in numerical computations.

2. Background

2.1. Performance measures

As mentioned above, we will assume that each process has a machine to itself, so there is no contention among processes. One hopes that applying more processes to a problem will allow an algorithm to solve it more quickly. We can quantify this hope by defining several notions.

- **Execution time** $T(p, n, A)$ is the time needed by algorithm A to compute a problem of size n on p processes. We will leave off the indices p, n, and A when it is clear what is meant. Execution time includes initialization and communication time, and is measured from the time the first process starts to the time the last one terminates.

- **Speedup** $S(p, n, A)$ is defined as

$$S(p, n, A) = \frac{\text{time required by the best serial algorithm}}{T(p, n, A)}.$$

Values of S range from 0 to ∞. By definition, $0 \leq S(1) \leq 1$. If we compare the distributed algorithm against $T(1, n, A)$, we get a value we will call the **rough speedup**, RS. It is a less honest measure of performance than the true speedup. If $RS(p) > p$, we say the algorithm exhibits a **speedup anomaly**. It shows that the distributed algorithm is poor when $p = 1$. If the distributed algorithm is the same as the serial algorithm for $p = 1$, a speedup anomaly implies that the serial algorithm is suboptimal.

- **Efficiency** $E(p,n)$ is defined as

$$E(p,n,A) = \frac{S(p,n,A)}{p}.$$

Values of E range from 0 to 1. We define the **rough efficiency** analogously to the rough speedup.

- The **useful-process point** $U(p,A)$ is the size n of problem that makes it worthwhile to use as many as p processes.

$$U(p,A) = \text{smallest } n \text{ such that } T(p,n,A) \le T(p-1,n,A)$$

In general, as the problem size increases, the expenses of distribution (which may be dependent on p) begin to be outweighed by its benefits.

2.2. Distributed-algorithm styles and methods

Many algorithms fit into one of the following classes.

- **Generate and solve.** A problem can be subdivided into subordinate problems, each of which can be solved independently of the others. A pool of **slave** processes stands ready to solve these problems as they are generated and distributed by a **master** process. This category includes tree-search algorithms such as alpha-beta search [12]. Our DIB example fits roughly into this class.

- **Iterative relaxation.** The data space can be divided into adjacent regions, which are then parcelled out to different processes. Each process carries out activities local to its region, communicating with neighbors when necessary. This category includes solution of numerical problems like PDEs and graph problems like finding a minimal spanning tree [13]. Termination is often difficult to determine in such algorithms. Both our palindrome example and our simplex example are in this class, but they are quite different from each other.

- **Passive data pool.** A large data space is managed by many processes, which support queries and updates on that space. Queries from **client** processes are directed to the appropriate data **server** processes. This category includes distributed file systems and other data structures such as hash tables [9]. These algorithms try to allow a high throughput of queries by letting non-interfering queries proceed simultaneously.

- **Systolic.** Data values flow through a set of processes, undergoing modification along the way. The processes are often arranged in a regular structure, such as a linear array or a square mesh. This category includes many numerical algorithms [17], pipeline algorithms, and multi-pass transformers such as compilers and scene analyzers.

Algorithms may engage in **restructuring** during the course of a computation. First, the allocation of data to processes may change. Data motion can be a result of attempts to balance load among processes or to bring values to where they are needed. The process structure may also change. New processes may be created as the size of the problem warrants, and new inter-process communication paths may be opened to fulfill new requirements. Restructuring often introduces bookkeeping difficulties as the algorithm tries to maintain enough information to find data and processes as they are needed. In general, such bookkeeping information includes **absolutes**, which are reliable data about the current situation, and **hints**, which are often correct but which must be verified before use. Hints are often far easier to maintain, but recovering from an inaccurate hint may be expensive.

A less dynamic form of restructuring comes from **quotient** schemes [14], in which a single **physical** process simulates the activities of several **virtual** processes. Quotient schemes are particularly useful in pipeline and iterative-relaxation algorithms when the grain of parallelism is too fine and communication is needed too often. All communication between virtual processes within the same physical process can be done cheaply, and communication that crosses the physical-process boundary can often be batched, so that one physical message represents many virtual messages. For example, systolic algorithms often perform only trivial calculation between computational steps. If the systolic array is arranged in a square lattice, a quotient in which one physical process represents a square cluster of 25 virtual processes reduces the amount of inter-process communication approximately 25-fold, since this scheme benefits both from removing internal communication and batching external communication. The boundary between physical machines is often a boundary between two representations of data as well. Within a physical machine, the fact that work is distributed often remains **implicit** in loops across arrays or traversals of graphs. Between physical machines, the fact that the data structures are distributed is **explicit**. Data often have two representations, therefore, depending on their proximity to this boundary.

Algorithms based on iteration often can be described either as synchronized or chaotic. **Synchronized** algorithms go through well-defined rounds, between which information is passed among the processes. The exchange of information becomes a bottleneck. One way to reduce that bottleneck is to use a **chaotic** algorithm, in which one process may start the next round before others have finished. The cost of such a scheme is often an increased number of rounds.

2.3. Sources of algorithm degradation

Speedup is typically less than p for several reasons:

- **Communication expense.** Messages use computational resources that cannot be simultaneously devoted to advancing toward a solution.

- **Communication delay.** Processes may have to wait for a message to arrive before they can continue to advance toward a solution.

- **Overlapped work.** To reach agreement on shared data, an algorithm can either broadcast the data, which incurs communication costs, or it can compute the data independently on several processes. Under this latter strategy, identical work is conducted by several processes, during which time the instantaneous speedup is 1 and the efficiency is $1/p$. If the same data appear in several places, we call the copies **carbons**.

- **Speculation.** There may be many ways to advance to a goal. A serial algorithm may sort them and try the best ones first. A distributed algorithm may try several at once. Although this strategy may occasionally be very lucky, leading to a speedup anomaly, it will often waste the efforts of those processes searching less useful paths when another process is pursuing the best path.

- **Reduced knowledge.** If performing calculations improves the knowledge base of the algorithm, simultaneous calculations cannot make use of the most recent knowledge, but must rely on information that is to some extent out of date.

- **Uneven allocation.** It is hard to allocate computation so that all processes have an identical amount. Those processes that finish a step earlier than the others may have to wait until the others complete before continuing.

2.4. Programming languages

Although we will not concentrate on how algorithms are represented in programs, it is worth pointing out some trends in programming languages intended for large-grain parallelism [5, 3, 22]. At the least, a reasonable language provides ordinary sequential operations and a way to send messages between processes. Inter-process communication can be abstracted as a form of remote procedure call, although there are rare situations in which this paradigm is not quite right (for example, requests that have an immediate answer and a delayed answer). Typically, the programmer must specify how work is divided among processes; the compiler does not attempt this division.

Several particularly useful features are coming into vogue.

- **Light-weight tasks.** Such tasks are relatively inexpensive to create and share memory with each other, although they may be subject to scoping restrictions for data access. Light-weight tasks are especially useful for maintaining the state of a server-client conversation while other conversations are taking place. Synchronization mechanisms are needed to prevent unwanted interference between tasks; these mechanisms include priority schemes, explicit conditional waiting, semaphores, and monitors.

- **Exceptions.** When a process sends a request that is misformed or illegal in some sense, it is possible to respond with an error indication. However, inspecting all responses for the presense of this indication is a heavy burden, and most programmers are not particularly careful about it. Instead, an exception mechanism such as is found in Ada [25] can be used to propagate errors.

- **Message-based type checking.** Wherever possible, the compiler can ensure that messages are properly formed. For complex algorithms built out of several compilation units that are compiled at different times, declaration libraries can be used. Inexpensive run-time checks are also possible [20].

- **Implicit and explicit message receipt.** Explicit receipt makes sense when the algorithm has reached a point where it knows that it cannot proceed unless a particular message arrives. However, messages that arrive during the course of other computation must also be dealt with. Implicit receipt starts a new lightweight task for each such message; the task begins its execution in whatever procedure the message is trying to call.

The Lynx language [21], which has been implemented both on the Crystal multicomputer [8] and the BBN Butterfly [23], has all of these features. Argus [19] includes the ideas of type checking and light-weight processes, and also has a well-developed notion of **transaction**, which is important in recovering from failures. Other important languages for distributed programming include CSP [15], NIL [24], and SR [2].

3. Palindromes

Our first example of a problem to be distributed is based on a puzzle. The following algorithm often leads to a number whose decimal representation is palindromic, that is, reads the same backwards and forwards:

> **while** n is not palindromic **do**
>> n := n + reverse(n);
> **end**;

Here, reverse(n) is the decimal representation of n written backwards. For example, if we start with the number 49, we iterate once to get 143, and once more to get 484, which is a palindrome. Many numbers reach a palindrome after only a few iterations. A few require many iterations. The number 196 seems to be special: The algorithm does not terminate within the first several million iterations. The first few **iterates** are 196, 887, 1675, 7436, 13783, 52514, and 94039. As the number grows longer, it becomes less likely that this algorithm will halt at any step, because any carry during the addition step is likely to result in a non-palindromic result, and the probability of carries increases with the length of the number.

It is not hard to write a serial program that implements the algorithm shown above to test whether 196 ever reaches a palindrome. At some point, it becomes worthwhile to attempt distributing the calculation, since iterations start taking significant amounts of time. We will express the useful-process point in terms of the number of digits in the current iterate. We will describe three distributed algorithms for this problem, the first of which was implemented by Bryan Rosenburg and William Kalsow at the University of Wisconsin on the Crystal multicomputer [8].

3.1. The wave algorithm

It does not seem feasible to distribute the work in iteration space, that is, have one process do some iterations and another process do others, since each iteration is strongly dependent on the previous one. Instead, we can divide the decimal representation of the current iterate into segments and give each segment to a different process. In order to perform one iteration, process i needs to add its segment with the reversal of the segment contained in process $p - i$. Each iteration therefore begins with a large swap of information. To avoid this data restructuring, we use a quotient technique. We will simulate $2p$ virtual processes with p physical ones. The ith physical process will simulate virtual processes i and $2p-1-i$, holding data from symmetric regions in the iterate. (We will always number processes starting at 0.) This arrangement is depicted in Figure 1, in which physical processes p_0 ... p_3 emulate virtual processes v_0 ... v_7. Each iteration requires reversing both segments held in the physical process, performing two additions, and preparing for the next iteration. This algorithm clearly fits in the iterative relaxation class.

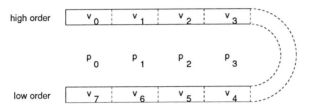

Figure 1: Quotient arrangement for the wave algorithm

Our algorithm is synchronous, in that all processes work on the same iteration at the same time. Between iterations, a **synchronization phase** accomplishes three tasks. First, any carries that leave a virtual process must be propagated to the next virtual process. We pass a wave of carry messages from the lowest-order virtual process to the highest-order one, with each virtual process waiting to see if the incoming carry will propagate out before sending its own value.

Second, the entire iterate may have grown. This occurs when there is a carry out of v_0. In this case, data must be restructured, because it is essential that the segments stored in any given physical process match each other in the reversal of the entire iterate. Approximately 40% of the iterations lead to a lengthening of the iterate. We pass a wave of restructure messages from the most significant virtual process down to the least significant one, indicating whether restructure is needed, and if so, possibly enclosing one digit that is being given to the next less-significant virtual process.[2]

Third, the iterate must be checked to see if it is now a palindrome, which would signal the termination of the algorithm. This check is performed independently by each physical process, and is usually quite rapid; only a few digits need to be scanned before a discrepancy is found. However, the entire computation needs to be informed about the results of this check, so we send a third wave (it does not matter in which direction) across the physical processes to see if all segments are palindromic. The end process knows whether the iterate is indeed a palindrome, so it sends a wave back telling the others either to stop or to continue with the next iteration. This part of synchronization, like the others, visits each physical process twice.

Each of our synchronization steps is systolic, in the sense that a small amount of information is passed in a pipeline fashion from process to process. Therefore, the entire structure of the algorithm can be described as a synchronous iterative relaxation, with a systolic synchronization step. We call this first algorithm the **wave algorithm** after its synchronization mechanism.

If there are p physical processes, it may be more efficient to let some sit idle instead of letting them join the computation, especially near the start of the computation, when the iterate is still quite small. The cost of synchronization is roughly $6(p-1)C$, where C is the cost of sending a single message. (The number 6 comes from the three passes during synchronization and the fact that each physical process simulates two virtual processes.) The computational cost of an iteration is roughly $2Ed/p$, where E is the cost of treating one of the d digits of the iterate. (The number 2 comes from the fact that each physical process simulates two virtual processes.) In our test implementation, the observed values of the constants were approximately $C = 50$ ms, $E = 6$ us.[3] Graph 1 shows the time per iteration as a function of d for various values of p.

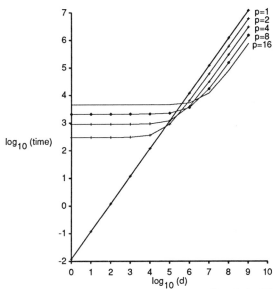

Graph 1: Time per iteration versus *d*

The point at which two processes can finish faster than one is at 50,000 digits, and three become better than two at 150,000 digits. The useful-process points are shown in Graph 2.

This algorithm was implemented. It was started on only one process. When the iterate size reached $U(2)$, a second process was brought into the calculation. (The program wrote periodic checkpoints so it could be restarted with any number of processes.) By the time the iterate was 1,000,000 digits long, four processes were in use. Together, they accumulated 142 machine days of calculation on VAX-11/750 computers, finishing in 47 real days. In contrast, a serial algorithm running on a single machine would have required about 90 machine days.

3.2. The exchange algorithm

The wave algorithm spends a significant amount of time accomplishing synchronization, and this amount of time is proportional to the number of processes. The second algorithm, while still synchronous, reduces the synchronization step to a pair of messages between adjacent physical processes; all exchanges are simultaneous. We therefore call it the **exchange algorithm**.

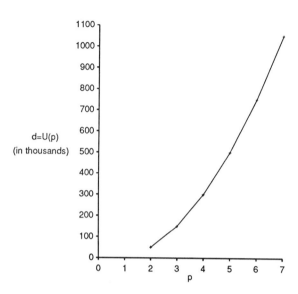

d=U(p)
(in thousands)

Graph 2: $U(d)$

The way to reduce the carry propagation is simple: We pass all carries simultaneously. In the unlikely event that an incoming carry propagates all the way out, the algorithm fails. The probability of failure is vanishingly small. (The probability of a carry propagating past d digits is 10^{-d}, assuming uniform distribution of digits.)

It is a little harder to avoid the digit redistribution wave. If each virtual process could predict which iterations lead to growth, it could send digits when appropriate to its less-significant neighbor. However, the pattern is not regular. Process p_0, which holds both the most and least significant segments, can predict this pattern by using just a few of the most and least significant digits. We let it predict whether the iterate will grow p iterations ahead of the current one. (This calculation usually requires less than p digits.)

After iteration i, p_0 passes information about iteration $i+p-1$ to p_1. During that same iteration, p_1 passes the information it received previously, that is, concerning iteration $i+p-2$, to p_2. In general, p_j passes information about growth $p-j$ iterations in the future to p_{j+1}. The last process receives data concerning growth on the very iteration it needs it. The other processes retain information for no more than p iterations, using it when needed. Along with this predictive information, each process also passes the moving digit, if any, to its neighbor. These digits constitute an exchange of messages between adjacent physical processes.

Finally, it is necessary to check whether the iterate is a palindrome. Instead of accumulating this information at each synchronization step, we let it pass slowly, one physical process at a time, each iteration. If a palindrome is actually found at iteration i, p_0 will hear from all the others by the end of iteration $i+p-1$, and it will inform the rest of the processes by the end of iteration $i+2p-2$. If it is important not only to discover the fact that a palindrome was found but also to print that palindrome, each process can remember the previous $2p-2$ values for its segment of the iterate. (Actually, only p_{p-1} needs to remember that many; p_0 can get by with only $p-1$ stored values.) Alternatively, since periodic checkpoints are valuable for such a long calculation to protect against failures, the calculation can be restarted from a checkpoint for just enough iterations to arrive again at the palindrome.

Each of these three details, carry propagation, data restructuring, and termination testing, involves an exchange of messages between each pair of adjacent physical processes. These messages can be batched together, which reduces the synchronization time per iteration to $2C$. This lower synchronization time improves the speed of the algorithm, as shown in Graph 3. Now $U(2) = 1000$, at which point any number of processes is worthwhile. In fact, $U(p>2) < 1000$.

The most important advantage of the exchange algorithm is that it replaces waves by simultaneous messages, reducing synchronization cost. The communication delay is no longer dependent on p. In order to accomplish this improvement, the exchange algorithm employs three different strategies. For carries, it attempts interdependent computations simultaneously. The cost of this strategy is the potential that a dependency will be missed, leading to an inaccurate result. In our case, the chance of failure is very small, and the failure can be detected. This strategy is a form of speculation.

For restructuring, it replaces communication with prediction. The cost of this strategy is the slight extra work required in p_0 to predict growth p iterations ahead and the slight extra memory required in all processes to remember predictions until they are needed. The fact that all processes use the same prediction at the same time is a form of overlapped work.

For termination, it replaces communication with delayed action. The cost of this strategy is the delay between terminating and discovering termination. In our case, approximately $2p$ iterations are wasted, which is negligible in comparison with the millions of iterations that may be needed before termination.

The fact that all three details of synchronization can be packaged in one message is a form of quotient: Three virtual messages are simulated by one physical message.

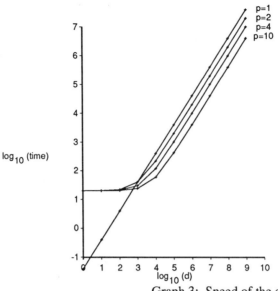

Graph 3: Speed of the exchange algorithm

3.3. The decoupled algorithm

The principal bottleneck of the exchange algorithm remains the synchronization step, although it is far more efficient than under the wave algorithm. Our third algorithm allows us to synchronize with neighbors far less often, completing many iterations between synchronization steps. It does this mainly by overlapping work.

Synchronization can be omitted if each process can predict the information that its neighbors would pass it. For example, p_i could predict a carry from its neighbor, p_{i+1}, if it had a carbon of both segments stored in p_{i+1}. Actually, this is far more information than is needed. Just a few digits from each of the segments in p_{i+1} are enough to predict carries for several iterations. Process p_i can manipulate the carbon to determine for itself whether a carry will occur, and it can even decide the amount of significant information remaining in the carbon.

For example, if we know that of the two segments held by p_{i+1}, one has high-order digits 6238[0-9] and the other has low-order digits [0-9]3423, we know that there will be no carry. (We are using [a-b] to mean that the digit in this position has a value between a and b.) After one iteration, we expect the high-order digits to be 6238[0-9]+3243[0-9]. So the new high-order digits will be 948[1-2], and the low-order ones will be [0-9]3423+[0-9]8326 = [0-9]1749. We have ignored the carry coming into the low-order digits; this carry is known to p_i as it simulates the high and low digits of p_{i+1}.) It takes twelve iterations in this case before we can no longer predict carries. Figure 2 shows what is known at each iteration.

```
    6238[0-9]   [0-9]3423
     948[1-2]   [0-9]1749
    1895[2-4]    [3-4]598
    108[4-5]    [0-9]579
    108[3-4]    [0-9]380
     19[1-2]    [0-9]181
     37[2-4]     [3-4]72
     64[5-9]     [6-9]45
    119[1-9]     [1-8]91
     31[0-8]     [1-8]02
      5[1-2]    [0-9]15
     10[2-4]     [3-4]0
      1[3-5]     [3-4]1
```

Figure 2: Predicting the carry

As this example shows, if the uncertainty is carefully represented, several iterations can be made with the same numbers.

Similarly, each physical process can predict when data restructuring is necessary if p_0 makes several of its highest- and lowest-order digits available to all the processes. Each of them can manipulate those carbon copies to predict carries out of the highest position. When an iteration that requires restructuring is finished, it is not necessary to actually transfer data from one virtual process to its lower-order neighbor, so long as that neighbor still has one significant digit in its carbon copy. Therefore, both carries and shifts deal with the same carbon copies, and both influence the number of significant digits remaining.

When a physical process has exhausted the carbon values from its neighbors, it can ask for new values, which it will then use for subsequent iterations. Some simple experimentation shows that j digits last about $3j$ iterations, although there is some variance in this figure. Instead of forcing a process to request more carbon values, we can supply enough values to be fairly sure (say with 99.9% probability) that they will last r iterations, and then after r iterations, automatically send the next batch of values. Alternatively, a double carbon method can be used, in which p_i simulates the same calculations that p_{i+1} is undertaking with the carbon values from p_i in order to determine when p_{i+1} will need new values. This strategy is an extreme form of overlapped computation, and would be fruitful only if messages are terribly expensive.

The decoupled algorithm may take even longer than the exchange algorithm to discover termination, since a message implying that an iteration resulted in a palindromic segment may be batched with carbon data, and only slowly wend its way to all the processes.

The decoupled algorithm is still synchronous, but with far fewer synchronization events. It is not chaotic, because the results do not depend on timings, and any process that consistently computes faster than its neighbors will eventually suffer communication delay due to the uneven allocation. At any instant, different processes may be undertaking different iterations, and messages may arrive with carbon data before they are needed. Such messages can be stored until their contents are required.

4. Distributed tree search

In contrast to the palindrome algorithms, which use iterative relaxation, our next example belongs in the general class of generate-and-solve algorithms. Also in contrast to the palindrome programs, which were intended to be used only once (for the special starting value of 196), our next example is a software package, intended to be used over a wide range of applications for a long time.

DIB, which stands for **d**istributed **i**mplementation of **b**acktracking, is a set of routines providing a simple interface for a wide range of tree-search applications such as recursive backtrack, branch and bound, and alpha-beta search [10]. DIB manages the computation, distributing work automatically and collecting results to be reported. The application program needs to specify only the root of the tree, the computation to be performed at each node, how to generate children, and how information is inherited (moving down the tree) and synthesized (moving back up the tree). In addition, the application program may optionally specify how to disseminate information (such as bounds) globally throughout the tree. DIB requires only minimal message-passing support from its environment and can recover from process failures even if they are not detected.

DIB currently runs on the Crystal multicomputer at the University of Wisconsin–Madison [8]. Many applications have been implemented quite easily, including exhaustive traversal (N queens, knight's tour, negamax tree evaluation), branch and bound (travelling salesman) and alpha-beta search (the game of NIM). Speedup is excellent for exhaustive traversal and quite good for branch and bound. Figure 3 shows how DIB fits into the Crystal environment.

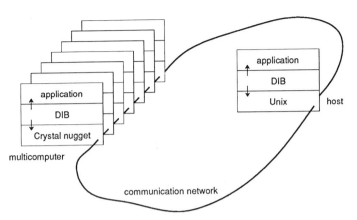

Figure 3: Organization of DIB

4.1. Why a software package?

Experience so far indicates that writing distributed programs is significantly more difficult than writing sequential programs. Programmers of distributed programs must deal with a variety of new issues, including synchronization, concurrency, communication protocols, and fault tolerance. Even seemingly simple tasks (for example, detecting termination) become complicated and error-prone in a distributed environment. Currently, distributed programming is almost always left to experts.

Two complementary approaches can make distributed programming easier. The first approach is to develop better programming languages, as touched on in the introduction. A second approach, typified by DIB, is to develop library packages in specific areas. These packages will be suitable only to restricted classes of applications and will lack the generality of a programming language. They will also probably be less efficient than direct implementations of specific algorithms. However, one can hope to make these packages relatively easy to use. In particular, the fact that the algorithm is distributed can be invisible to the programmer. Such tools enable a novice programmer to write distributed programs in a basically sequential manner. As a result, programmers who are unable or unwilling to master the techniques of distributed programming can still use the full power of multicomputers. DIB achieves this goal by presenting a sequential interface to the programmer in which tree-search programs can be written.

4.2. The application interface

The applications for which DIB is designed all have one feature in common: They all generate and evaluate trees. Typically, the structure of the tree is not known in advance, but is elaborated as the tree is traversed. Each node in the tree goes through several states: Creation (when the node is created), active life (when its children are going through their life cycles), and termination (when the last child has terminated). When a node is active, some of its children have already terminated, some are active, and others are yet to be created. As each one is created, it receives **inherited** information from its parent. When it terminates, it reports its **synthesized** information back to its parent. Children that are created after some of their siblings have terminated may inherit information from their terminated siblings (by way of their parent); we call this feature **sibling inheritance**. Terminated nodes no longer participate in the calculation.

Computation may take place at various times during the life of a node. Typically, a node engages in calculation each time it creates a child and each time a child terminates and reports its synthesized information. Leaves of the tree may perform calculations when they become active, even though they create no children.

DIB recognizes two cases in which information is disseminated other than by the parent-child exchanges described above. The first is broadcast information, which as soon as it is generated by some active node must be sent to all active nodes. Bound information in a branch-and-bound search [18] fits this case. The second case is when the synthesized information reported by a child is of such value that the other active children of the parent will benefit from immediate notification. In this case, DIB allows active children to receive data updates, and they have the opportunity to pass information to their active children, as well. This technique is important in implementing alpha-beta search [11].

4.3. Sample application: Permutations

We will demonstrate the sort of problem for which DIB is intended with a simple task: For all permutations of the numbers from 1 to n, count how many satisfy some property P. All the permutations of n can be described as leaves of a permutation tree, where a node at depth d in the tree represents a choice for the first d digits of the permutation, and its children represent the choice for digit $d + 1$, as well. All leaves are at level n; there are $n!$ of them. Figure 4 shows the tree for $n = 3$.

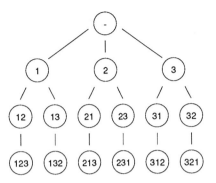

Figure 4: Tree for permutations on 3 numbers

The interface between the application and DIB is a set of procedures, most of which reside in the application and are called by DIB when necessary. We will describe them informally here; a more formal definition can be found elsewhere [10]. These procedures deal with a data type known as ProblemType, which is used to represent nodes in the search tree. For our application, ProblemType might be a record with these fields:

inherited constants:
  ProblemSize — the number of digits in the permutation
inherited variables:
  Depth — the depth of the current node
  Permutation — an array [1:ProblemSize] indicating the choices
    so far (only the entries from 1 to Depth are in use)
synthesized variables:
  Count — number of leaves in our subtree satisfying property P.

The first application procedure is called by DIB to create the root of the entire tree.[4]

```
procedure FirstProb(in Size : integer) : ProblemType;
begin
        return ProblemType'(
                ProblemSize := Size,
                Depth := 0;
                Permutation := empty;
                Count := 0);
end FirstProb;
```

The central application procedure creates children. It is used both for first children and for later ones; in the latter case, it accepts the previous child as a parameter so it can modify it slightly to form the next child.

```
1      procedure Generate(
2              out Done : Boolean;
3              in First : Boolean;
4              in out Parent : ProblemType;
5              in out Child : ProblemType;
6              );
7      begin
8              if First then
9                      with Parent do
10                             if ProblemSize = Depth then — Parent is a leaf
11                                     Done := true;
12                                     if P(Permutation) then
13                                             Count := 1
14                                     else
15                                             Count := 0;
16                                     end;
17                             else — create first child
18                                     Child := Parent;
19                                     Child.Depth +:= 1;
20                                     Child.Permutation[Child.Depth] := first value;
21                                     return;
22                             end;
23                     end;
24             else — not the first child
25                     with Child do
26                             — Child is the older sibling
27                             if Permutation[Depth] = last value then
28                                     — no more children
29                                     Done := True;
30                             else
31                                     — create next child
32                                     Permutation[Depth] = next legal value;
33                     end; — with
34             end — not first child
35      end Generate;
```

The construction of the Generate procedure is usually the most time-consuming part of
programming a DIB application. It must cover several cases, which are often handled
quite differently: (1) The parent is a leaf, (2) the parent is creating its first child, and (3)
the parent is creating a subsequent child.

The Generate procedure allows us to initialize the child with information inherited
from the parent, as we do in line 18. We also create subsequent children on the mold of
their predecessors (line 25), so they all share the same inheritance from their common
parent.

Synthesized information is passed up the tree when a child terminates. At that time, DIB calls Combine:

```
procedure Combine(
        in Child : ProblemType;
        in out Parent : ProblemType;
        out Update : Boolean
        );
begin
        Parent.Count +:= Child.Count;
        Update := false;
end Combine;
```

Here we use Combine to increment the parent's running count of the number of leaves where the property P holds. If we had set Update to true, then another procedure, Update, would have been called for all the active children of the parent to allow this synthesized information to affect their computations. In our case, their computation is completely independent of ours, so we no not need this facility.

The rest of the DIB-application interface is composed of (1) output and debugging procedures, provided by the application, and (2) a broadcast facility, comprised of one procedure invoked by the application to start a broadcast and another invoked by DIB when a broadcast arrives. Our application needs only one output routine:

```
procedure PrintAnswer(in P : ProblemType);
begin
        write(P.Count);
end PrintAnswer;
```

4.4. Distributing the work — first attempts

Since DIB is ignorant of the calculation needed to create children (it just calls the Generate function, provided by the application), there is no hope of distributing the calculations performed within any one node of the tree. Instead, the nodes of the tree themselves must be distributed. We can imagine a separate process for each node of the search tree. As a node creates children, new processes are created. An initial message transmits any inherited information to the child, and a final message from the child transmits synthesized information. Update information is passed from parents to active children when necessary.

This form of distribution has several problems.

(1)   Creating a process is likely to be an expensive operation. The parent cannot continue until its child processes are formed, leading to synchronization delay.

(2)   There may be a limit on how many processes may be formed.

(3)   At least $2t$ messages are required if the tree has t nodes. Although not as expensive as creating a process, sending a message is still relatively costly.

(4)   During most of their active period, processes representing nodes high in the tree are idle, waiting for their children to report their results. That is, work is very unevenly distributed. The overall efficiency of the algorithm is therefore quite low.

(5)   This distribution suffers from information loss, since the entire tree is calculated simultaneously. There is little chance for global broadcast or sibling inheritance to have much effect. If the application has been written properly, information loss will not cause inaccurate results, but it can lead to a significant loss of efficiency. In our permutation example, however, neither global broadcast nor sibling inheritance is used, so this inefficiency does not arise.

We can improve the situation greatly by applying quotients in various ways. The easiest is to place entire subtrees within one physical process. For example, we can decide that all tree nodes of depth d are roots of subtrees that are governed by a single process. We don't know how many such processes there will be, so we still need to create processes dynamically; objections (1) and (2) still hold. However, the physical processes that manage the subtrees are active as long as the subtree is, mitigating objection (4) somewhat. Objection (5) is not quite so severe, since sibling inheritance can proceed within a subtree, although not between subtrees, and global broadcast has an effect, since leaves are not all evaluated simultaneously.

Most importantly, objection (3) has been answered in two significant ways. First, a large number of messages has been avoided. Second, the data structures needed inside the subtrees are much more efficient, both in space and time, than those needed above the subtrees. In the part of the tree above depth d, we need an explicit tree structure. This structure will include pointers both to the processes governing the children (it may be necessary to send them an Update message) and to the process governing the parent (in order to transmit synthesized data). Starting at depth d, a much simpler structure suffices: the one used by the serial algorithm, namely, a recursion stack. The stack represents the currently active nodes in a path from the root of the subtree down to the node that has just been created. The subtree is searched in strict depth-first order. After a node terminates, its space on the stack can be reused for its next younger sibling. The boundary between the explicit and implicit structures is at depth d; above that depth, we use a separate process for each node, with explicit tree representation, but below that depth, there is only one physical process for each subtree, with implicit tree representation. We can call the processes above the boundary **supervisors** and the ones below **calculators**.

Having introduced a quotient, we have also introduced new problems.

(6)   Different subtrees will require different amounts of computation, so some physical processes terminate first and are left idle. This is a new form of uneven work allocation.

(7)   The number of physical processes available may not match the number needed by this distribution. There may be too many physical processes, in which case a larger d should have been chosen, or there may be too few, in which case a

smaller d may be preferable. However, even if d is selected correctly, there will very likely be more subtrees than there are physical processes to govern them.

We can answer these new objections by introducing dynamic data restructuring into our distribution scheme. Those subtrees at level d for which no calculator process is available will remain dormant for a while. As a calculator finishes its subtree, it not only transmits its synthesized information, but also (implicitly) requests more work. Some dormant subtree is then given to the idle calculator, which can then continue working, albeit for a new master. (We pass over a slightly more static strategy, which is to allocate to the i th calculator each subtree numbered $i \bmod c$, where there are c calculators.)

Our policy should be to compute left-most subtrees first. The application writer can sort children of a node to optimize a left-right traversal of the search tree. That is, if parts of the search can be pruned based on sibling inheritance or global broadcast, it is most efficient to place the prunable branches to the right and the data-generating branches to the left. Since we will compute the left subtrees before the right ones, sibling inheritance and global broadcast will have a chance to prune computations far to the right in the search tree.

The supervisory processes can also benefit from forming a quotient. In fact, reducing all the supervisory work to one physical process will not only reduce the amount of communication and idleness in that community, but it will also make it easier to assign new work to calculator processes that have just terminated, since their physical parent will not change.

There will still be some imbalance in the amount of work demanded of calculator processes. As the last subtrees are being calculated, newly terminated calculator processes will become truly idle. We can reduce this effect by increasing d (which has the effect of generating more computational tasks, each one smaller), but we must pay for this change in increased communication expense and delay (more subtrees, therefore more messages). However, increasing d while leaving c fixed may reduce the effects of information loss. Without knowing the application, it is difficult to predict the correct value for d.

4.5. An elegant distribution algorithm

The previous algorithm suffers from a lack of elegance, which we may characterize as inhomogeneous processes (supervisor and calculator) and unbalanced work. The algorithm chosen for DIB discards the distinction between supervisors and calculators. It balances work by yielding to the temptation to dynamically restructure data whenever necessary.

DIB assumes a fixed number of processes. A process is either **satisfied** (that is, it has tree frontiers to expand) or **hungry** (it has finished its portion of the search tree and has nothing to do). At the outset, process p_0 has charge of the root of the entire search and is satisfied. All the others are hungry, having been given none of the search at all.

A satisfied process searches the tree until it is finished, listening all the while for cries of hunger from colleagues. A cry for hunger elicits a handout from the satisfied process: It gives away part of its tree in an attempt to satisfy the hungry process. Eventually, it will finish whatever part of the search it has not given away, and will become hungry. At this point, it cries in hunger to one or more peers (we will discuss soon how we choose them) and waits for handouts.

The elegance of this method is the way it manages to balance work. In general, all processes are satisfied. When one finishes its assigned task, it quickly finds more work to satisfy it for a while longer. Experimenting with this way to distribute work shows that after the initial flurry, as all the hungry processors converge on p_0 for work, almost no communication takes place. Later, perhaps half-way through the entire computation, processes begin to get hungry. The work they are given in response to their cries is smaller than the original handout (they are given deeper subtrees than before), but enough to satisfy them for a while. Communication increases rapidly near the end, when handouts are only crumbs that are quickly gobbled down and leave the process hungry again. But when that happens, the computation is almost over. Soon it terminates, and all the processes become quiet. (We announce termination by sending a message to all processes. This message is generated by p_0 when it discovers that it has finished the entire tree.)

Although elegant and extremely efficient, this algorithm proved to be surprisingly tricky to implement. We can distinguish problems of mechanism and problems of policy.

Mechanism complexities

The hardest mechanism problem was maintaining the frontier between explicit and implicit tree representations. We need to use implicit (stack) representation whenever possible, because it is much more efficient, both in space and time. However, an explicit tree structure is needed to record the fact that some siblings of a node have been generated but given away, whereas others have been generated and are still untreated. Upward-facing pointers are needed to report synthesized values back up the tree, possibly traversing process boundaries as they travel.

The explicit representation needs to have a unique identifier for each node so that when a synthesized result comes back long after a problem was handed out, the originator can find the part of its tree that was waiting for that result. We chose a Dewey-decimal notation, where each node is named by a sequence of integers, one for each node in the path from the root to that node, indicating the sibling number of each node in the path. For example, the name 3 4 6 belongs to a level-3 node, the sixth child of the fourth child of the third child of the root. Whenever answers come back from handouts, the originating process searches its tables for the Dewey name carried by the returning value in order to account for it properly in the explicit part of the tree.

Unfortunately, the frontier between explicit (at the top) and implicit (at the bottom) is not static. It grows deeper in response to cries for work. In general, DIB prefers to give away subtrees at the highest level possible. If there are no remaining children of the local root (that is, the root of the subtree the satisfied process is working on when it hears a cry for help), perhaps the child of that root on the currently active path has unexplored children. If so, that child can be placed in the explicit structure, along with all its unexplored children. The frontier has thereby grown deeper.

The frontier must not grow so deep that it collides with the currently active point of calculation, because that would violate data-structure assumptions made by the tree evaluator. We allow a satisfied process to refuse to give away work if the frontier will move too close to the active point.

A related problem is one of *concurrency control*. The process wants to carry on two activities simultaneously: searching the tree and responding to cries for work. It can only respond to cries if that will not destroy sensitive data structures. Therefore, DIB only responds to messages at a clean point in the recursive search (after a node has been finished and its synthesized values have been combined with its parent). As the search activity climbs back up the tree, it will eventually cross the frontier from implicit to explicit representations. It must deal with that frontier carefully, ignoring the now-stale values in its recursion stack and using the explicit structures instead.

Policy complexities

Several matters of policy must be decided:

(1)   How much work should be handed out in response to a request?

(2)   What should be done if no work can be handed out?

(3)   How many peers should be asked for work?

(4)   How are those peers to be selected?

We experimented to find answers to these questions. In brief, here are some of the results we found.

(1)   Work is balanced best if we hand out about one half of all highest-level unexplored nodes. Unless there are many high-level siblings, it is almost as good to hand out only one node. This simpler choice has the advantage that the handout can be specified in a shorter message.

(2)   A request can be forwarded to another peer if it cannot be satisfied at the process to which it is first directed. An alternative is to remain silent and let the hungry process ask again after a while. We use this alternative when dealing with fault-prone machines.

(3)   One peer suffices unless there is a danger of machine failure, which we will discuss below.

(4)   At the outset, every process should bother p_0, since it is known to be satisfied. Afterward, a random choice works fine, as does a circular choice.

4.6. Performance

Our implementation of DIB demonstrates very high efficiency for problems that do not require sibling inheritance or broadcast. An example of such an application is the n-queens problem: Find all possible arrangements of n (chess) queens on an $n \times n$ board such that no queen can attack another queen. A count of solutions is passed up the tree. Graph 4 shows timings (in seconds) for the eleven-queens problem. The time curve is measured in seconds. We scale the rough efficiency curve by 100 to make it visible. The rough efficiency is generally above 90% and degrades quite slowly as extra machines are added.

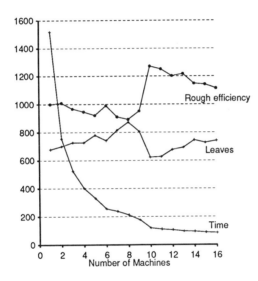

Graph 4: 11 Queens

DIB is not so efficient for problems that require broadcast for pruning. Degradations are generally due to speculative loss: Parts of the tree are searched in the hope they will be useful, but they would be pruned by a serial algorithm before reaching that part of the tree. Occasionally, speculation leads to an efficiency anomaly, as one process working in the middle of the tree serendipitously finds a very good path that prunes subtrees that a serial algorithm would search. A secondary source of degradation is reduced knowledge, because it takes some time for a broadcast to reach all the processes. During that delay, they may continue to search parts of the tree that will be pruned as soon as the message arrives.

A good example of an application that needs broadcast is the travelling salesman problem: Find the cheapest Hamiltonian path through a complete graph of n vertices with given edge costs. We implemented a DIB solution to this problem using branch and bound, broadcasting each newly discovered minimum path cost so that a search that has already exceeded the cost of the best path can be pruned. Graph 5 shows timings for the 11-city travelling salesman problem for 1 to 16 machines. The times are far greater than more sophisticated algorithms can achieve; we do not claim that branch and bound is the best way to solve this problem. The rough efficiency curve, which is scaled by 1000, shows an anomaly with more than 10 machines; the order DIB searched the tree was better than the serial algorithm. The number of leaves reached is shown to explain this anomaly; when the search found a good path early, fewer leaves were reached, and the rough efficiency was higher.

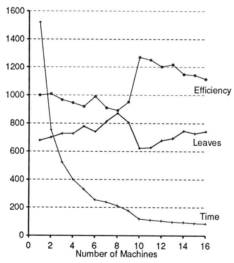

Graph 5: Travelling salesman

The cost of speculation can be reduced substantially by coding the Generate procedure of the travelling-salesman application so that it sorts children of a node in increasing order of distance. When this is done, the entire problem is solved approximately four times as fast, and the speculation anomalies disappear. This is demonstrated in Graph 6.

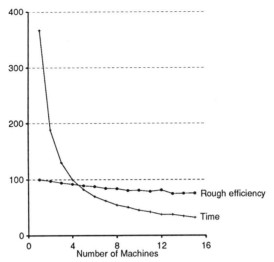

Graph 6: Sorted travelling salesman

Cases where sibling inheritance are needed show the poorest efficiency. In alpha-beta search [16], results synthesized by a node are used to prune the trees for their younger siblings. If the siblings are already active, the Update feature of DIB can be used to transmit new information. We found that this feature helps substantially, but that speculation and reduced knowledge combine to cause severe degradations. This situation is shown in Graph 7, where alpha-beta search was used to evaluate a tree with randomly chosen values at the leaves. The graph shows the efficiency and speedup curves and the number of leaves searched (reduced by a factor of 100) for a tree of fanout 3, depth 11.

4.7. Fault tolerance

So far, we have not worried about recovering from the effects of processes failing unexpectedly. However, the way DIB distributes work lends itself to an elegant recovery mechanism. Whenever a process hands a subtree to another process, it must remember that the subtree has been given away so that it can later accept the synthesized results and so that it does not accidentally give the subtree away again. If the result never comes back because of a failure, the original process can activate the subtree itself or give it away again.

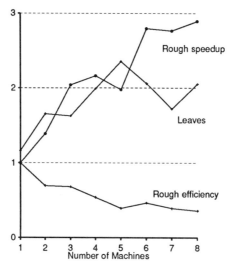

Graph 7: Alpha-beta search

It is not always easy to distinguish whether a result is delayed because of failure or because the computation is slow. In some environments, failed computers are detected immediately, and this information can be returned to the process that is waiting for an answer. In other environments, repeated probes by the problem originator (pinging) or repeated assurances by the recipient (heartbeats) can be used to achieve the same effect.

For DIB, we decided to ignore information about process failure, because it can be expensive to generate such information. Instead, we assumed that failures are silent: A process starts working on a problem and never responds with the answer. When should a subtree that was given away be reactivated by its originating process?

The DIB solution is simple: When a process becomes hungry and cries for help, it occupies itself until work arrives by activating subtrees that it has given to others. We call this executing **redundant** work. If its cry for help is rewarded, it can discard the redundant work. Similarly, it discards redundant work if the process dealing with it reports the results. That process might report results after the subtree has been finished redundantly, in which case the superfluous results are just discarded. If a process hears a cry for help while doing redundant work, it gives away some of that work, properly labelled as redundant. If an answer to that work does not arrive soon, it may itself be retried, leading to doubly redundant work.

This algorithm is very robust. Unless p_0 fails, the work will eventually get done. There are inefficiencies, however, in the way redundant work is executed. One problem involves the mechanism of crying for help. If failures are common, and if unanswerable cries are forwarded, there is a remarkably large chance that a hungry process will be ignored, as its cry eventually is sent to a failed process. We avoid this danger by sending another cry of hunger periodically until the process is satisfied.

A more subtle inefficiency comes from a chain of processes ancestral to a failing process. Assume that p_0 gives some work to p_1, which in turn gives some of it to p_2, which fails before reporting the result back to p_1. When their other work is finished, both p_0 and p_1 will redo the work they gave away. What p_1 is redoing is worth the effort; p_0 however is redoing both work already finished but not yet reported by p_1 and work being redone by p_1, both of which are unnecessary. This problem becomes more serious with longer chains. Experimentation found that the problem can be alleviated by having any process that is about to embark on redundant work tell this fact to the process that was supposed to be doing that work. If that process is still alive, it treats this message as a request for work.

4.8. Summary

Within its limited scope of applications, DIB manages to distribute work remarkably well (with almost no uneven allocation), whether it is distributing original work or redundant work. Very little time is spent on messages, and very few messages are sent throughout a fairly large computation. The principal degradation that is seen is due to speculation and reduced knowledge, which are likely to plague any distribution of tree search.

5. The simplex method

The simplex method is an important linear programming tool [7]. It is used to solve a numerical problem that consists of variables, constraints on the values that those variables may assume, and an objective function. The objective function is a linear combination of constrained variables. The numerical problem to assign values to all the variables in order to maximize the value of the objective function while satisfying the constraints.

For example, consider two variables, y and z, both of which are constrained to be nonnegative. In addition, let us pose two additional constraints:

$$2y+3z \leq 6$$

$$3y+4z \leq 10$$

Let us set our objective function to

$$6y+4z$$

One legal set of values for the variables is (0,2); it yields an objective function value of 8. A better set of values is (3,0), yielding 18. The goal is to find the best set of values for y and z.

The initial step of the simplex method consists of adding "slack" variables to the constraints to make them equalities. In our example, we add u and w, both of which are constrained to be nonnegative.

$$2y+3z+u \quad = 6$$

$$3y+4z+ \quad w = 10$$

If our original variables have constraints other than simple nonnegativity, "artificial" variables may be added to achieve such a constraint. For our discussion, we will assume that we will always start with problems to which slack and artificial variables have already been added if necessary.

5.1. The serial algorithm

We can represent the system in the matrix form as follows.

solve $Ax = d$ while maximizing cx
where

A is an $m \times n$ matrix	— representing the constraints
x is a column of n elements	— the variables to be given values
d is a column of m elements	— the right hand side of the constraints
c is a row of n elments	— the coefficients of the objective function

In our case,

$$n = 4 ; \quad m = 2$$

$$A = \begin{bmatrix} 2 & 3 & 1 & 0 \\ 3 & 4 & 0 & 1 \end{bmatrix} \quad d = \begin{bmatrix} 6 \\ 10 \end{bmatrix} \quad c = \begin{bmatrix} 6 & 4 & 0 & 0 \end{bmatrix} \quad x = \begin{bmatrix} y \\ z \\ u \\ w \end{bmatrix}$$

The simplex method builds a slightly larger matrix, B, as follows:

$$B = \begin{bmatrix} A & d \\ c & -z \end{bmatrix}$$

In our case,

$$B = \begin{bmatrix} 2 & 3 & 1 & 0 & 6 \\ 3 & 4 & 0 & 1 & 10 \\ 6 & 4 & 0 & 0 & 0 \end{bmatrix}$$

The variable z denotes the value of the objective function. In our case, we have started with

$$x = \begin{bmatrix} 0 \\ 0 \\ 6 \\ 10 \end{bmatrix},$$

so the initial value of z is 0. The matrix B always represents a solution to the formula $Ax=d$, but not necessarily the optimal one. The columns that have only a single 1 in the A part are the **basis** columns, and they indicate which variables are set to non-zero values. Their values are taken from the d column at the row where the 1 appears. In our case, u and v are basis variables, and are set to 6 and 10, respectively.

The simplex method consists of repeated iterations of the following steps.[5] After each iteration, B is a better solution than was formed previously.

(1)   Select the **pivot** column j such that $c_j > 0$. If there is no such column, the optimal solution has been found.

(2)   Select the **pivot** row i with the smallest positive ratio $d_i / A_{i,j}$ for $A_{i,j} > 0$. If there is no such row, stop; the objective function is unbounded above.

(3)   Perform row operations on matrix B to achieve

$B_{i,j} = 1$

$B_{row,j} = 0, \quad 0 \leq row \leq m+1, \; row \neq i$

A typical way to perform the row operations is as follows:

```
for row := 1 .. m+1 do
        ratio := B[row,j] / B[i,j];
        for col := 1 .. n+1 do
            if row ≠ i then
                    B[row,col] −:= ratio * B[i,col]
            else
                    B[row,col] /:= B[i,col];
            end;
        end;
end;
```

At each iteration, the simplex method transforms the problem, increasing the objective function while observing the constraints. In our case, we have a choice during the first iteration to choose either $j = 0$ or $j = 1$. (We number both rows and columns starting from 0.) If we pick column 0, then at step 2 we pick $i = 0$, and the row operations result in this matrix:

$$B = \begin{bmatrix} 1 & 3/2 & 1/2 & 0 & 3 \\ 0 & -1/2 & -3/2 & 1 & 1 \\ 0 & -5 & -3 & 0 & -18 \end{bmatrix}$$

At this point, column 3 has left the basis and column 1 has entered it. There are no positive c components left, so the current situation is optimal. The optimal value of x is [3 0 0 1].

If we had picked column 1 instead of column 0, we would again have $i = 0$, but this time B would turn out this way:

$$B = \begin{bmatrix} 2/3 & 1 & 1/3 & 0 & 2 \\ 1/3 & 0 & -4/3 & 1 & 2 \\ 10/3 & 0 & -4/3 & 0 & -8 \end{bmatrix}$$

This corresponds to the solution [0 2 0 2], which has objective value only 8. It would take several more iterations to achieve the same result we got earlier.

As this example shows, the order in which pivot columns are chosen can make a large difference in the number of iterations needed. Unfortunately, there is no perfect rule for picking a good order. One easy heuristic is to pick the column with the largest value of c. In our case, that rule works well, but in general, it does not seem preferable to other rules. There are pathological cases in which it is possible to enter a cycle, choosing a column with positive c in every iteration but never improving the objective function. Luckily, these cases are quite rare and almost never occur in practice.

The simplex method works remarkably well. Although theoretically exponential in its worst case and polynomial in the expected number of iterations [4], practice has shown that it performs quite rapidly, requiring about m to $3m$ iterations to finish (for $m < 50, n < 100$). The principle cost of this method is performing the row operations, which must inspect and modify every entry in B. We will investigate several ways to distribute this work.

5.2. Distribution by rows

Let us assume that the task is to be done by p processes such that p divides m. (These processes are numbered 0 ... $p-1$.) We will distribute B by assigning several contiguous rows to each **calculator** process. To be precise, we will give process k the m/p rows starting with km/p.

We could have a **controller** process hold the c row. It would start each iteration by performing step 1, that is, choosing the pivot column, based on the values of c and some heuristic. It would then broadcast its choice to all calculators. We can avoid this broadcast by letting each calculator keep a carbon copy of the c row, and let them overlap work by each applying the same heuristic and choosing the same pivot column. This organization is shown in Figure 5.

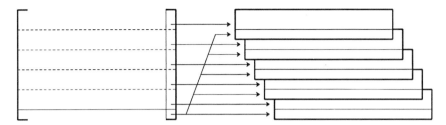

Figure 5: Distribution by rows

Performing step 2 requires comparing the ratios for all rows. Each calculator can decide the best row (that is, the one with the minimal positive ratio) of those it controls. This work is carried out in parallel by all calculators. In order to choose a single pivot row, however, these p locally best ratios must be compared and the globally best one chosen. This subproblem is known as a **voting** problem. We can imagine several ways to perform the vote.

- **Centralized**. Let a controller process receive data from each calculator, decide which is best, and tell all the calculators the result. Each calculator needs to know the entire pivot row chosen in order to perform row operations in step 3. Each calculator could send its entire best row to the controller, which would pick the best and send it to all calculators. Alternatively, the calculators could send their ratios, and the controller would ask the winning calculator to broadcast the pivot row. These two alternatives are diagrammed in Figure 6.

  We can estimate the amount of time needed to perform step 1 under these alternatives. Let the time required to send a message with q numbers be $a+bq$, where a is the per-message cost and b is the per-number cost. We will assume that two messages with different senders and receivers do not interfere with each other, but that a single sender or receiver serializes the messages with which it is involved. The first alternative would require $2p(a+b(n+1))$ time and the second would require $p(a+b)+(a+b)+p(a+b(n+1))$ time. (We are assuming that computing the winner takes insignificant time.) If the per-message cost is significantly higher than the per-number cost, the first method is better.

- **Binary tree**. Arrange the calculators in a logical tree, and collect votes from the bottom up. Specifically, we will use $\lceil \log_2 p \rceil$ rounds to collect the votes and the same number to distribute the winning row to all calculators. During round 0, each odd-numbered p_i sends its vote to p_{i-1}, which selects a winner between its own row and the one it receives. During round r, only processes p_i such that

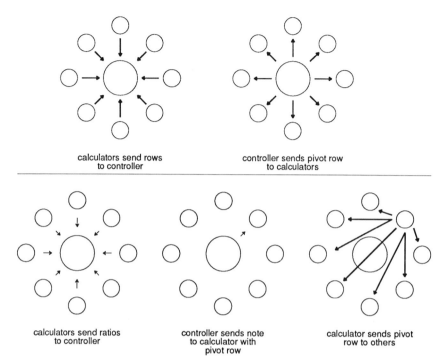

Figure 6: Centralized voting

$2^r \mid i$ are active. For such processes, let $v = i / 2^r$. If v is odd, the process sends its vote to v_{v-1}. The root of the tree is p_0. Assuming we send entire rows and not ratios, this method takes $2 \lceil \log_2 p \rceil (a + b(n+1))$ time.

- **Perfect broadcast.** A clever method based on Latin squares is especially effective when $p = 2^i$ for some i [1]. Each calculator p_i has a list L_i of all other calculators. These lists are arranged so that $L_i[k] \neq L_j[k]$ for any slot k and any pair of calculators p_i and p_j. During round r, p_i sends its best row (either its own, in round 0, or the best it has heard so far, in later rounds) to calculator $p_{L_i[r]}$. During that round, it also receives a message from some other calculator. The lists can be arranged so that after $\log_2 p$ rounds, each calculator knows the best row. This method takes the same amount of time as the previous one, because each round requires the time for two messages. If a process can send and receive at the same time, then the total time is reduced to $(\log_2 p)(a + b(n+1))$.

Once each calculator has been told the winning row, it can proceed to perform row operations independently of the other calculators. All calculators overlap the row operation applied to the c row.

At the end of the iteration, the process starts again. If a calculator is slightly slower (for example, if p does not divide m exactly, so work is not evenly distributed), all must wait at this stage.

Step 2 of the algorithm, the synchronization step, is the bottleneck. We would prefer synchronizations less frequently than every iteration of the simplex method. One way to approach this goal is to let processes perform some iterations using only their locally best row and ignoring the globally best row. Unfortunately, this policy immediately leads the B matrix into a state known as the "infeasible region", in which the values for x no longer satisfy the constraints. An intuitive explanation is that each calculator is trying to maximize the objective function while ignoring some constraints. The amount of work required to bring the system back into feasible region is high (approximately equal to the work done in the wrong steps). This approach therefore introduces at least as much work as it saves. Instead, we can rearrange the distribution of data to decrease the synchronization requirements.

5.3. Distribution by columns

We will now explore the possibility of distributing the B matrix by columns instead of rows. For simplicity, we will assume that p divides n, and we will give each process n/p columns of B. Every process will be given a carbon of the d column. This organization is shown in Figure 7.

This arrangement of data makes step 1 of the algorithm hard. How can all processes agree on the pivot column? We would like to avoid synchronizing the calculators at every iteration, so we need some method other than voting to select the pivot column.

A method of overlapping will do nicely. Assuming we have selected a few likely-looking columns for eventual pivot, we can put carbon copies of those columns in each process. We restrict the choice of pivot column to this set of **globally known** columns. Every calculator has the same set of globally known columns and uses the same selection algorithm to pick a pivot column, so all will agree in their selection. Once they have chosen the pivot column (step 1 of the simplex algorithm), they can all decide on the pivot row (step 2) independently, as well, since that is a calculation dependent only on the pivot column and the d column, both of which exist as carbons in each process. They can then undertake the row operations (step 3) simultaneously, each performing the operations both on its local columns and on the carbons of the globally known columns and the d column.

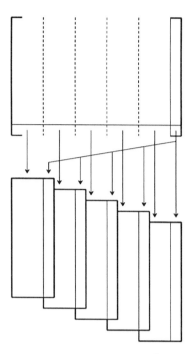

Figure 7: Distribution by columns

At the end of an iteration, some globally known column may still have a positive c value, and the next iteration can proceed immediately without any pause for synchronization. It is only necessary to communicate when the globally known columns have been depleted. We will call the intervals between synchronization **rounds** to distinguish them from **iterations**.

At the start of each round, each process chooses any k columns from the set of private columns having a positive c value. Of these kp columns, any s columns ($s \leq kp$) are chosen by a voting algorithm to broadcast to all processes. (We allow k to be greater than p/s to cover the case in which one process has very few acceptable columns, but another has many.) Each process then has its own n/p private columns, plus the s globally known columns, at most k of which are duplicates of its private ones. A round continues until either a fixed limit on number of iterations has been reached,[6] or no globally known column can be used (all c values are nonpositive).

Selection of the values of k and s involve interesting tradeoffs. If k is too small, it may not be possible to collect s globally known columns with positive c. If k is too large, collecting those columns involves more data transmission than is needed. If s is too small, only one or two iterations might be possible in a round. If s is too large, the cost of each iteration rises (since each process must perform overlapped work on the s globally known columns), and there is no guarantee that even a very large s will allow the round to progress beyond a single iteration.

5.4. Analysis of the columnwise distribution method

We assume that the time to select k columns for broadcast and the time to choose a column for an iteration are both negligible. In each iteration, for some constants c and d,

$$\text{time to choose column} = 0 \qquad \text{— by assumption}$$
$$\text{time to choose row} \quad = dm \qquad \text{— overlapped work}$$
$$\text{time for row operations} \quad = c\ (n/p+s+1)\ (m+1)$$

Let us represent the number of iterations in a round by f. There is no way to predict f, but it is roughly dependent on s. The cost of synchronization between rounds can be represented as $2\lceil \log_2 p \rceil\ (a+bsm)$, using a binary-tree voting algorithm, where a is the per-message cost and b is the per-number cost, as before. We will assume that $1.5m$ iterations suffice to solve the problem, independently of whether we are using the serial algorithm or the distributed one. (This assumption is, of course, not really justifiable; the number of iterations may be quite different.) The number of rounds is $1.5m/f$. The total time taken by the distributed algorithm is then

$$1.5\frac{m}{f}f\ (dm+c\ (\frac{n}{p}+s+1)(m+1))+2\lceil \log_2 p \rceil\ (a+bs\ (m+1))$$

ignoring time for initialization and final output.

In contrast, for the serial method,

$$\text{time to choose column} = 0 \text{— by assumption}$$
$$\text{time to choose row} \quad = dm$$
$$\text{time for row operations} \quad = c\ (n+1)\ (m+1)$$

The total time needed by the serial simplex method is therefore

$$(1.5m\ )(dm+c\ (n+1)(m+1)).$$

The speedup for p processes is the ratio of these two quantities.

We can predict speedup by assigning values to the parameters a, b, c, and d. Typical values for a Lynx program in the Crystal environment are shown here in milliseconds:

$a = 30$ — per-message cost
$b = 0.011$ — per-number (4 bytes) cost
$c = 0.04$ — per-cell cost in row operations
$d = 0.04$ — per-row cost to find pivot row

If we set $k=3$ and $f=6$, we get the expected efficiency curves shown in Graph 8.

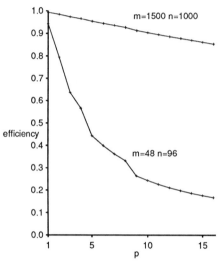

Graph 8: Efficiency of distribution by columns

The problem needs to be substantial in size before it makes sense to use many processes. The steep decline in efficiency for the smaller problem at $p=5$ and $p=9$ are due to the increase in synchronization cost when p exceeds a power of 2; it takes one extra step to vote.

We can see the effect of varying s in Graph 9, in which $p=5$, $m=48$, and $n=96$.

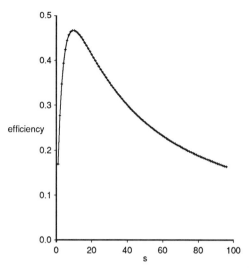

Graph 9: Effect of varying *s*

This algorithm was implemented by Chandrashekhar Bhide, using Lynx on Crystal. His program used a centralized voting strategy, so synchronization was more expensive than our estimates. Table 1 shows the behavior observed with randomly generated problems for $p=2$, *and* $s=k=n/7$. These results are not very promising. The speedup was seldom better than 1, so the efficiency was seldom above 0.5. Part of the reason is that the distributed method often performed more iterations than the serial method. Increasing the value of s generally reduced the number of rounds, with an attendant reduction in running time. However, the effect of s depends on the particular problem being solved and the way s columns are chosen for global replication. Graph 10 shows how the number of iterations, rounds, and total running time varied with s in an experiment run with $m=48$, $n=96$, $k=5$, *and* $p=2$. In comparison, the serial algorithm on the same problem required 144 iterations, which took 109 seconds.

m	n	iterations serial	iterations distributed	speedup
3	6	3	3	.03
5	10	9	8	.08
10	20	18	16	.34
15	30	21	20	.65
20	40	39	50	.69
25	50	63	77	.85
30	60	66	76	1.00
35	70	84	117	.84
40	80	104	113	1.11
45	90	131	164	1.01
48	96	144	199	.92

Table 1: Speedup for $p=2$

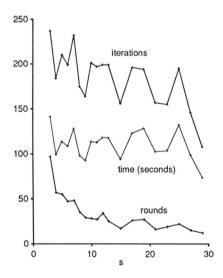

Graph 10: Effect of s

The only clear tendency is that the number of rounds decreases with increasing s, which means that synchronization is not needed so often.

Distribution by columns is a good demonstration of a general phenomenon: In numerical problems, if the order of operations is changed, it is not always possible to predict what effect that will have on the rate of convergence. Our distributed algorithm chooses a different list of pivot columns from that chosen by the serial algorithm. Often, it seems that our choice leads to worse behavior because it increases the number of iterations. Occasionally we get better behavior. We hope that each iteration is faster under the distributed algorithm, because synchronization is needed only between rounds. However, to make rounds last a long time, each iteration is more expensive (since processes overlap work on the s globally known columns).

6. A final word

It is not easy to design distributed algorithms. The methods shown in this chapter often help, as we have seen, but often lead to less than perfectly efficient programs. For numerical problems, it may well be impossible to predict the effect of distribution on convergence rate, so experimentation is necessary to discover where the bottlenecks are and where to make tradeoffs. On the other hand, high-quality algorithms can often be found, as shown by the DIB example.

This chapter has just scratched the surface of the area of large-grain algorithms. Many problems have been explored already, and others are under active investigation. We can look forward to the time when DIB-like packages will provide efficient parallel implementations for a wide variety of problem classes.

Acknowledgements

Research reported in this chapter was supported in part by the National Science Foundation under grants MCS-8105904, MCS-8303134 and DCR-8451397, and by DARPA contract N00014-82-C-2087.

The palindrome study was undertaken by Bryan Rosenburg and William Kalsow. DIB is a joint project of the author and Udi Manber. Chandrashekhar Bhide suggested the two distributed approaches to simplex algorithm and implemented the distribution-by-columns approach in Lynx [13]. I would also like to thank Marvin Solomon, the Charlotte group, and the Crystal group for ideas and suggestions.

Footnotes

[1] The fastest implementations of remote-procedure call require on the order of a millisecond [6]; some of the slower ones require on the order of 80 milliseconds.

[2] The algorithm for distributing digits is somewhat complex. If the iterate has an odd number of digits, the middle virtual process must have the odd digit. In general, virtual processes i and $p+1-i$ must have the same number of digits.

[3] The high value for C was due to some debugging code; communication is not nearly that slow on Crystal.

[4] DIB is written in Modula [26], but we will present the interface in an informal language that bears resemblance to both Modula and to Ada [25].

[5] For simplicity, we assume that our problem has a unique solution.

[6] The number of iterations per round may be limited to prevent infinite cycles.

References

1. N. Alon, A. Barak, and U. Manber, "On disseminating information reliably without broadcasting," Computer Sciences Technical Report #621, University of Wisconsin–Madison (December 1985).

2. G. R. Andrews, "The Distributed programming language SR — mechanisms, design and implementation," *Software — Practice and Experience* **12** pp. 719-753 (1982).

3. G. R. Andrews and F. B. Schneider, "Concepts and Notations for Concurrent Programming," *ACM Computing Surveys* **15**(1) pp. 3-44 (March 1983).

4. K. H. Borgwardt, "Some distribution-independent results about the asymptotic order of average number of pivot steps of the simplex method," *Mathematics of Operations Research* **7** pp. 441-462 (1982).

5. P. M. Cashin, "Inter-process communication," Technical Report 8005014, Bell-Northern Research (June 1980).

6. D. R. Cheriton and W. Zwaenepoel, "The Distributed V Kernel and its Performance for Diskless Workstations," *Proceedings of the Ninth ACM Symposium on Operating Systems Principles*, pp. 128-139 (In *ACM Operating Systems Review* 17:5) (10-13 October 1983).

7. G. B. Dantzig, *Linear Programming and Extensions,* Princeton University Press, Princeton, NJ (1963).

8. D. DeWitt, R. Finkel, and M. Solomon, "The Crystal multicomputer: Design and implementation experience," Technical Report 553 (To appear, IEEE Transactions on Software Engineering) , University of Wisconsin–Madison Computer Sciences (September 1984).

9. C. S. Ellis, "Distributed data structures: A case study," *Proc. 5th International Conference on Distributed Computing Systems*, pp. 201-208 (May 1985).

10. R. Finkel and U. Manber, "DIB — A distributed implementation of backtracking," *ACM TOPLAS*, (to appear) ().

11. R. A. Finkel and J. P. Fishburn, "Parallelism in Alpha-Beta Search," *Journal of Artificial Intelligence* **19**(1) (September 1982).

12. R. A. Finkel and J. P. Fishburn, "Improved speedup bounds for parallel alpha-beta search," *IEEE Transactions on Pattern Analysis and Machine Intelligence* **PAMI-5**(1) (January 1983).

13. R. A. Finkel, B. Barzideh, C. W. Bhide, M-O Lam, D. Nelson, R. Polisetty, S. Rajaraman, I. Steinberg, and G. A. Venakatesh, "Experience with Crystal, Charlotte, and Lynx: Second Report," Computer Sciences Technical Report #649, University of Wisconsin–Madison (July 1986).

14. J. A. Fishburn and R. A. Finkel, "Quotient networks," *IEEE Transactions on Computers* C–**31**(4) pp. 288-295 (April 1982).

15. C. A. R. Hoare, "Communicating sequential processes," *CACM* **21**(8) pp. 666-677 (August 1978).

16. D. E. Knuth and R. W. Moore, "An analysis of alpha-beta pruning," *Artificial Intelligence* **6**(4) pp. 293-326 (Winter 1975).

17. H. T. Kung and C. E. Leiserson, "Systolic Arrays (for VLSI)," *Symposium on Sparse Matrix Computations*, pp. 256-282 (Duff, I. S. and Stewart, G. W., editors) (November 1978).

18. E. L. Lawler and D. Wood, "Branch and Bound methods: a survey," *Operations Research* **14**(4) pp. 699-719 (1966).

19. B. Liskov and R. Scheifler, "Guardians and actions: Linguistic support for robust, distributed programs," *ACM TOPLAS* **5**(3) pp. 381-404 (July 1983).

20. M. L. Scott and R. A. Finkel, "A simple mechanism for type security across compilation units," Computer Sciences Technical Report #541, University of Wisconsin–Madison (May 1984).

21. M. L. Scott and R. A. Finkel, "LYNX: A dynamic distributed programming language," *1984 International Conference on Parallel Processing*, (August, 1984).

22. M. L. Scott, "A framework for the evaluation of high-level languages for distributed computing," Computer Sciences Technical Report #563, University of Wisconsin–Madison (October 1984).

23. M. L. Scott, "Lynx reference manual," BPR 7, Computer Science Department, University of Rochester (March 1986).

24. R. E. Strom and S. Yemini, "NIL: An integrated language and system for distributed programming," *Proceedings of the SIGPLAN '83 Symposium on Programming Language Issues in Software Systems*, pp. 73-82 (In *ACM SIGPLAN Notices* 18:6 (June 1983)) (27-29 June 1983).

25. United States Department of Defense, "Reference Manual for the Ada Programming Language," ANSI/MIL-STD-1815A-1983 (February 1983).

26. N. Wirth, "Modula: A language for modular multiprogramming," *Software — Practice and Experience* **7**(1) pp. 3-35 (1977).

Characterizing Parallel Algorithms

Leah H. Jamieson

School of Electrical Engineering
Purdue University
West Lafayette, Indiana 47907

One of the significant problems that must be addressed if we are to realize the computing potential offered by parallel architectures has to do with developing a better understanding of the relationship between parallel algorithms and parallel architectures. In this paper, research on the mapping of algorithms to reconfigurable parallel architectures is presented. The thrust of this work is in identifying those characteristics that have the greatest effect on the execution of parallel algorithms and in identifying a correspondence between those characteristics and the characteristics of parallel architectures.

1. Introduction

In both parallel architectures and parallel algorithms, there exist many design choices for which there are no direct counterparts in conventional serial processing. In architectures, examples of such choices include number of processors, local versus global memory organization, synchronous versus asynchronous execution, and interconnection topology. In parallel algorithms, issues that do not arise in serial programming include determination of the number of processors needed/useful for a task, data allocation across memories, synchronization beneficial or necessary, and interprocessor communication requirements. The move to parallelism has introduced new degrees of freedom to both the architecture and algorithm design process. For effective use of parallel systems, it is essential to obtain a good match between algorithm requirements and architecture capabilities.

Information that captures the relationships between parallel algorithms and parallel architectures can be of use in a number of different ways:

- Algorithm-to-architecture mapping information bears directly on the algorithm design process. General knowledge about what constitutes an effective match between a parallel algorithm and a parallel architecture can accelerate the process of developing new parallel algorithms for a given machine.

- An understanding of the relation between algorithms and architectures is a prerequisite for the fast, efficient design of algorithmically-specialized systems (e.g., see [26-28, 48, 61, 75, 82]). Given a fixed set of algorithms, architectures tailored for the execution of those algorithms can be developed if the architectural requirements of the algorithms are understood. A set of algorithms, characterized by the union of their attributes, will guide the design decisions for the algorithmically-specialized parallel architecture.

- A means of measuring the "goodness" of the match between an algorithm and an architecture can provide the capability to predict the performance of a particular algorithm on a particular architecture.

- A general method of relating algorithms and architectures will allow efficient use of reconfigurable parallel systems. Systems that allow the machine configuration to vary as a function of the current algorithm and the current system state (e.g., resources available) are being designed (e.g., PASM [68], CHiP [74]). Integral to the effective use of these flexible parallel systems is the ability to select machine configurations based on knowledge about the algorithms to be executed. In order to accomplish this automatically, the operating system will need to use information about the characteristics of the algorithms to select successive configurations of the parallel architecture.

- A classification of algorithms based on characteristics that span the range of structures typically seen in parallel algorithms can be used to construct and evaluate benchmark sets [29]. The characterization of algorithms can therefore play an important role in devising ways to evaluate parallel architectures.

There are a number of approaches to the mapping of parallel algorithms to parallel architectures. The work by Li, Wang, and Lavin [50] and O'Dell and Browne [56] provide examples of language-based approaches to the problem of relating algorithms to architectures. Two powerful (and related) approaches that have been applied to a variety of architectures are mappings based on graph transformations (e.g., work by Bokhari [9], Berman and Snyder [8], Browne [11], Chiang and Fu [15], Fishburn and Finkel [31], Kuhn [44], Schwartz and Barnwell [65]) and geometric approaches based on linear transformations (e.g., work by Cappello and Steiglitz [14], Fortes and Moldovan [32]). The strength of these methods is that they employ the formalisms of graph transformations or geometric representations to address specific aspects of the mapping problem: process-to-processor assignment, minimization of communications, folding of large problems onto smaller arrays of processors. Much recent work has addressed the specific mapping problem having to do with the design of systolic arrays (e.g., work by Cappello and Steiglitz [14], Dewilde, Deprettere, and Nouta [24], Fortes and Moldovan [32], Frison, Gachet, and Quinton [33], Jagadish et al. [38], Kung [46]). For the application domain of image processing, relationships between broader characteristics of parallel algorithms and architectures have been explored by Swain, Siegel, and El-Achkar [77], Cantoni and Levialdi [13], Etchells and Nudd [29], and Aggarwal and Yalamanchili [2]. These approaches, and the work presented here, focus on identifying a wide range of characteristics of algorithms and relating these characteristics to classes of architectures or architecture configurations.

The rationale behind the "characteristics-based" approaches is that many algorithms do possess an identifiable structure. For example, at the data dependency level, many algorithms share similar communications patterns. At the process level, algorithms based on the same paradigm -- e.g., divide-and-conquer -- may exhibit similar communications requirements. Algorithms that operate on similar data structures may lend themselves to execution on similar architectures. Allen has observed [3] that, for example, in the area of digital signal processing, it is possible to identify a set of canonical algorithm structures (e.g., second order section, FFT, autocorrelation, convolution) and that algorithms that can be expressed in terms of these structures can be constructed from a set of basic building blocks (e.g., multiplications, complex multiplications, butterflies, sums-of-products, address arithmetic). A concise description of the algorithm in terms of a basic set of features allows selection of an appropriate machine configuration and can facilitate the mapping process by relating the characteristics of the current algorithm to known layout patterns.

In order to identify patterns within a specific characteristic (e.g., in communications) a number of techniques, including graph-based and geometric, can be applied. In the graph-based mapping systems mentioned above, graph transformations are used to map a graphical representation of the algorithm onto a graphical representation of the architecture. In the characteristics-based approach, graph transformations can be used to recognize patterns in the graphical representation of the algorithm, for example, by searching for isomorphisms between the data dependencies in the current algorithm and a stored library of communications patterns. For each such pattern, the library provides guidance for mapping the algorithm onto possible architecture configurations. The graph-based approach is in some sense more general in that it does not depend on the existence or recognition of patterns in the algorithm (e.g., in the algorithm's data dependency/communications topology structure). However, it does not necessarily lend itself to exploiting such patterns when they do exist. In the absence of identifiable patterns, the characteristics-based approach will in fact fall back on mappings such as those provided by the graph-based or geometric approaches. Conversely, it may be that information gained from the characteristics analysis can be incorporated into language-, graph-, or geometric-based systems. The approaches therefore complement each other.

The focus of this paper is on examining the characteristics that have the greatest effect on the execution of parallel algorithms. In the next section, we define a framework in which parallel algorithms can be discussed. We then identify a set of algorithm characteristics that relate to the mapping of parallel algorithms to parallel architectures. Finally, we present two examples that demonstrate various aspects of the relations between algorithm characteristics, parallel architectures, and performance.

2. The Virtual Algorithm Model

Although there is no formal notion of the scope of an algorithm in terms of how "complicated" a problem a single algorithm solves, it will be convenient to assume that an *algorithm* performs a single well-defined function, and that a *task* is performed by execution of a collection of algorithms. From an applications point of view, the objective will always be to perform a task, with the selection/design of algorithms and architectures being the means by which the task is accomplished. Thus a task might be "Identify an object in a noisy image," with component algorithms including filtering, edge detection, contour tracing, scale and orientation normalization, computation of a Fourier descriptor, and template matching. Clearly there are many issues relating to the interaction of the algorithms that comprise a task; however, our focus here will be on the individual algorithm. Future work will attempt to take into account the aggregate requirements of a set of algorithms.

In order to establish at what point an algorithm is in some sense "bound" to an architecture, the "algorithm development life cycle" shown in Fig. 1 is assumed. Using the area of digital signal processing for the purpose of illustrating the components of the graph, an example of a *problem statement* might be "Obtain a spectral estimation of a given signal." An *algorithm approach* might be the decision to use an FFT to obtain the spectral estimation, without specifying any details of the FFT implementation. The *virtual algorithm* represents the point at which the computational steps to be performed have been determined, but the mapping of those steps to an architecture has not yet been committed. In the signal processing example, this might correspond to the decision to use a radix-2 decimation-in-time formulation of the FFT. The virtual algorithm might be represented in terms of a dataflow or signal flow graph, or an operator net [4]. The virtual algorithm representation contains information from which a data dependency graph can be derived and from which the theoretical maximum parallelism can be determined. The *ideal algorithm* will correspond to the parallel algorithm that would be selected if no

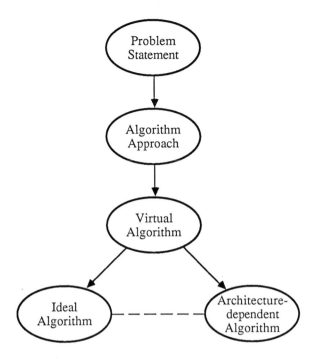

Fig. 1. Algorithm development life cycle.

constraints were placed on the architecture configuration. In general, this will be the best known parallel implementation of the virtual algorithm. In the example, for computation of an N-point FFT and assuming speed as the performance criterion, this might be a pipelined $N/2$-processor, $\log_2 N$-step SIMD algorithm using a perfect shuffle or cube-type interconnection network [40, 58, 76], with internal parallelism and pipelining in the butterfly computational units [78]. The ideal algorithm may in fact never be used. However, it may suggest strategies for implementation of the algorithm and it provides a reference point against which architecture-dependent algorithms can be compared. This potential (but possibly indirect) relationship between the ideal algorithm and architecture-dependent algorithms is indicated by the dashed arc in Fig. 1. The *architecture-dependent algorithm* represents the binding of the algorithm steps to a particular architecture. In the signal processing example, this would include decisions about the number of processors, synchronization, data allocation, and inter-processor transfers to be used.

The model shares some basic ideas with other work that has dealt with the algorithm-to-architecture mapping problem. Browne [10, 11] has characterized the process of mapping an algorithm (a program for an "abstract machine'') to an architecture in terms of the action of binding values to parameters, and distinguishes between types of architectures in terms of when this binding is performed. In the context of silicon compilation, Denyer makes the distinction between "behavioral specification" and "structural specification" [22]. Dewilde and Annevelink describe the mapping of an algorithm onto a hardware architecture as the translation from a semantic level, which constitutes an input/output specification of the function to be performed, to a structural level, in which a signal flow graph is used to represent the operations performed and the timing relations between the operations [23]. In terms of the virtual algorithm model, the problem statement and algorithm approach represent behavioral or semantic descriptions, while the virtual algorithm (and certainly the architecture-dependent algorithm) represent structural descriptions.

In order to solve the "task-to-architecture" mapping problem in its entirety, all five points in the life cycle graph must be addressed. Thus, in the context of performing a task, a number of different algorithm sets may be possible, corresponding to a number of distinct *problem statements* for the possible component algorithms. Given a *problem statement,* the selection of an *algorithm approach* and *virtual algorithm* will necessarily depend largely on the specific application area. A true representation of the *task* solution space replicates the algorithm development life cycle for each algorithm in a number of candidate sets of algorithms (representing the alternative approaches to performing the task), but also allows branching at the point of choosing the *algorithm approach* (e.g., alternative algorithms for obtaining a spectral representation) and at the point of specifying the *virtual algorithm* (e.g., alternative FFT formulations). This branching within a given problem statement is shown in Fig. 2. For solution of a task, such a graph would be replicated for each of the many different problem statements that could serve as components to the solution of the task. In addition, when considering a sequence of algorithms that comprise a task, not only the basic architecture, but the current system state, plays a part in determining an appropriate architecture-dependent algorithm. Factors such as the allocation of data at the junctures between individual algorithms and the combined resource requirements for the set of algorithms that will execute concurrently may influence decisions made about the mapping of an individual algorithm onto the architecture. The full "task-to-architecture" problem is therefore a very complex one.

In this paper, we focus our attention on the transition from a *virtual algorithm* to an *architecture-dependent algorithm*. At the point of specification of the virtual algorithm, the computational steps to be performed have been determined, but the specific structure of the implementation -- the allocation of data, the assignment of computational steps to processors -- has not yet been defined. It is the mapping from the virtual algorithm to the architecture-dependent algorithm that will determine the actual execution

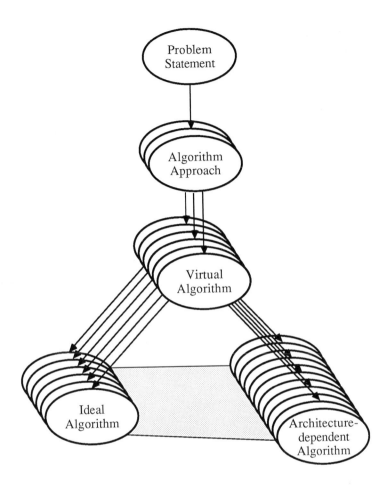

Fig. 2. Algorithm development life cycle:
branching from a single problem statement.

profile. Using the virtual algorithm as our starting point, we identify characteristics of algorithms and architectures that bear on the expected performance of architecture-dependent algorithms.

3. Architecture Model

We assume a target parallel architecture with the general attributes
listed below. The purpose of the assumptions is to define a very general architecture framework, so that no undesirable restrictions are imposed on the

architecture by the assumptions. Rather, by examination of the algorithm attributes, the subset of capabilities required in the architecture can be selected. Thus the framework is more general than most existing parallel systems, but many architectures, including PASM [68, 71], Ultracomputer [35], RP3 [59], the Cosmic Cube [66], the Butterfly [18], the Connection Machine [36], MPP [5], Clip 4 [25], CHiP architectures [74], and pyramid architectures [79, 81], can be characterized in terms of subsets of these design attributes. The principal classes of architectures not covered by the assumptions are dataflow systems and heterogeneous architectures.

Assumptions for the target architecture are as follows:

- The system consists of a large number of homogeneous processors. In mapping the algorithm to an ideal parallel architecture, the number of processors available is considered to be unlimited.

- The system can be organized with processors accessing a shared global memory or with each processor having an associated local memory, or a hybrid of the two approaches.

- The system is partitionable into independent submachines of various sizes. The partitioning can be changed dynamically at execution time. For execution of a single algorithm, this assumption merely states that the machine size can be selected to meet the needs of the algorithm. In the more general case of considering a set of algorithms comprising a task, the partitioning assumption means that a number of algorithms can be executing simultaneously, each on a different submachine.

- Each partition of the system (and the entire system itself) is capable both of SIMD and MIMD operation, and can dynamically switch between modes during execution.

- The system has a flexible interconnection network that can provide a wide variety of communications patterns within each partition. In any specific instance of mapping a particular algorithm onto a particular architecture, details of the interconnection network will be of importance. For example, in a multistage network (e.g., [30, 47, 58, 69, 70]), any single point-to-point data transfer can be done in one pass through the network, regardless of the "distance" between the source and the destination, and many permutations can be performed in a single network pass (e.g., see [1]). In such a network, communications distance is largely irrelevant, but the *number* of data transfers may be a significant factor in the algorithm's performance. In contrast, in systems in which communications time is proportional to communications distance (e.g., [5, 66, 74]), both number of transfers and distance will be important. When a system with specific communications capabilities is being considered, such information can be used in conjunction with the algorithm characteristics in performing the algorithm-to-architecture mapping [41].

Given the above assumptions, the problem becomes one of selecting the architecture configuration -- memory organization, partition size, mode of operation, network configuration -- that best matches the attributes of the algorithm.

4. Algorithm Characteristics

Parallel algorithms can be characterized in many ways, along a number of dimensions (e.g., see [29, 45, 77]). Similarly, parallel architectures can be described in terms of a number of attributes (e.g., see [19, 37, 73]). Ideally, a set of orthogonal characteristics would describe parallel algorithms and a corresponding set of orthogonal characteristics would describe parallel architectures, with a unique bijection performing the mapping from one to the other. Experience shows that the relationship between parallel algorithms and parallel architectures is clearly too complex to conform to such a desirable model. In the absence of independence, completeness therefore becomes the relevant goal. Using the application areas of image, speech, and signal processing as our frame of reference, we have identified a set of characteristics that appear to capture the salient attributes of parallel algorithms in their relationships to parallel architectures [39]. These features were derived from the design and analysis of over 50 specific algorithms from the selected application domains (see [71] for a bibliography), and refined by the iterative process of applying the preliminary characteristics sets to new algorithms. Table 1 summarizes some of the relationships between these algorithm characteristics and a set of parameters that describe the principal components of reconfigurable parallel architectures. A "1" entry in the table indicates a probable primary dependence between the algorithm and architecture characteristics, an entry of "2" indicates a likely secondary dependence and "3" denotes a less strong dependence. Other relationships can certainly hold; however, for a wide range of algorithms, these appear to characterize many of the algorithm-architecture dependencies.

1. *Nature of the parallelism.* Under this heading come a number of attributes having to do with the "kind" of parallelism that is used and the the way in which the algorithm and/or data can be decomposed.

    a. *Data parallelism versus function parallelism.* Parallelism can be achieved by dividing the data among the processors, by decomposing the algorithm into segments that can be assigned to different processors, or by pipelining (which is a special case of decomposition of the algorithm into functional segments). The type of parallelism will

Table 1. Relationship Between Algorithm and Architecture Characteristics

Architecture Characteristics

Algorithm Characteristics	Number of PEs	Memory Organization	Memory Size	Mode (SIMD/MIMD/Pipe)	Network	Synchronization	Processor Capability	Data Types	Addressing Modes	Data Structures	I/O
Type of Parallelism	3	2	3	1		3					
Data Granularity		1	2	1	3					2	
Module Granularity		1		1	1	1					
Degree of Parallelism	1	2		2							
Memory	2	2	1								
Uniformity		2		1	3						
Synchronization		2		1	2	1					
Static/Dynamic		1		1	2	3					
Data Dependencies		2			3	1					
Fundamental Ops.		2			2		1				
Data Types					3			1	2		1
Data Structures		2		3	2				3	1	
I/O		3	3	2	2						1

affect the allocation of data, the assignment of processes to processors, and the basic decision as to what mode of parallelism (SIMD/MIMD/pipeline) to use. Function parallelism will almost always imply MIMD operation. Data parallelism will often be amenable to SIMD implementation; however, data parallelism alone is not sufficient to guarantee good performance in SIMD mode. At some level of decomposition, an algorithm must exhibit both data and function parallelism in order for pipelined operation to be applicable. Memory organization can be tied to the type of parallelism in that data parallelism is often well implemented using local memories. The type of parallelism will also act as a broad indicator of the number of processors. With data parallelism, the number of processors will typically be proportional to the data set size, and utilization of large numbers of processors will not be uncommon. Algorithms based on function parallelism will more typically use numbers of processors counted in the tens rather than the thousands.

b.   *Data granularity.* Data granularity deals with the "size" of the data items processed as a fundamental unit, and will have a bearing on the data allocation, communications requirements, processor capability, and memory requirements. Fine-grain algorithms will often be suitable for SIMD or pipelined implementation using local memories. The data granularity generally will not affect the overall memory requirements, but may bear on the amount of memory that must be readily accessible to each processor. The data granularity will provide an indication of the bandwidth needed to communicate a single data item.

c.   *Module granularity.* Module granularity [45] quantifies the amount of processing that can be done independently, either of other processes or of operations being performed in other processors. It is essentially a measure of the frequency of synchronization, and will affect the choice of SIMD versus MIMD operation, the assignment of processes to processors, the memory organization, the communications requirements, and the likelihood of equalizing the execution times of component parts of the algorithm. Algorithms characterized by fine-grain module granularity will require frequent synchronization. If possible (based on the other attributes of the algorithm), SIMD or pipelined execution, in which communications can be performed with less overhead than in MIMD operation, will be preferred, as will a local memory organization. Because of the frequent communications, a fast network capability is imperative. Large-grain algorithms typically have less of a need for efficient communications. The large amount of processing done between synchronization points often suggests MIMD execution.

2. *Degree of parallelism.* This will be related to both the data granularity and the module granularity. Its most direct impact will be on the choice of machine size and on the maximum speedup attainable. In addition, the degree of parallelism will in practice often be related to the mode of operation (massive parallelism will rarely be associated with MIMD execution) and the memory organization (with massive parallelism, global memory organization can lead to significant contention in accessing memory).

3. *Uniformity of the operations.* If the operations to be performed are uniform (e.g., across the data or feature set), then SIMD or pipeline processing may be feasible. Uniformity will generally be associated with data parallelism. It will be possible to talk about different levels of uniformity, depending on the granularity or resolution at which operations are being examined. For example, Bayesian classification of the pixels of an image is a highly uniform operation in that the same computations are performed on each pixel of the image. However, implementations of the floating point arithmetic may operate in a data-dependent fashion, producing non-uniform execution at this finer level. If the operations are not uniform, then MIMD processing will be chosen and strategies to equalize the computational load across the processors may come into play. These strategies may be applied statically at compile time or dynamically at execution time.

4. *Synchronization requirements.* In addition to the synchronization requirements implied by the module granularity, consideration of precedence constraints is implicit in characterizing the synchronization requirements. This will affect the assignment of processes to processors and the scheduling of various components of the algorithm.

5. *Static/dynamic character of the algorithm.* The pattern of process generation and termination will affect the processor utilization, the scheduling of sub-processes, the mode of processing, the memory organization, and the communications requirements. Identification of an algorithm as being dynamic generally rules out SIMD or pipelined execution. A dynamic algorithm will need to be supported by either a global memory organization or a communications network and/or an I/O system capable of providing a fast means of loading the local memories with the data and code needed for a new process.

6. *Data dependencies.* The data dependencies in an algorithm will play the largest role in dictating data allocation patterns and communications characteristics. They will also have a major part in the decision to use a global versus local memory organization. Preparata discusses the desirability of classifying algorithms in terms of their communications patterns [60]. Work on characterizing communications with respect to a number of

parallel architectures has been reported by Levitan [49]. Existing algorithms provide a wealth of information about specific mappings. Much of this experience can be incorporated into a library of known mappings.

In the characteristics-based approach, the handling of data dependencies is conceptually a graph isomorphism problem. The library contains known data-dependency structures and (potentially multiple different) mappings of these structures onto architecture configurations. The purpose of the stored information is to make available mappings which experience has shown to be useful but which might be difficult or prohibitively time-consuming to derive directly. Although this approach uses the intermediary of the stored database of patterns, the major steps involved in mapping the data dependency graph for the current algorithm onto one of the stored structures are similar to those employed by Berman and Snyder for mapping an algorithm directly onto an architecture [8]. We will assume that both the data dependencies of the algorithm and the communications topology of the architecture are represented as graphs. The first step in the mapping process deals with the size of the problem. In some algorithms, e.g., the neighborhood operation of convolution with a 3×3 window, the data dependency structure is effectively independent of the problem size: the data needed to compute a particular output pixel is independent of the size of the input image. Such structures can be represented in terms of the basic data dependency patterns. In other algorithms, e.g., an FFT, the communication required is size-dependent: a 16-point FFT needs 16 input values to compute one output value; a 1024-point FFT need 1024 input values. The structure of the communications may be regular, but the specific communications required will depend on the problem size. In this case, the library may contain construction rules for the communications graph rather than storing a single pattern, as is sufficient for the size-independent structures, or storing the communications graphs for all possible sizes. The second step in the mapping process deals with the communications topology, and maps the nodes of the algorithm's data dependency graph to nodes in the architecture graph. In the characteristics-based approach, this step is performed by comparing the data dependency structure of the algorithm to the library of structures. A broad outline of one possible approach is as follows:

1. Classify the data dependency structure of the algorithm as size-independent or size-dependent.

2. If the structure is size-dependent, measure the size S of the input problem.

3. For each stored graph structure:

   a. If the algorithm's data dependency structure is size-dependent, generate a template graph of size S using the construction rules for the stored structure.

    b.   Test for isomorphism with the algorithm's data dependency graph.

4.   Once an isomorphism is found, say for pattern P, use the associated mapping information stored in the library. This information may include specific assignment of algorithm nodes (operations, processes) to architecture nodes (processors) and strategies for handling any remaining mismatches between problem size and architecture size (e.g., [31]).

5.   If no isomorphism is found, then graph transformations such as those presented by Berman [7], Bokhari [9], Chiang and Fu [15], DeGroot [20], Fortes and Moldovan [32], and Kuhn [44] can be used. An alternative approach in the absence of an exact isomorphism may be to explore techniques for identifying the "closest" match.

Although the identification of a communications pattern can be formulated conceptually as a graph isomorphism problem, the complexity of graph isomorphism will in most cases make the direct and exhaustive test for isomorphism infeasible. Therefore an issue of interest is the organization and representation of the stored structures so that the searching/matching process can be performed efficiently. Different graph isomorphism algorithms have been shown to be preferable depending on whether it is likely that the two graphs being compared are or are not isomorphic [63]. The search process may be assisted by the inclusion of auxiliary information with each of the stored patterns. Characterization in terms of information such as minimum/maximum/average degree of each node in the graph, measures of "distance" of communication or data dependencies, and classification into broad categories (e.g., neighborhood dependencies, global dependency) may allow faster convergence to the correct stored pattern (or determination that no such pattern exists in the library). In some cases it may also be possible to employ linear-time algorithms that have been developed for special classes of graphs (e.g., see [53]). Characterizing the data dependencies lies at the heart of the algorithm-to-architecture mapping problem, and there are many approaches to be explored.

7.   *Fundamental operations.* The basic operations performed in the algorithm will dictate the processor capabilities needed. To the extent that the operation identified as being the basic unit of processing also determines some communications requirements, this characteristic will also have a bearing on the network and memory organization.

8.   *Data types and precision.* The atomic data types and data precision will bear most directly on the individual processor capability and on the memory requirements, but may also imply requirements for

communications bandwidth.

9. *Data structures.* Many algorithms can be characterized as having a "natural" data structure (or structures) on which operations are performed. The ability of an architecture to support the needed access patterns, to exploit possible regularity in the structures, and to allow the needed interactions between parts of the structures will affect algorithm performance.

In using the algorithm characteristics, it is necessary to make a distinction between an attribute that is required for a particular architecture implementation and one that is preferred. For example, uniformity is a requirement for SIMD processing: although it is possible to construct an SIMD implementation of a non-uniform algorithm (e.g., by cycling through the processors and executing one instruction from each while disabling all of the other processors), performance will be so bad that it is not a reasonable implementation to consider. In contrast, if an algorithm has a high degree of uniformity, then, depending on its other attributes, it may be that an SIMD implementation is preferable, but an MIMD implementation may run only slightly more slowly.

5. Examples

In this section, two examples are presented. Both are aimed at providing, in an informal way, an affirmative answer to the question *"Do the algorithm characteristics capture the aspects of the virtual algorithm that will dictate the performance of architecture-dependent algorithms?"* The first example illustrates the range of effects that a single virtual algorithm characteristic (in the example, data granularity) can have on the choice of architecture configuration. The second example examines the complete set of algorithm characteristics for a particular virtual algorithm, and looks at how the implications of these characteristics relate to a particular architecture-dependent algorithm.

5.1. Data Granularity and Dynamic Time Warping

Dynamic time warping (DTW) is a technique widely used in word recognition to eliminate the effects of nonlinear time fluctuations in speech patterns [64]. It forms a part of a template-matching operation in which an input utterance, represented by a time sequence of parameter vectors, is compared to stored templates that represent the training utterances for the system. DTW uses dynamic programming to obtain an optimal time alignment between the input utterance and a stored template. The matching process is performed for each word in the stored vocabulary; the input utterance is identified as being the vocabulary word corresponding to the template for which the minimum distance is obtained (or the input utterance is rejected if all distances exceed some threshold).

Fig. 3a sketches the major aspects of the DTW computation. The horizontal axis represents time in the vocabulary word (template), with $\vec{a_i}$ being the feature vector that represents the i-th time segment of the utterance, $1 \leq i \leq I$. If each time segment (typically 10 to 40 ms) is represented by f features (typically 4 to 16), then $\vec{a_i} = a_{1,i}, a_{2,i}, ..., a_{f,i}$. The vertical axis represents time in the input utterance, with $\vec{b_j}$ representing the j-th time segment, $1 \leq j \leq J$, and $\vec{b_j} = b_{1,j}, b_{2,j}, ..., b_{f,j}$. Dynamic programming is used to find a path from $(1,1)$ to (I,J) for which the cumulative distance C is minimized. Such a path is shown by the line that meanders from $(1,1)$ to (I,J) in Fig. 3a. Portions with slope greater than 1 (e.g., in the first quarter of the word) roughly correspond to sounds that "lasted longer" in the input word than in the template; portions with slope less than 1 (e.g., at the end of the word) correspond to sounds that were more prolonged in the template. The cumulative distance $C(i,j)$ is a weighted sum of the local distance $d(i,j)$ between the feature vectors $\vec{a_i}$ and $\vec{b_j}$ combined with the cumulative distance that represents the minimum-cost path that leads from point $(1,1)$ to point (i,j). $C(i,j)$ will therefore typically be a function of $C(i-1,j)$, $C(i,j-1)$, $C(i-1,j-1)$, and $d(i,j)$; computation of the $C(i,j)$'s will proceed in diagonal waves, from the lower left corner $(1,1)$ to the upper right corner (I,J), with the C's computed in wave k being a function of the C's computed in wave $k-1$. Typical values for the number of waves range from 20 to 60.

The above description outlines the principal components of DTW algorithms. A number of additional constraints derived from the characteristics of speech can be applied to the DTW computation. One such constraint is the use of an adjustment window, spanned by w in Fig. 3b. The window restricts the domain of the time warp to those (i,j) that lie within the width of the window, prohibiting time differences between the input word and template that are "too large." Typical values for w range from 12 to 20. A second technique, pruning, aborts a DTW match as soon as the minimum cumulative distance exceeds a preset threshold.

Myers et al. have reported that dynamic time warping accounts for from 50 to 90% of the computation time in word recognition on a serial computer [55]. Because the number of DTW matches performed is proportional to the number of stored templates, the time needed for DTW matching can limit the size of the vocabulary that can be used in a word recognition system. Parallel algorithms and architectures for DTW are therefore of interest for real-time speech recognition applications (e.g., [12, 84]). The DTW algorithm has three natural data granularities: *large-grain,* in which the basic data unit is a complete stored template, *medium-grain,* in which the basic unit is one wave of one DTW match, and *fine-grain,* in which the basic unit is one vector distance. We explore the effects that this single virtual algorithm characteristic

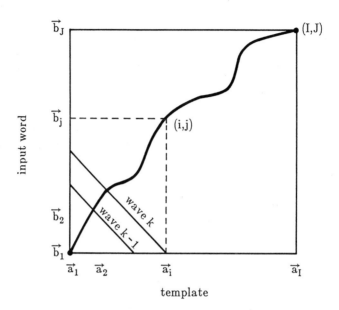

Fig. 3(a) Dynamic time warping.

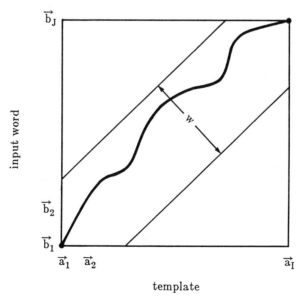

(b) DTW adjustment window w.

can have on the ideal and architecture-dependent DTW algorithms.

5.1.1. Large-Grain Implementation

In the large-grain algorithm, the basic data unit is a complete stored template, representing one vocabulary word. The natural implementation associates one template with each processing element (PE) and performs one DTW match in each PE, executing a serial DTW algorithm to compare the input word to the template. After all of the DTW matches have been performed, the template yielding the minimum cumulative distance is selected. This virtual algorithm dictates a number of architecture requirements. The processing implied by this choice of data granularity also has an effect on some of the other attributes of the virtual algorithm, which in turn affect the architecture requirements.

- *Number of processors:* The number of PEs will be less than or equal to the size of the vocabulary (i.e., the number of templates): for a V-word vocabulary, $\#PEs \leq V$. It is not possible to use more PEs.

- *Processor capability:* Each processor must be able to execute the complete dynamic programming algorithm, including the computation of the local vector distances ($d(i,j)$'s) and the cumulative distances ($C(i,j)$'s).

- *Memory organization:* The logical association in this algorithm is of one template with each processor. This is most naturally implemented using local memories, but can be implemented either using local memories or a global memory. However, if a global memory is used, contention in accessing the memory may degrade performance. If the logical global memory is implemented by a number of physical memory modules with access from processors to the memories accomplished through a routing network (e.g., Ultracomputer [35]), the memory access time may be longer than with a direct association of a local memory with each processor. If local memories are used, then communications will be required at the end of the processing in order to collect the distance scores and select the template yielding the minimum distance.

- *Memory requirements:* In a local memory implementation, each PE memory must hold one complete template and the input word. In a global memory implementation, total memory requirements are for the set of templates and the input word. In both cases, an additional small work space is needed.

- *Communications:* In a local memory implementation, it will be necessary to collect results from all of the PEs. Recursive doubling [43] provides the canonic structure for this. Since this step is performed only once per input word, network capability to support optimal collection of results may not be critical.

The use of the large grain size does not directly dictate the mode of parallelism in this case. However, the large grain size does have a potential effect on the uniformity of the operations performed in the algorithm, and the uniformity characteristics combined with the sparse synchronization requirements do impact the choice of mode.

- *Uniformity:* In a broad sense, the algorithm is highly uniform: each of the processors is performing the same serial DTW algorithm, using a different reference template. Much of the structure of the algorithm is data-independent, i.e., the processors perform the same steps, regardless of the values of their individual data elements. There are three major sources of potential non-uniformity in the algorithm: (1) Both fixed and floating point implementations of DTW are possible. In floating point implementations, the time to compute the local distances ($d(i,j)$'s) may vary, depending on the individual $\vec{a_i}$ and $\vec{b_j}$ values. (2) Computation of the cumulative distance for point (i,j) involves selecting the path from one of the preceding points, typically $(i-1,j)$, $(i,j-1)$, or $(i-1,j-1)$, basing the selection on the minimization of a weighted cost function. For the same (i,j) values, matches to different templates, being performed concurrently in different processors, may select different predecessor points. Simulation has shown that this local non-uniformity can have a significant effect on execution time if masking is used so that only processors selecting the same predecessor point are active at the same time [83]. (3) In serial DTW algorithms, pruning of DTW matches as soon as the minimum cumulative distance exceeds some threshold can sometimes provide significant savings. In parallel implementations, at least one of the matches (the "successful" match) must go to completion. If V matches are performed in V processors, pruning will therefore have no effect on the execution time. However, if fewer processors, say N, are used, with $\lceil V/N \rceil$ matches performed sequentially in each processor, then pruning may reduce the overall execution time. In this case, it may be useful to examine strategies for distributing the templates among the processors in such a way that pruning occurs with approximately the same frequency in all of the processors. Such strategies will depend on the speech characteristics of the vocabulary.

- *Synchronization:* The only synchronization required in the large-grain parallel DTW algorithm is for the collection of results (i.e., selection of the minimum distance) at the end of processing.

- *Mode:* Because of the possible non-uniformities and the minimal synchronization, MIMD implementation will potentially be preferred.

5.1.2. Medium-Grain Implementation

In the medium-grain implementation, the basic data unit is one wave of one DTW match. All of the cumulative distances along the diagonal wave k can be computed concurrently, and must follow computation of the distances in wave $k-1$. Most systolic array algorithms for DTW (e.g., [12, 84]) have been of this type.

- *Number of processors:* If an adjustment window of width w is used, then w processors can be used to compute the w $C(i,j)$ values in wave k. This can be combined with the large-grain approach for a maximum degree of parallelism of wV for a V—word vocabulary.

- *Processor capability:* Each processor must be able to compute local vector distances ($d(i,j)$'s) and cumulative distances ($C(i,j)$'s).

- *Memory requirements:* At any point in the algorithm, each processor must have immediate access to two f-element feature vectors plus the cumulative distances computed for the predecessor points.

- *Communications:* The cumulative distances for the predecessor points (e.g., $C(i-1,j)$, $C(i,j-1)$, $C(i-1,j-1)$) must be communicated to the processor in which $C(i,j)$ is being computed. Typically this will entail local near-neighbor communications. If the feature vectors are not stored locally in the PEs, communications to access or distribute these vectors will be needed. In addition, at the end of processing, it will be necessary to collect the final distance scores from all of the PEs, as in the large-grain case.

- *Synchronization:* Synchronization is required for each wave, to ensure that the values needed from the previous wave are available. Specific synchronization requirements accompany the communications requirements. Synchronization is therefore much more frequent that for the large-grain case.

- *Mode:* Because of the frequent synchronization and the regularity of the computations, SIMD or pipeline processing is attractive. The potential non-uniformities are the same as for the large-grain case. However, these must now be weighed against the much more stringent synchronization requirements.

5.1.3. Fine-Grain Implementation

In the fine-grain implementation, the basic unit of computation is viewed as a single element of a feature vector, and the basic computation performed is the computation of one term in the vector inner product for local distance $d(i,j)$. The parallelism achieved by dealing with the individual feature values can be combined with the wave-level and word-level parallelism of the medium- and large-grain approaches.

- *Number of processors:* For the vector computations, the degree of parallelism is f, the number of elements in the feature vectors. Combining the fine-grain parallelism with the wave-level parallelism gives degree fw; combining it with both the wave-level and word-level parallelism gives fwV.

- *Memory requirements:* Each PE must hold the two current vector elements. Storage is needed to accumulate the inner product and, once the $d(i,j)$ value has been computed, to compute the cumulative distance $C(i,j)$.

- *Communications:* Communications will be needed to accumulate the terms of the inner product. In addition, the communications required for the medium-grain approach will be needed.

- *Synchronization:* Synchronization is frequent. Specific requirements will depend on how the inner product terms are accumulated, but the canonic order of operation will be as follows:

   compute $a_{\ell,i} * b_{\ell,j}$ for $1 \leq \ell \leq f$
   accumulate product terms to form $d(i,j)$
   compute $C(i,j)$

   Synchronization is needed at every step.

- *Mode:* The frequent synchronization (basically the fine module granularity imposed by the fine data granularity) dictates SIMD or pipelined operation.

The DTW example demonstrates the effect that a single algorithm characteristic can have on the design of an architecture-dependent algorithm. As the data granularity changed, so the architecture most suitable for execution of the virtual algorithm changed. In some cases (e.g., number of processors) this change was a direct result of the change in data granularity. In others (e.g., mode), it was a change brought about by the effect of the data granularity on other algorithm characteristics (uniformity, synchronization, module granularity).

5.2. Virtual Algorithm Characteristics of a 2-Dimensional FFT

In the second example, the complete set of algorithm characteristics for a single algorithm is derived. The *algorithm approach* is computation of a 2-dimensional discrete Fourier transform (DFT):

$$F(v,w) = \sum_{k=0}^{M-1} \sum_{\ell=0}^{M-1} S(k,\ell) e^{-2\pi jvk/M} \, e^{-2\pi jw\ell/M} \quad \text{for } 0 \leq v,w < M$$

where S is the input array and F is the DFT of S; both are $M \times M$ arrays. The *virtual algorithm* that we will consider is the "row-column" method for computing the DFT. $F(u,v)$ is computed in two steps: first, calculation of 1-dimensional DFTs on the rows of S:

$$G(k,w) = \sum_{\ell=0}^{M-1} S(k,\ell) \, e^{-2\pi jw\ell/M} \quad \text{for } 0 \leq k,w < M$$

where row k of G is the 1-dimensional DFT of row k of S; followed by calculation of 1-dimensional DFTs on the columns of G:

$$F(v,w) = \sum_{k=0}^{M-1} G(k,w) \, e^{-2\pi jvk/M} \quad \text{for } 0 \leq v,w < M$$

where column w of F is the 1-dimensional DFT of column w of G. The computations performed are outlined in Fig. 4. For the *virtual algorithm,* we will assume that the 1-dimensional DFTs are computed using conventional fast Fourier transform (FFT) algorithms (e.g., [62]).

5.2.1. Virtual Algorithm Characteristics

The characteristics of the virtual algorithm are described below.

- *Type of parallelism:* The type of parallelism characteristic of the 2-D FFT algorithm is data parallelism.

- *Data granularity:* A number of different data granularities are possible. A large-grain algorithm will have the row/column as the basic element on which operations are performed; the basic operation is a 1-D FFT on a row or column. A finer-grain algorithm will operate on the basic element of the 1-D FFT: the FFT butterfly operation. In a very fine-grain algorithm, the arithmetic operations that comprise the the 1-D butterfly can be viewed as the basic elements. In the discussion that follows, the row/column is taken as the grain size.

- *Module granularity:* For the large-grain algorithm, the module granularity is a 1-D FFT on one row or column.

- *Degree of parallelism:* For the large-grain algorithm, the degree of parallelism is M: the number of rows and columns in the arrays.

- *Memory requirements:* Exact memory requirements depend on the serial algorithm used to perform the 1-D FFTs. In general, requirements will be $3M^2 + M$: M^2 for each of S, G, and F; M for the weighting (twiddle) factors used in computing the M-point 1-D FFTs. The memory needed can be reduced, e.g., by using "in-place" algorithms that use the same memory locations for the input and output vectors in the 1-D algorithm, and by computing the weighting factors as they are needed rather than storing them (e.g., see [62]). (However, since the same weighting factors are used for each of the $2M$ 1-D FTTs, this will add considerable redundant computation.)

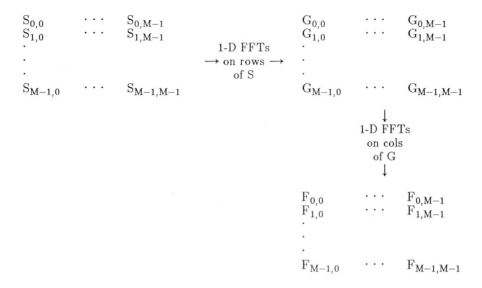

Fig. 4. Computations in row-column 2-D FFT.

- *Uniformity:* The operations performed are uniform across the rows of S and across the columns of G. At the finest level (and depending on the data types and implementation), data-dependent non-uniformities may occur within the floating point arithmetic operations.

- *Synchronization requirements:* The synchronization requirements can be specified at a number of different levels of refinement. The large-grain row-column algorithm consists of two steps:

  compute G; compute F

  The basic synchronization requirement is that computation of G must be completed before computation of F can begin. A more detailed statement of the synchronization requirement is that column w of G must be completed before computation of column w of F can begin. Further refinement requires examination of the particular 1-D FFT algorithm used, and requires only that all of the inputs to a butterfly operation be available before that butterfly can be computed. In the discussion that follows, the coarse synchronization requirement is assumed.

- *Static/dynamic character:* The algorithm structure is static.

- *Data dependencies:* The data dependencies can be observed from the definition of the 2-D DFT: each element $F(v,w)$ of the output array is a function of $S(k,\ell)$ for all k and ℓ, $0 \leq k, \ell < M$. In the row-column algorithm, the data-dependencies can be associated with the two steps of the algorithm. Each element of row k of G depends on all of the elements of row k of S:

$$\text{for fixed k,} \quad G(k,w) = f\left(S(k,\ell)\right) \quad \text{for all } \ell$$

Each element of column w of F depends on all of the elements of column w of G:

$$\text{for fixed w,} \quad F(v,w) = f\left(G(k,w)\right) \quad \text{for all } k$$

- *Fundamental operations:* For the large-grain algorithm, the fundamental operations performed are M-point 1-D FFTs.

- *Data types and precision:* These will be application dependent. The most general case uses complex data and results.

- *Data structures:* The basic data structures for the 2-D DFT are $M \times M$ arrays. For the row-column algorithm, operations are performed on M-point vectors.

- *I/O:* I/O will be application dependent.

Table 2 summarizes the implications of the virtual algorithm characteristics most directly related to the parallel processing aspects of the algorithm. The outcome is typical of parallel algorithms: a single ideal algorithm and an associated ideal architecture are not clearly indicated. Rather, some attributes of the virtual algorithm imply one machine type for the ideal implementation (e.g., the type of parallelism and data granularity favor a local memory organization) while others suggest a different architecture configuration (the global data dependencies would be well supported by a global memory organization). And in fact, a wide range of parallel DFT/FFT algorithms have been developed for a wide range of architectures (e.g., [6, 16, 17, 34, 40, 42, 51, 52, 54, 57, 72, 80]). Both the conflicting conclusions that can be drawn from the different algorithm characteristics and the large number of different architecture-dependent algorithms that have been proposed are indicative of the complexity of the algorithm-to-architecture mapping problem. Since we cannot identify a single ideal solution for this algorithm, we will consider one "good" parallel algorithm for the 2-dimensional DFT and examine how it relates to the implementation attributes implied by the virtual algorithm characteristics.

5.2.2. A Parallel Algorithm for the 2-Dimensional FFT

A row-column algorithm for the 2-dimensional FFT is presented in [72]. The algorithm uses M PEs organized as processor-memory pairs. PE i holds row i of the $M \times M$ input array S. The algorithm consists of three steps:

Table 2. Relationship Between Algorithm and Architecture Characteristics:
2-Dimensional FFT

Architecture Characteristics

Algorithm Characteristics

	Number of PEs	Memory Organization (Local/Global)	Memory Size	Mode (SIMD/MIMD/Pipe)	Network
Type of Parallelism		L		S/P	
Data Granularity		L		M/S	
Module Granularity				M/S	
Degree of Parallelism	0/(M)				
Memory			$0(M^2)$		
Uniformity		L		S/P	
Synchronization				S/M	
Static/Dynamic				S/P	
Data Dependencies		G			F*
Fundamental Ops.					F*
Data Structures				S/P	

*F = full communication capability

> *compute 1-D FFTs on the rows of S to obtain G*
> *transpose G*
> *compute 1-D FFTs on the columns of G to obtain F^T*

This is shown in Fig. 5. The resulting matrix F^T is the transpose of the 2-dimensional DFT of S. F^T can be transposed to give F; however, this may not be necessary depending on what further processing is done on F.

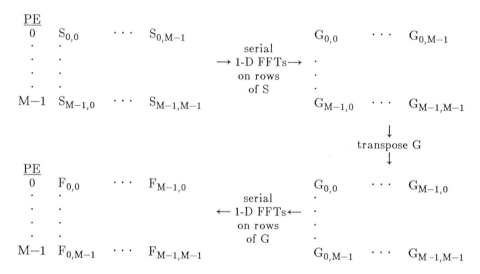

Fig. 5. Computation of 2-dimensional DFT of $M \times M$ array S using M PEs.

To form the transpose, G^T, of G such that each row of G^T is in a different PE, the basic operation performed is the transfer of array element $G(k,w)$ from PE k to PE w. This is done for M $G(k,w)$'s in parallel, using an interconnection function that sends data from PE k to PE $(k+i) \bmod M$ for all of the $G(k,w)$ for which $(w-k) \bmod M = i$. The parallel transfer operation is performed once for each i, $1 \le i < M$. For each i value, the element that PE k sends is the w-th element of the row of G held in PE k, i.e., $G(k,w)$, where $w = (k+i) \bmod M$. That element, received in PE w, is stored as the k-th element of the row of G^T being created in PE w, i.e., $G^T(w,k)$, where $k = (w-i) \bmod M$. This is shown in Fig. 6. The transfers can be expressed in terms of *uniform shift* interconnection functions:

$$shift_d(x) = (x+d) \bmod N$$

for a given d, $0 < d < N$, and $0 \le x < N$ where N is the size of the network

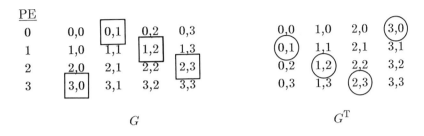

Fig. 6. Example of data transfers to transpose an array. Boxes indicate
$G(k,w)$'s for which $(w-k)\ mod\ M = 1$. These elements are
transferred simultaneously using a $shift_1$ transfer, and are stored in
the circled locations in G^T.

(in this case the number of PEs, M). From [47] and [67], a number of interconnection networks can perform each of these shifts in a single pass through
the network, including the omega [47], generalized cube [70], indirect binary
n-cube [58], data manipulator [30], and ADM [69]. Performing the transpose
therefore requires $M-1$ parallel data transfers for such networks.

The serial complexity of $2M$ 1-D FFTs (i.e., an $M \times M$ 2-D DFT) is
$M^2\log_2 M$ butterflies. The parallel implementation of the 2-D DFT executes
two serial FFT algorithms and has a complexity of $M\log_2 M$ butterfly steps.
Thus, an ideal speedup of M is achieved for butterfly operations with an overhead cost of $M-1$ data transfer steps. A counting argument can be used to
show that the transpose algorithm achieves a lower bound on the number of
transfers needed to transpose G. Element $G(i,j)$ must be transferred from PE
j, for all $0 \le i,j < M$, and $i \ne j$. The total number of elements that must be
transferred is therefore $M(M-1)$. At most M elements can be moved in one
data transfer step, so at least $M-1$ data transfer steps are needed. The space
used by the algorithm is $5M^2$: each PE holds one row of S, one row of G, one
row of G^T, one row of F^T, and M weighting factors for computing the 1-D
FFTs. With "in-place" algorithms, this can be reduced to $3M^2$ (by overlapping the storage for S and G and for G^T and F^T). Further reductions can be
achieved at the expense of execution time if the weighting factors are either
computed as they are needed or are broadcast to the PEs rather than stored
locally.

5.2.3. Virtual Algorithm Characteristics Versus Architecture-Dependent Algorithm Characteristics

In order to consider the relation between the attributes of the FFT algorithm and the attributes derived from the virtual algorithm characteristics, we

consider the columns of Table 2.

- *Number of processors:* The degree of parallelism of the virtual algorithm is M; the algorithm presented uses M PEs. As discussed in [40], the algorithm can be generalized to use $N \leq M$ PEs. When implemented on N PEs, $N \leq M$, the arithmetic complexity will be derived directly from the the serial algorithm used to perform the 1-D FFTs. If the complexity of the serial 1-dimensional algorithm is C, then the arithmetic complexity of the parallel 2-D FFT algorithm is $2\lceil M/N \rceil C$. If N divides M then the speedup over the serial arithmetic complexity will be N, which is ideal. An $O(M^2 \log_2 M)$ serial algorithm will be reduced to a parallel algorithm that has arithmetic complexity $O(2^r M \log_2 M)$ when $M/(2^r)$ PEs are used. The cost of performing the transpose will be $O(2^r M)$, so the communications overhead will never dominate the cost of the algorithm.

- *Memory organization:* The algorithm uses a local memory organization, as called for by most of the virtual algorithm characteristics. The need for global communications derived from the data dependencies of the DFT is handled at a relatively low cost by the transpose operation.

- *Memory size:* The algorithm uses $5M^2$ space as compared with the memory requirements of $3M^2 + M$. The memory used is distributed across M memories, $5M$ per memory.

- *Mode:* The virtual algorithm characteristics yield no definitive decision about the preferred mode for the algorithm. The highly uniform data parallelism, static nature of the algorithm, and predominance of vector operations point to SIMD or pipelined implementations. The large data granularity and the infrequent synchronization implied by the large data and module granularity are most typical of MIMD algorithms, although, coupled with high uniformity, they are not incompatible with an SIMD implementation. The algorithm presented can be executed either as an SIMD or MIMD algorithm. In an MIMD implementation, fine-grain non-uniformities in the arithmetic operations will not slow down the algorithm; in an SIMD implementation, some time savings may be achieved by overlap of the control operations with the computations performed in the PEs. The synchronization-intensive transpose is best implemented in SIMD mode.

- *Network:* The data dependencies and use of 1-D FFTs call for a full communications capability: an output $F(v,w)$ computed in PE i must receive information from all other PEs. In the algorithm, this is reflected in the use of the uniform shifts to perform the transpose. In order to achieve the stated performance, the communications network must be able to support the uniform shifts.

6. Conclusions

In this paper we have described a characteristics-based approach to relating parallel algorithms and parallel architectures. One application of this approach is being explored in the design of an operating system for the PASM [71] architecture. The operating system will select an algorithm implementation and machine configuration from a database of known implementations, using information about the algorithm characteristics and the current state of the system [21, 41].

The problem of gaining an understanding of the relationships between algorithms and architectures is a critical one. Toward this end, we have presented the model of virtual algorithms and architecture-dependent algorithms. This model allows us to specify the point at which architectural considerations enter into the algorithm design process, and therefore allows us to trace relationships between virtual algorithm characteristics and architecture features that support these characteristics. The algorithm characteristics described here represent one step in the development of an effective means of relating parallel algorithms to parallel architectures.

Acknowledgments

Discussions with many colleagues are gratefully acknowledged: George Adams, Fran Berman, Ed Delp, Bob Douglass, Art Feather, Dennis Gannon, Jack Goldberg, Dave Mizell, H. J. Siegel, Phil Swain, Len Uhr, Bob Voigt, and Frank Weil. This research was supported by the United States Army Research Office, Department of the Army, under grant number DAAG29-82-K-0101, by the United States Air Force Systems Command, Rome Air Development Center, under contract number F30602-83-K-0119, and by SRI International. Facilities for preparation of this document were provided by RIACS and Purdue University's School of Electrical Engineering.

References

[1]   G. B. Adams III and H. J. Siegel, "On the Number of Permutations Performable by the Augmented Data Manipulator Network," *IEEE Transactions on Computers*, Vol. C-31, April 1982, pp. 270-277.

[2]   J. K. Aggarwal and S. Yalamanchili, "Algorithm Driven Architectures for Image Processing," *Workshop on Algorithm-Guided Parallel Architectures for Automatic Target Recognition*, July 1984, pp. 1-31.

[3]   J. Allen, Plenary Address, *1986 IEEE Workshop on VLSI Signal Processing*, November 1986.

[4]   E. A. Ashcroft and J. Jagannathan, "Operator Nets," in *Fifth Generation Computer Architectures*, J. V. Woods, editor, North Holland, Amsterdam, 1986, pp. 177-201.

[5]   K. Batcher, "Design of a Massively Parallel Processor," *IEEE Transactions on Computers*, Vol. C-29, September 1980, pp. 836-840.

[6]   G. D. Bergland and D. E. Wilson, "A Fast Fourier Transform for a Global, Highly Parallel Processor," *IEEE Transactions on Audio and Electroacoustics*, Vol. AU-17, June 1969, pp. 125-127.

[7]   F. Berman, "Experience with an Automatic Solution to the Mapping Problem," *The Characteristics of Parallel Algorithms*, L. H. Jamieson, D. B. Gannon, and R. J. Douglass, editors, MIT Press, 1987.

[8]   F. Berman and L. Snyder, "On Mapping Parallel Algorithms into Parallel Architectures," *1984 International Conference on Parallel Processing*, August 1984, pp. 307-309.

[9]   S. H. Bokhari, "On the Mapping Problem," *IEEE Transactions on Computers*, Vol. C-30, March 1981, pp. 207-214.

[10] J. C. Browne, "Characterization of Parallel Architecture," *1985 International Conference on Parallel Processing*, August 1985, p. 665.

[11] J. C. Browne, "Formulation and Programming of Parallel Computations: A Unified Approach," *1985 International Conference on Parallel Processing*, August 1985, pp. 624-631.

[12] D. J. Burr, B. D. Ackland, and N. Weste, "Array Configurations for Dynamic Time Warping," *IEEE Transactions on Acoustics, Speech, and Signal Processing*, Vol. ASSP-32, February 1984, pp. 119-128.

[13] V. Cantoni and S. Levialdi, "Matching the Task to an Image Processing Architecture," *6th International Conference on Pattern Recognition*, October 1982, pp. 254-257.

[14] P. R. Cappello and K. Steiglitz, "Unifying VLSI Array Design with Linear Transformations of Space-Time," *Advances in Computing Research*, Vol. 2, JAI Press, 1984, pp. 23-65.

[15] Y. P. Chiang and K. S. Fu, "Matching Parallel Algorithm and Architecture," *1983 International Conference on Parallel Processing*, August 1983, pp. 374-380.

[16] R. A. Collesidis, T. A. Dutton, and J. R. Fisher, "An Ultra-High Speed FFT Processor," *1980 IEEE International Conference on Acoustics, Speech, and Signal Processing*, April 1980, pp. 784-787.

[17] M. J. Corinthios, "A Fast Fourier Transform for High-Speed Signal Processing," *IEEE Transactions on Computers*, Vol. C-20, August 1971, pp. 843-846.

[18] W. Crother et al., "Performance Measurements on a 128-Node Butterfly Parallel Processor," *1985 International Conference on Parallel Processing*, August 1985, pp. 531-540.

[19] P. E. Danielsson and S. Levialdi, "Computer Architectures for Pictorial Information Systems," *Computer*, November 1981, pp. 53-67.

[20] D. DeGroot, "Partitioning Job Structures for SW-Banyan Networks," *1983 International Conference on Parallel Processing*, August 1983, pp. 106-113.

[21] E. J. Delp, H. J. Siegel, A. Whinston, and L. H. Jamieson, "An Intelligent Operating System for Executing Image Understanding Tasks on a Reconfigurable Parallel Architecture," *1985 IEEE Computer Society Workshop on Computer Architectures for Pattern Analysis and Image Database Management*, November 1985, pp. 217-224.

[22] P. B. Denyer, "System Compilers," in *VLSI Signal Processing, II*, S. Y. Kung, R. E. Owen, and J. G. Nash, editors, IEEE Press, New York, 1986, pp. 3-13.

[23] P. Dewilde and J. Annevelink, "Hierarchical Design of Processor Arrays Applied to a New Pipelined Matrix Solver," in *VLSI Signal Processing, II*, S. Y. Kung, R. E. Owen, and J. G. Nash, editors, IEEE Press, New York, 1986, pp. 106-116.

[24] P. Dewilde, E. Deprettere, and R. Nouta, "Parallel and Pipelined VLSI Implementations of Signal Processing Algorithms," *VLSI and Modern Signal Processing*, S. Y. Kung, H. J. Whitehouse, and T. Kailath, editors, Prentice-Hall, Englewood Cliffs, NJ, 1985, pp. 257-276.

[25] M. J. B. Duff, "Parallel Algorithms and their Influence on the Specification of Application Problems," in *Multicomputers and Image*

Processing, K. Preston, Jr. and L. Uhr, editors, Academic Press, New York, 1982, pp. 261-274.

[26] *Computing Structures for Image Processing*, M. J. B. Duff, editor, Academic Press, London, 1983.

[27] *Intermediate-Level Image Processing*, M. J. B. Duff, editor, Academic Press, London, 1986.

[28] *Languages and Architectures for Image Processing*, M. J. B. Duff and S. Levialdi, editors, Academic Press, London, 1981.

[29] R. D. Etchells and G. R. Nudd, "Software Metrics for Performance Analysis of Parallel Hardware," *DARPA Image Understanding Workshop*, June 1983, pp. 137-147.

[30] T. Feng, "Data Manipulating Functions in Parallel Processors and Their Implementations," *IEEE Transactions on Computers*, Vol. C-23, March 1974, pp. 309-318.

[31] J. P. Fishburn and R. A. Finkel, "Quotient Networks," *IEEE Transactions on Computers*, Vol. C-31, April 1982, pp. 288-295.

[32] J. A. B. Fortes and D. I. Moldovan, "Parallelism Detection and Transformation Techniques Useful for VLSI Algorithms," *Journal of Parallel and Distributed Computing*, Vol. 2, August 1985, pp. 277-301.

[33] P. Frison, P. Gachet, and P. Quinton, "Designing Systolic Arrays with DIASTOL," in *VLSI Signal Processing, II*, S. Y. Kung, R. E. Owen, and J. G. Nash, editors, IEEE Press, New York, 1986, pp. 93-105.

[34] B. Gold and T. Bially, "Parallelism in Fast Fourier Transform Hardware," *IEEE Transactions on Audio and Electroacoustics*, Vol. AU-21, February 1973, pp. 5-16.

[35] A. Gottlieb, et al., "The NYU Ultracomputer -- Designing an MIMD Shared Memory Parallel Computer," *IEEE Transactions on Computers*, Vol. C-32, February 1983, pp. 175-189.

[36] W. D. Hillis, *The Connection Machine*, MIT Press, Cambridge, MA, 1985.

[37] R. W. Hockney and C. R. Jesshope, *Parallel Computers: Architecture, Programming and Algorithms*, Adam Hilger Ltd., Bristol, 1981.

[38] H. V. Jagadish, R. G. Mathews, T. Kailath, and J. A. Newkirk, "A Study of Pipelining in Computing Arrays," *IEEE Transactions on Computers*, Vol. C-35, May 1986, pp. 431-440.

[39] L. H. Jamieson, "The Mapping of Parallel Algorithms to Reconfigurable Parallel Architectures," *Intermediate-Level Image Processing*, M. J. B.

Duff, editor, Academic Press, London, 1986, pp. 53-63.

[40] L. H. Jamieson, P. T. Mueller, Jr., and H. J. Siegel, "FFT Algorithms for SIMD Parallel Processing Systems," *Journal of Parallel and Distributed Computing*, Vol. 3, March 1986, pp. 48-71.

[41] L. H. Jamieson, H. J. Siegel, E. J. Delp, and A. Whinston, "The Mapping of Parallel Algorithms to Reconfigurable Parallel Architectures," *ARO Workshop on Future Directions in Computer Architecture and Software*, May 1986, pp. 147-154.

[42] C. R. Jesshope, "The Implementation of Fast Radix 2 Transforms on Array Processors," *IEEE Transactions on Computers*, Vol. C-29, January 1980, pp. 20-27.

[43] P. M. Kogge and H. S. Stone, "A Parallel Algorithm for the Efficient Solution of a General Class of Recurrence Equations," *IEEE Transactions on Computers*, Vol. C-22, August 1973, pp. 786-793.

[44] R. H. Kuhn, "Efficient Mapping of Algorithms to Single-Stage Interconnections," *7th Annual Symposium on Computer Architecture*, May 1980, pp. 182-189.

[45] H. T. Kung, "The Structure of Parallel Algorithms," in *Advances in Computers*, Vol. 19, Academic Press, New York, 1980, pp. 65-112.

[46] S. Y. Kung, "On Supercomputing with Systolic/Wavefront Array Processors," *Proceedings of the IEEE*, Vol. 72, July 1984, pp. 867-884.

[47] D. H. Lawrie, "Access and Alignment of Data in an Array Processor," *IEEE Transactions on Computers*, Vol. C-24, December 1975, pp. 1145-1155.

[48] *Integrated Technology for Parallel Image Processing*, S. Levialdi, editor, Academic Press, London, 1985.

[49] S. P. Levitan, "Evaluation Criteria for Communication Structures in Parallel Architectures," *1985 International Conference on Parallel Processing*, August 1985, pp. 147-154.

[50] H. Li, C-C. Wang, and M. Lavin, "Structured Process: A New Language Attribute for Better Interaction of Parallel Architecture and Algorithm," *1985 International Conference on Parallel Processing*, August 1985, pp. 247-254.

[51] W. T. Lin and C. Y. Ho, "A New FFT Mapping Algorithm for Reducing the Traffic in a Processor Array," in *VLSI Signal Processing, II*, S. Y. Kung, R. E. Owen, and J. G. Nash, editors, IEEE Press, New York, 1986, pp. 328-336.

[52] W. Liu, T. Hughes, and W. T. Krakow, "A Rasterization of Two-Dimensional Fast Fourier Transform," in *VLSI Signal Processing, II*, S. Y. Kung, R. E. Owen, and J. G. Nash, editors, IEEE Press, New York, 1986, pp. 281-292.

[53] G. S. Lueker and K. S. Booth, "A Linear Time Algorithm for Deciding Interval Graph Isomorphism," *Journal of the ACM*, Vol. 26, April 1979, pp. 183-195.

[54] V. Milutinovic, J. Fortes, and L. Jamieson, "A Multimicroprocessor Architecture for Real-Time Computation of a Class of DFT Algorithms," *IEEE Transactions on Acoustics, Speech, and Signal Processing*, Vol. ASSP-34, October 1986, pp. 1301-1309.

[55] C. Myers, L. R. Rabiner, and A. E. Rosenberg, "Performance Tradeoffs in Dynamic Time Warping Algorithms for Isolated Word Recognition," *IEEE Transactions on Acoustics, Speech, and Signal Processing*, Vol. ASSP-28, December 1980, pp. 622-635.

[56] R. O'Dell and J. C. Browne, "Parallel Structuring of Control and Resources Management Systems for Parallel Programs," *1986 International Conference on Parallel Processing*, August 1986, pp. 153-156.

[57] M. C. Pease, "An Adaptation of the Fast Fourier Transform for Parallel Processing," *Journal of the ACM*, Vol. 15, April 1968, pp. 252-264.

[58] M. C. Pease, "The Indirect Binary n-Cube Microprocessor Array," *IEEE Transactions on Computers*, Vol. C-26, May 1977, pp. 458-473.

[59] G. F. Pfister et al., "The IBM Research Parallel Processor Prototype (RP3): Introduction and Architecture," *1985 International Conference on Parallel Processing*, August 1985, pp. 764-771.

[60] F. P. Preparata, "VLSI Algorithms and Architectures," *11th Symposium on the Mathematical Foundations of Computer Science*, September 1984, pp. 149-161.

[61] *Multicomputers and Image Processing: Algorithms and Programs*, K. Preston, Jr. and L. Uhr, editors, Academic Press, New York, 1982.

[62] L. R. Rabiner and B. Gold, *Theory and Application of Digital Signal Processing*, Prentice-Hall, Englewood Cliffs, NJ, 1975.

[63] R. C. Read and D. G. Corneil, "The Graph Isomorphism Disease," *Journal of Graph Theory*, Vol. 1, Winter 1977, pp. 339-363.

[64] H. Sakoe and S. Chiba, "Dynamic Programming Algorithm Optimization for Spoken Word Recognition," *IEEE Transactions on Acoustics, Speech, and Signal Processing*, Vol. ASSP-26, February 1978, pp. 43-49.

[65] D. A. Schwartz and T. P. Barnwell III, "A Graph Theoretic Technique for the Generation of Systolic Implementations for Shift-Invariant Flow Graphs," *1984 IEEE International Conference on Acoustics, Speech, and Signal Processing*, March 1984, pp. 8.3.1-8.3.4.

[66] C. Seitz, "The Cosmic Cube," *Communications of the ACM*, Vol. 28, January 1985, pp. 22-33.

[67] H. J. Siegel, *Interconnection Networks for Large-Scale Parallel Processing: Theory and Case Studies*, Lexington Books, D. C. Heath, Lexington, MA, 1985.

[68] H. J. Siegel et al., "PASM: A Partitionable SIMD/MIMD System for Image Processing and Pattern Recognition," *IEEE Transactions on Computers*, Vol. C-30, December 1981, pp. 934-947.

[69] H. J. Siegel and R. J. McMillen, "Using the Augmented Data Manipulator Network in PASM," *Computer*, Vol. 14, February 1981, pp. 25-33.

[70] H. J. Siegel and R. J. McMillen, "The Multistage Cube: A Versatile Interconnection Network," *Computer*, Vol. 14, December 1981, pp. 65-76.

[71] H. J. Siegel, T. Schwederski, J. T. Kuehn, and N. J. Davis IV, "An Overview of the PASM Parallel Processing System," in *Tutorial: Computer Architecture*, D. D. Gajski, V. M. Milutinovic, H. J. Siegel, and B. P. Furht, editors, IEEE Computer Society Press, Washington, DC, 1986, pp. 387-407.

[72] L. J. Siegel, "Image Processing on a Partitionable SIMD Machine," *Languages and Architectures for Image Processing*, M. J. B. Duff and S. Levialdi, editors, Academic Press, London, 1981, pp. 294-300.

[73] B. W. Smith and H. J. Siegel, "Models for Use in the Design of Macro-pipelined Parallel Processors," *12th Annual Symposium on Computer Architecture*, June 1985, pp. 116-123.

[74] L. Snyder, "Introduction to the Configurable, Highly Parallel Computer," *Computer*, Vol. 15, January 1982, pp. 47-56.

[75] *Algorithmically Specialized Parallel Computers*, L. Snyder, L. H. Jamieson, D. B. Gannon, and H. J. Siegel, editors, Academic Press, Orlando, FL, 1985.

[76] H. S. Stone, "Parallel Processing with the Perfect Shuffle," *IEEE Transactions on Computers*, Vol. C-20, February 1971, pp. 153-1661.

[77] P. H. Swain, H. J. Siegel, and J. El-Achkar, "Multiprocessor Implementation of Image Pattern Recognition: A General Approach," *5th International Conference on Pattern Recognition*, December 1980, pp. 309-317.

[78] E. E. Swartzlander, Jr., *VLSI Signal Processing Systems*, Kluwer Academic Publishers, Boston, 1986.

[79] S. L. Tanimoto and A. Klinger, editors, *Structured Computer Vision: Machine Perception through Hierarchical Computation Structures*, Academic Press, New York, 1980.

[80] C. D. Thompson, *Fourier Transforms in VLSI*, Electronics Research Laboratory, University of California, Berkeley, Memorandum No. UCB/ERL M80/51, October 1980.

[81] L. Uhr, "Pyramid Multi-computer Structures, and Augmented Pyramids," in *Computing Structures for Image Processing*, M. J. B. Duff, editor, Academic Press, London, 1983, pp. 95-112.

[82] *Evaluation of Multicomputers for Image Processing*, L. Uhr, K. Preston, Jr., S. Levialdi, and M. J. B. Duff, editors, Academic Press, Orlando, FL, 1986.

[83] M. A. Yoder and L. H. Jamieson, "Simulation of a Highly Parallel System for Word Recognition," *1985 International Conference on Acoustics, Speech, and Signal Processing*, March 1985, pp. 1449-1453.

[84] M. A. Yoder and L. J. Siegel, "Dynamic Time Warping Algorithms for SIMD Machines and VLSI Processor Arrays," *1982 IEEE International Conference on Acoustics, Speech, and Signal Processing*, May 1982, pp. 1274-1277.

Measuring Communication Structures
in Parallel Architectures and Algorithms

Steven P. Levitan

Department of Electrical and Computer Engineering
University of Massachusetts, Amherst, MA 01003

This work focuses on analyzing metrics for communication structures in parallel computer architectures. We review several measures or "metrics" of the interconnection networks of parallel architectures and we evaluate these metrics as predictors of machine/algorithm performance. We have done this by simulating six tasks on each of eight architectures and comparing the metric-predicted ranking of the machines/algorithms with the actual time complexity of the algorithms. We show that, in general, accepted metrics of machine architectures do not perform well as predictors of run-time performance. However, the performance of some parallel algorithms running on parallel machines correlates very well with the performance of other parallel algorithms on those same machines. We propose a set of tasks to be used as the beginning of a performance suite for evaluating the communication structures of parallel architectures.

1. Introduction

Parallel processing is not new. However, only recently have the interprocessor communication structures necessary to support parallel processing been recognized as a key issue in the design of parallel machines [50]. While there have been many attempts to classify and evaluate different communication structures [20, 36, 30], a real theory of parallel, or communication, complexity has not yet emerged [38].

Such a theory would allow computer architects to evaluate new ideas in parallel computer organization and perform comparative analysis between existing and proposed designs. It would give programmers a method for predicting the performance of parallel algorithms running on different parallel machines and a way to judge the performance of algorithms running on different architectures, in terms of optimality for each particular architecture. Just as the current theory of algorithmic complexity is a basis for algorithm design and analysis, an extension of this theory to parallel algorithms and parallel machines will form a basis for parallel algorithm design and analysis techniques.

Many models for parallel computation have been proposed. Borodin and Hopcroft [9] give a hierarchy of models which could be extended to include The parallel comparison model of Valiant [54], Schwartz's Paracomputer [44] and Thompson's [51] and Vuillemin's [55] models of VLSI. Galil and Paul [21] also give a comparative analysis of models.

Models, however, are not real machines. Computer architects and programmers are concerned with how accurately models predict the abilities of real machines. Machines that can be built would approach the abilities of any model with a time penalty which is some function of the size of the system. How much of a time penalty is not always clear. For example, Schwartz [44] claims the Ultracomputer will be able to simulate the Paracomputer (and all simpler models) with only a factor of $\log(N)$ time penalty. He is using about $N * \log(N)^2$ hardware to perform routing between a set of N processors and N memory registers. In the notation we describe below, this is a $O_i(\log(N))$ instruction time penalty. This penalty is, itself, more than the running times of some of the algorithms we want to discuss. This makes it difficult to predict the running times of real algorithms on real machines.

The problem is that a model of a parallel computer for a theory of parallel complexity must account for the time penalties involved in communication. However, these time penalties are different for different parallel architectures. Since we clearly do not want a different theory of complexity for every parallel architecture, we must develop a theory which explicitly addresses the differences in communication structures of different architectures.

In an attempt to solve these and other problems, researchers have proposed several measures of communication structures. Some possible measures that have been proposed are: Diameter, Bandwidth and the Average Distance between every two processors [35], Message Capacity [26], Average Message Delay, Average Message Density, and Connections per Processor [58], the time to perform a Bitonic sort, the number of mapping functions performed by the network, the number of bijection functions performed by the network, and the time it takes for a network to simulate other networks [46,47].

Not all of these measures have been used in the context we are proposing, as a tool for complexity analysis, and not all of them would be appropriate for that purpose. More importantly, rather than some measure of the "raw power" of the structures, we are concerned with characterizations of the structures that will give us some insight into the way they will support the communication needs of real algorithms. Therefore, the best measures are the ones that help us predict algorithm performance. Measures (or metrics) which predict algorithm performance could then be used in a general theory of parallel complexity.

There is no reason to believe that any of the proposed metrics are better than any others. Moreover, very little has been done to verify the quality of the metrics themselves. To address this issue, we have chosen a set of metrics to evaluate on the basis of how well they can be used to predict the abilities of different multiprocessors. The metrics we choose to evaluate attempt to capture those aspects of the multiprocessor communication structures which will directly affect algorithm runtime performance.

The research reported here consisted of several steps [33,34]. The organization of this paper follows our work. We first present the assumptions or "ground rules" for analyzing parallel machines and algorithms used throughout the work. We next select a set of multiprocessor architectures based on the communications structures of those architectures. We then consider a set of metrics for those structures which might be useful to capture their communications abilities. Next, we examine the runtime performance of several classic tasks on the different architectures. At the same time, we use the metrics to predict the performance of the machines on those different tasks. We then compare the runtime performance of the algorithms and the predictions of the metrics.

The rest of the paper evaluates how well the metrics can be used to capture the abilities of the architectures, and therefore predict their relative merits on different algorithms. The result is an evaluation of the metrics themselves, as tools to capture the nature of of machine/algorithm interactions. Besides simply evaluating the metrics as predictors, this work is also an examination of the way algorithms "fit" on different parallel architectures. We have tried to reveal the relationship of the communication needs of algorithms to the abilities of machine architectures. Finally, we conclude with a proposal for a set of benchmark routines or a performance suite of algorithms which we believe are more suitable for evaluating multiprocessor architectures.

Thus, the goal of the work is twofold: To characterize parallel algorithms in terms of their communication needs and to select and verify useful measures that can aid in the analysis of parallel algorithms. These characterizations and metrics are viewed as necessary precursors to a valid theory of parallel algorithm complexity.

2.   The Model

We have chosen to use a version of the network model [41] for our research. We are trying to be "architecturally accurate." When we discuss machines, we would like to eliminate as much as possible the hidden costs of converting these ideals to practical systems. We are doing this because we are seeking to evaluate metrics that will be used on real machines. As much as possible, we should consider architectures that can be realized directly. In particular, we are not considering any common-memory based architectures.

Our model assumes that the time for messages to travel along the arcs connecting processors is very small compared to the time to put the message on the arc, or to take it off at the other end. However, we are very restrictive about the abilities of the processors to handle messages. We charge one unit of time equal to the execution of an instruction for each "send" or "receive" operation.

More importantly, we do not assume that the processors can detect messages on their input arcs "for free." Each input must be tested which also takes a unit of time. In a hardware sense this means we are not allowing interrupts. Although hardware for arbitrating interrupts is well understood, the arbitration mechanism and the vectorization mechanism require hardware and gate delays, which grow as the system grows. We have decided to remove that complication from our analysis. As we show later, this means that our Fully Connected machine is weaker in our model than in other models. Routing is not a trivial problem since arriving messages are not immediately known at each processor.

2.1   Time Complexity of Communication and Computation

In addition to the instruction level delays for communication, we are also interested in considering the time for gate delays and wire propagation delays in a certain restricted sense. We are concerned with comparing operations that might take different amounts of time on different architectures. For instance we might need to contrast a single bit boolean OR operation on one architecture, with an arithmetic ADD on another. To allow for such "apples and oranges" comparisons, we need to have a consistent measure of how long operations take.

For serial machines the actual time to perform operations is not an issue and it is often consciously ignored by theorists. "An add is an add is an add," at least asymptotically. However, in large systems the time to move data around cannot be ignored. More importantly, when analyzing parallel algorithms we must contrast the time to move data between processors with the time to compute. Since we are talking about different degrees of sophistication of our processing elements, we must also separate out the difference in complexity between a "Gate Level" operation and a "Fetch Execute Instruction" level operation. As part of the model, there must be a way to compare different notions of computation with each other and with equally different implementations of communication.

There is no agreement about uniform time measures for communication costs

in parallel models of computation. Even the propagation delays in different types
of wires have been dealt with differently in the past. For instance, fabricators of
silicon circuits claim that the delay in silicon wires is intrinsically proportional
to the length of the wire squared; wires to them are delay lines. Designers of
integrated circuits however, use length as a rule of thumb. Using repeater circuits,
Thompson [53] and Mead and Conway [38] claim a logarithmic delay. With
other assumptions, researchers have different results. G. Bilardi, M. Pracchi, and
F. P. Preparata [7] propose three different models for three special cases of signal
propagation in VLSI silicon: $O(1)$, $O(L)$, and $O(L^2)$. They conclude that for
most current technology we can use $O(1)$.

Of course, things are no better when we try to decide gate delay time. Is it a
function of the number of inputs to a gate? Is it a function of the driven load? Is
it a function of the technology? The answer to all these questions is "Yes." For
operations, or instructions, we have no better solution. After all, machines are
made out of wires and gates.

We propose the following resolution to the above problem. If we are considering CPU operations or instructions we use a subscript $O_i()$ (or none); if we
are considering gate delays we use a subscript $O_g()$ and wire delays call for a
subscript of $O_w()$. We use a similar convention for the lower bounds notation:
$\Omega_i()$, $\Omega_g()$, $\Omega_w()$. We will use the three subscripts where appropriate in the
analysis of our algorithms. This frees us to pursue our analysis without regard
to technology-dependent conditions of the relationships between these domains.
Also note, that we use $\log(x)$ to mean logarithm to the base 2 of x, $\ln(x)$ to
mean the natural logarithm of x, and $x * y$ to mean $x \times y$.

3.   Machines, Metrics, and Algorithms

For our work in testing metrics, we have chosen a representative sample of current or proposed parallel architectures. We consider one Single Instruction stream
Multiple Data stream (SIMD) machine, the Content Addressable Parallel Processor. We consider five "graph-structured" machines of increasing complexity: the
Star, Linear, Tree, Shuffle, and Full networks. We also include two unique designs
of our own, the Broadcast Protocol Multiprocessor, and the Fully Interconnected
with Content Addressable Parallel Processors machine. The machines have been
selected on the basis of their broad range of communication characteristics.

We have chosen the metrics of Diameter and Bandwidth based on Lint [35],
Path Count based on Horowitz and Zorat [26] Thickness based on Gannon [22],
and our own measure, Narrowness [33]. These metrics span a range of complexity
from simple measures to more sophisticated graph theoretic properties of the
network.

The algorithms have been selected by three criteria. First, we have taken familiar, well studied, and well understood serial algorithms and considered them in the parallel environment. Second, we have limited the domain of our algorithms to combinatoric algorithms. These are integer arithmetic non-numerical algorithms generally concerned with decisions rather than numbers. Third, we have been concerned with sub-algorithms or "kernels." These are useful algorithms which occur often in larger programs.

3.1 Machines

The architectures we have chosen span the taxonomy presented by Anderson and Jensen [1]. We give the Anderson-Jensen classification for each machine along with a brief description.

3.1.1 The Content Addressable Parallel Processor (CAPP)

Single Instruction Multiple Data (SIMD) architectures [17] are based on the concept of a single central controller broadcasting instructions to all the processors in the machine. Each processor has its own data which it operates on in parallel with all the other processors of the machine. The Content Addressable Parallel Processor (CAPP) architecture is like a SIMD machine with simple processors [18]. The CAPP does not directly fit into the Anderson-Jensen classification (Figure 1).

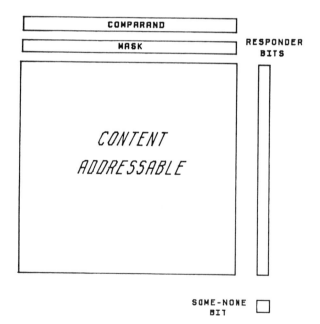

Figure 1. Content Addressable Parallel Processor (CAPP)

In a CAPP both instructions and data are broadcast. The processors, or processing elements, can operate on both their own internal data and data broadcast by the central controller. This allows for such operations as comparison of values, global updates of values, and arithmetic operations where one of the operands is a broadcast value. Each processor of the CAPP can save the results of these operations and conditionally execute future operations based on these results. The results are kept in boolean flag registers called "responder bits."

Another important feature of the CAPP is the ability of the processors to report their results back to the central controller. There are two types of summary reports. One is a simple existence result: Are there any processors with responder bits set? The second is a count of the number of processors with their responder bits set.

A design for a VLSI CAPP is proposed by Weems [56]. A survey of CAPPs and related architectures is provided by Foster [19] and Weems [57]. The CAPP is interesting primarily because of the broadcast and response circuitry. Also, it has reasonably little processing power at each "processor," but it is still a powerful parallel processor.

3.1.2 Star Network

The Star machine is made up of a group of processors (Spokes), each connected to a single central, or Hub, processor. The distance between any two processors is short (two) but the number of simultaneous messages that can be transferred is only one. This is an Indirect, Centralized Routing, Dedicated Path (ICDS) machine (Figure 2).

3.1.3 Broadcast Protocol Multiprocessor (BPM)

This machine is like a Star Network, except that the Hub is not a processor. Rather, the Hub is a single register that has the property that all processors can read its contents at once. All Spoke processors can attempt to write at once into the register, but only one (chosen at random) will succeed. This structure also models the communication abilities of networks like the Ethernet [39]. Details of this architecture are in [31, 32]. Depending on the central resource, this is a Direct, Shared Path Global Bus (DSB) or an Indirect Bus with a Central Switch (ICS) machine (Figure 3).

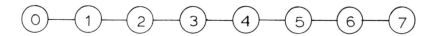

Figure 4. Linear Network

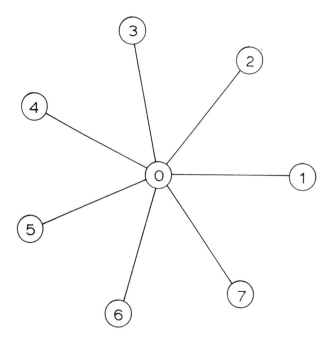

Figure 2. Star Network

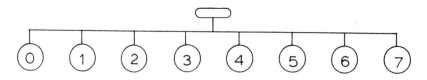

Figure 3. Broadcast Protocol Multiprocessor (BPM)

3.1.4 Linear Network

This machine is simply a group of processors interconnected in a linear fashion. There is a bidirectional link between a processor and its left and right neighbors. There is no "end around wrap." This architecture is interesting because it has a large distance between the two end processors, but the number of simultaneous messages in the network can also be large. This is a Direct, Dedicated Path Linear (DDL) machine (Figure 4).

3.1.5 Tree Network

This Machine is a group of processors connected in a binary tree. There is a distinguished Root and Leaves. This architecture has been proposed by several authors since the communication distance is shorter than it is for the Linear network and the number of simultaneous messages it can support is better than the Star network. This is an Indirect, Decentralized, Dedicated path, Irregular (IDDI) machine (Figure 5).

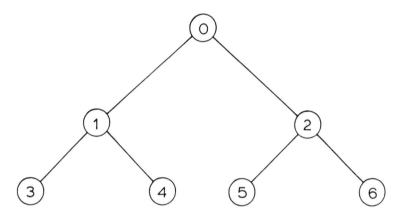

Figure 5. Tree Network

3.1.6 Shuffle Network

This interconnect pattern was originally proposed by Stone [49] as a method for interconnecting dynamic memories. In our machine, each processor $1 \ldots N - 1$ is connected by four unidirectional links. Processor I can send to processor $(2 * I) \bmod N$, and processor $(2 * I + 1) \bmod N$. N must be a power of two. This can be thought of as a network where each processor is the root of a binary tree, connecting all processors. This is an Indirect, Decentralized, Dedicated path, Regular (IDDR) machine (see Figure 6 for numbering conventions).

3.1.7 Fully Interconnected Network (Full)

In this machine, every processor can communicate directly with every other processor. There are N^2 bidirectional links. However, each processor can send and receive only one message at a time. This is often considered the best possible interconnection scheme, even though it is expensive. This is a Direct, Dedicated path, Complete (DDC) machine (Figure 7).

3.1.8 Fully Interconnected with CAPP (Full with CAPP)

This machine has the feature that each processor has its own CAPP as part of

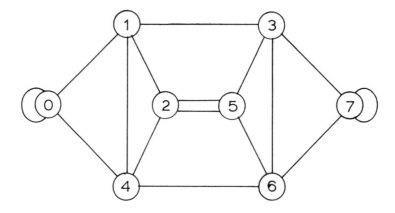

Figure 6. Shuffle Network

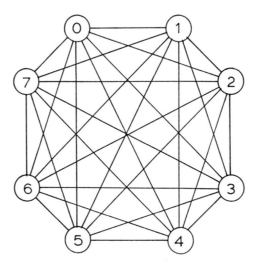

Figure 7. Full Network

its own memory. Furthermore, we let each processor's communication links end in a word of the CAPP of every other processor. This allows each processor to broadcast to every other processor in "unit" time (Figure 8).

The Full with CAPP machine is constructed by having each processor share a different word of CAPP with every other processor. This means that when one processor broadcasts to all words of its CAPP, the value appears in one word of

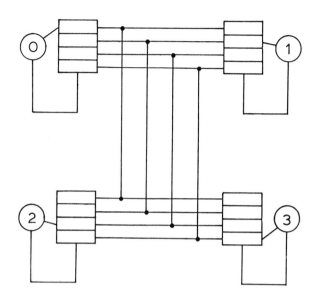

Figure 8. Full with CAPP Network

each of the other processor's CAPPs. Conversely, when one processor searches its CAPP it is really examining one word from each other processor's CAPP. Of the N^2 words $(W[i,j])$, each is written into by processor i and read and searched by processor j. This could be extended to allow both processors i and j to read and write the same word. However, to allow both processors to search the same word would imply N response bits for each word. Our algorithms do not need that capability.

The Full with CAPP architecture is useful in cases where a regular CAPP would need to serialize operations. CAPPs, being SIMD machines, need to serialize at every decision point in an algorithm. There is only one controller. If there is a branch point based on a data value in the words of memory, the controller needs to "turn off" some words, perform one case, then go back and deal with the other case. Our CAPP architecture has a stack of response store bits to facilitate that operation. This simplifies programming, but does not necessarily speed up the algorithms. For instance, in the sorting task a traditional CAPP needs to find the K-th largest value for $K = 1 \ldots N$. The operation is serialized since the CAPP cannot find both the 17th and 31st largest values at the same time. The Full with CAPP machine, having N controllers, can perform the ranking operation in parallel.

3.2   Metrics

Two relatively simple metrics that have been proposed for measuring the communications abilities of parallel architectures are Diameter and Bandwidth [35]. Diameter can be informally described as the worst case time for one message to go from one node of the network to some other node. Bandwidth is the maximum number of messages that can be sent and received in the network at once.

We want our definitions of Diameter and Bandwidth to capture our intuitive notions of "How fast can we get a message from one end of the network to the other?" and "How much communication can be going on at once?" We would also like the definitions to be both precise and general enough that we can apply them to different architectures than the ones we are considering in this work, without resorting to intuition for every special case. While these goals are easily realized for Diameter, we will see that it is more difficult to formulate a satisfactory definition of Bandwidth.

3.2.1   Diameter

We define Diameter empirically in terms of a set of experiments. These experiments proceed as follows: We have each processor in the network send a message to every other processor, one at a time. The worst case delay among the $N * (N-1)$ trials is the Diameter of the network. Delay is measured in units of receive-send (forwarding) time $O_i()$, gate delays $O_g()$, or wire length $O_w()$, as appropriate.

3.2.2   Bandwidth

There are two problems that arise when we attempt to formalize our intuitive notions of Bandwidth. The first problem is that there is a difference between the number of messages that can be active in a network and the number of messages that the processors can handle. The second problem is that we must decide whether to count broadcast messages with the same content as one message or several. Examining the behavior of infinite size networks with different size branching factors aids us in resolving both problems.

We first address the problem of the difference between the number of messages that could exist in the communication pathways of a network and the number that can be processed by the system. Consider the case of a message originating at the root of an infinite binary tree and passed down from parent to children such that it is broadcast to all nodes of the tree. The number of nodes, N, which have seen the message at any given step time, S, is the sum of the first S Fibonacci numbers, $F(S)$. The sum is itself a Fibonacci number: $F(S + 2) - 1$ [29]. The reason for this growth pattern is the forwarding rule of our model which implies that a processor can only forward one message at a time. In a binary tree the number of new nodes at each step is the number of nodes which are one step old, plus the number of nodes which are two steps old. Three-step old nodes have sent messages to both their children and therefore do no further work. The growth is exponential. $F(S)$ is about equal to $\phi^S/\sqrt{5}$, where ϕ is the "golden ratio" defined to be $(1+\sqrt{5})/2$.

For a trinary tree the nodes stay active for three time steps, for a quaternary tree for four time steps, etc. For an infinitely branching infinite tree all nodes stay active. Therefore, each time step doubles the number of active nodes. For an infinitely branching tree, it is our forwarding rule which prevents us from concluding that an infinite number of messages could be sent throughout the network in the first few time steps. The number of active nodes $= 2^{S-1}$. Even in the fully connected architecture it takes eight time steps to broadcast a single number to 128 processors.

The preceding argument demonstrates why we believe that Bandwidth ought to be defined in terms of the number of messages that processors can handle rather than simply the number of messages that can be active in a network. In fact, it substantiates our intuitive notion that Bandwidth is not simply the maximum number of messages that could "live" in the network at any given time; but rather it must be some function of the number of messages that can be generated or received by the processors in the network at a given time. Under the definition of Bandwidth that we give below, the maximum possible value of the Bandwidth of an N processor machine is N.

We now turn to the second problem that arises when we attempt to formalize our definition of Bandwidth. Do we count copies of identical messages as one message or as individual messages? Put another way, do we count the number of messages sent, or the number received?

Again we turn to the extreme case of an infinite number of processors. If all processors are listening to an infinitely long broadcast bus, and one processor sends a message, then they all receive a copy of the message "immediately." We can say that there are an infinite number of messages received, or multiple copies of one message, or that the message moves down the wire with speed $O_w()$ and therefore takes infinite time to propagate. If we allow infinite buses to hold more than one message, we could claim that the messages need not even be multiple copies of the same original, and thus we could use the $O_w()$ propagation delay as a virtual buffer. We could even pipeline messages. The infinite broadcast bus does not resolve our dilemma.

If, however, we look at an infinite receive bus, a "wired OR" for example, only one message can be carried on the bus. It can be a message of consensus, a Some/None response, but it is only one datum of global information. The propagation of the message on this bus must be at least $\Omega_w()$, but it is only one

message. Therefore, for the purpose of symmetry (and simplification) we have
decided to take the Bandwidth of both broadcasting and receiving buses as one
message per unit time, $O_i()$.

Having considered the two problematic issues regarding Bandwidth, we are
now able to state precise definitions for Diameter and Bandwidth. We define
Diameter as the worst-case time to get a message from one processor of a network
to another. We measure Diameter with the "time trials" experiment given above.
We assume there is no other computation or communication in progress in the
network at the time of the experiment.

We define Bandwidth as the total number of messages that can be sent or
received by processors in the system in one unit of time ($O_i(1)$). We make the
further restriction that a communication on a broadcast or receive bus counts
as one broadcast, or receive, message respectively. Therefore, Bandwidth is the
maximum number of unique messages that can be generated or consumed in unit
time in a network.

The first two columns of table 1 gives the Diameter and Bandwidth of the
architectures we are discussing. We can see that the machines are not completely
differentiated based on these two metrics. The difference between the Star and the
BPM is that the BPM uses broadcast messages, while the Star does not. The Tree
and Shuffle networks differ in that the Shuffle has no root node with its associated
bottle-neck. The difference between the Full and the Full with CAPP machines is
that the Full with CAPP uses broadcast and report buses for its communications.
In the next section we examine more sophisticated metrics which hope to capture
these distinctions.

3.3   Message Capacity Metrics

Next we examine metrics for those structures that characterize how well the
parallel machines support multiple message transfers between arbitrary processors
in the network. We want to judge the ability of the network to handle the "typical"
communication patterns that will be encountered during the execution of "typical"
programs. Even though these metrics seem more general and less precise than
Diameter and Bandwidth they do capture important properties of the networks
and for some problems they can be used for deriving Lower Bounds. These
"message capacity" metrics are based on an examination of message density along
each of the interconnection paths between processors in the machine. They give
a figure of merit for the message capacity of the network. There are several ways
of calculating message capacity.

For comparison, we have chosen three different measures of message capacity.
The measures are Path Count, Narrowness, and Thickness. Path Count can be

Table 1. Metrics of the Architectures

Machines			Metrics				
			. . . . . Paths				
	Diam.	Bndwdth.	Total	Avg.	W − C	Thick.	Narr.
Serial	1	1	1	1	1	1	1
Linear	N	N	N^3	N^2	N^2	1	N
Star	1	1	N^2	N	N^2	N	1
Tree	$\log(N)$	N	$N^2 \log(N)$	$N \log(N)$	N^2	1	N
Shuffle	$\log(N)$	N	$N^2 \log(N)$	$N \log(N)$	$N \log(N)$	N	1
BPM	1	1	N^2	N	N^2	1	N
CAPP	1	1	N^2	N	N^2	1	N
Full	1	N	N^2	N	N	N^2	$1/N$
Full/CAPP	1	N	N^2	N	N	N^2	$1/N$

thought of in terms of the number of paths through each node of the network, or equivalently the number of messages each node must handle in the task of each processor sending one message to every other processor. Narrowness measures the worst-case bottleneck which occurs when one partitions the network in all possible ways. Thickness is a refinement of Bandwidth; it is based on the task of each processor sending one message to one other processor in the "other half" of a partitioned network. The bisection width is minimized for all partitionings of the network into two parts, with half of the processors in each partition.

3.3.1 Path Counting Message Capacity

Our first measure is an extension of the ideas in Horowitz and Zorat [26]. They use the number of communication paths which each processor in a network is responsible for as a measure of the communication overhead in the system.

We define a Path in terms of the particular task of every processor sending one message to every other processor in the network. This generates $N * (N - 1)$ original messages. We make a distinction between original and total messages. For most machines, messages must be forwarded several times to get to their destinations. Each time a message is forwarded it is counted again.

We say there is a Path through a processor, X, if processor X is the source or destination of a message, or if processor X forwards a message from some processor, Y, to some processor Z. The number of paths through X is the number of messages (one to every other processor) that X creates, plus the number of messages (one from each other processor) that X terminates, plus the number of messages that X forwards for other processors, which are also sending one message to every other processor.

Using these rules we can calculate the total number of messages involved in the task of each processor sending to every other processor, and therefore calculate the number of paths that each processor must handle. We calculate the total number of messages, the average number of messages each processor must handle, and the number of messages the "worst-case" processor must handle.

3.3.2   Congestion based Message Capacity

Another measure, "Narrowness" [33] gives a more direct measure of the congestion to be expected in a network. The idea here is that for algorithms with lots of message traffic, the congestion dominates all other interactions in the system. Therefore, a measure which abstracts out that single property of the network is a good metric.

We calculate the Narrowness of the network as follows. We partition the network into two groups of processors A and B where the number of processors in each group is a and b respectively and assume $b \leq a$. Now we count the number of interconnections between A and B, call this i. We find the maximum value of (b/i) for all partitionings of the network. We call this measure the Narrowness of the network.

3.3.3   Worst-Case Bandwidth as Message Capacity

Gannon [22] uses a modified version of Bandwidth to give a figure of merit to networks. His work is an extension of the ideas used in VLSI complexity theory [52]. The basis of the measure is again that we partition the network, this time into two equal halves. Now we assume the underlying task is for each processor in one group to send a message to another processor in the other group. We compute the normalized time for this task. Gannon calls this the "Bisection Bandwidth." To avoid confusion we will call it the "Thickness" of the network.

Gannon shows how we can use this measure to give lower bounds for certain problems. Assume P is a problem for which parallel algorithm A provides a solution on a given architecture. If A needs to send k messages between two halves of the network during its execution and the Thickness of the network is T, then the communication part of the execution of A is bounded by k/T. If, in addition, the Numerical part of the execution of A has S steps and the network has N processors, then the entire execution time is bounded by:

$$\text{Time} \geq S/N + k/T.$$

Gannon examines a class of algorithms based on "Divide and Conquer" with a Shuffle Exchange communication pattern. Of course, not all algorithms can be characterized by this exchange operation and determining k for other solution techniques is somewhat tricky. This does not negate the value of Thickness as a measure of machine communication capabilities. It is a valid measure of the different interconnection structures' abilities to support message passing.

We should note that for our machines, Thickness is proportional to $N/$ Narrowness. This is a consequence of the fact that the (b/i) value is maximized when the networks are partitioned in half. This is not generally true for all networks. Networks made up of tightly interconnected sub-networks, loosely interconnected, generally will not have their narrowest point at the middle of the network. Rather, it will be between some sub-network and the rest of the network.

3.4   Algorithms

The algorithms have been chosen to represent a broad range of complexity. They are also important "kernel" algorithms which occur as sub-problems in larger programming tasks. We explain our choice of algorithms and the algorithms themselves in the following sections. broadcasting, reporting, extrema, and packing are discussed first. Then we discuss sorting and the minimum spanning tree. We briefly give our motivation for selecting these algorithms and review some related work in the following sections.

We have deliberately chosen communication intensive problems for use in studying the metrics. This is despite the widely accepted belief that multiprocessors, in general, are more efficient when the ratio of communication to computation is low [30, 27, 24]. Jones, for example, calls the amount of computation done per communication the "grain size" of the tasks. She recommends that programmers try to keep the grain size of their parallel algorithms high. In other words, one would like to have the "semantic content" of the communications as high as possible. That way each message would be maximally useful to the completion of the total algorithm.

For our purposes in assessing the metrics, however, it is more informative to consider small grain size algorithms. There are two reasons for this. First, since our interest is on the effect of different network organizations on system performance, we want to "overload" the network as much as possible, i.e., stress the multiprocessor on the side of communication rather than computation. We have done this in order to clearly expose the impact of organization on performance.

Second, it is often possible to trade communication time for computation time, and in general that is the kind of tradeoff that will be done by creators of parallel algorithms. But in our analysis, we do not want the possible tradeoffs to complicate the issues of communication behavior. The simple algorithms were chosen expressly to keep such tradeoffs from obscuring the issues of communication behavior that are our primary concern.

The first four computational problems that we address in this section are relatively simple. Broadcasting is the problem of sending one value to all processors in the network. Reporting is the problem of determining if any processor has a particular value. Broadcasting and reporting are essentially pure communication algorithms; there is virtually no computation component to the algorithms. Maximum finding is finding the largest value from a set of items distributed one to a processor in the network. Packing is the problem of squeezing null values out of a sparse array of item values. The items are distributed throughout the network, and the algorithm must preserve the relative order of the non-null items. Maximum finding and packing involve slightly more computation than broadcasting and reporting, but they are still predominantly communications problems, with very minor computational components.

Sorting and computing the minimum spanning tree of a graph have been chosen as more complex tasks which also need a large amount of communication. For both these tasks our algorithms are based on initial conditions of each processor having one element of the input set. This might not always be the fastest way to compute the result, however we still want to use communication rather than computation whenever possible.

3.4.1 Broadcasting

Broadcasting is the task of sending a single message to all processors in the multiprocessor system. This is a basic operation for all multiprocessors. It is often provided by the hardware architecture itself, as in the case of the CAPP and the BPM machines. Even when it is not provided in hardware, it is a necessary function and must be somehow implemented to allow the processors to at least initiate and terminate tasks. Many parallel algorithms use broadcasting as a subalgorithm in their larger computation. For example, in several of the minimum spanning tree algorithms we need to tell every processor which node of the tree was chosen after each iteration of the algorithm.

For those machines that do not support broadcasting directly in hardware, the time that it takes to perform the operation is critical. As we see in later sections, it is often the most limiting aspect of the computation. Jordan [28] and Nassimi and Sahni [40] present algorithms for broadcasting in the Finite Element Machine and a SIMD square grid machine respectively.

3.4.2 Reporting

The purpose of reporting is the opposite of that for broadcasting. It is to gather information about the state of the network to a central location. In some cases the result is available to all processors as a side effect of the algorithm. The algorithm gathers one bit of information, reflecting existence of a particular value in the network. The demands the algorithm makes on the network are different than for broadcasting since there is some computation involved, logically OR'ing partial results before sending a new result on to its next destination. Even though this is a rather trivial amount of computation it imposes a synchronization constraint on the algorithm.

It is also worth noting that, aside from buses, Full interconnect structures are optimal for broadcasting problems. However, they are not optimal for fan-in tasks like reporting. For fan-in, binary (or trinary) tree structures are optimal [33]. Of course, the Full interconnect can just use the subset of its connections which make up a binary tree.

For broadcasting, the order that data flows around the network is irrelevant. Even redundant copies of the data value were not a problem. For reporting things are different. Each processor must wait for any partial values it is responsible for combining to arrive before it sends its results on. The delay due to synchronization is a coefficient on the time of the algorithm. Although it has no effect on the asymptotic running times of the algorithms, the difference is important.

3.4.3 Extrema (Maximum or Minimum) Finding

Extrema, or maximum, finding is the first "non-trivial" algorithm we explore. It is generally not available as a hardware operation in multiprocessors. It is an extension of reporting and can often be implemented as multiple calls to a report operation. It is also a useful sub-algorithm, particularly in algorithms where the multiprocessor is performing a distributed search operation. Each processor can speed up its local processing by having the best value found anywhere available for its local processing.

Maximum finding on multiprocessor architectures has been widely discussed. Hirschberg [25] and Chang and Roberts [12] discuss circular, or ring, architectures and give $O(N * \log(N))$ algorithms. Kung [30] describes algorithms on systolic arrays and trees. In trees, extremum can be found in $O(\log(N))$. Bokhhari [8] presents an $O(n^{2/3})$ algorithm for a square, n by n array augmented by a global bus. For a common memory model, Valiant [54] and Shiloach and Vishkin [45] present algorithms that have the surprisingly low running time of $O(\log(\log(N)))$. However, they do not discuss memory contention problems.

3.4.4 Packing

Packing is the task of moving data from higher numbered processors to lower numbered processors that have space in such a way that the relative order of the data among processors is unchanged. This is often used as a sub-algorithm to sorting and searching problems. As an example, parallel distribution sorts use this sub-algorithm. Packing is also a way to increase the locality of data in distributed processing algorithms. The multiprocessor can "shift down" related data so that the processors which have the data are physically near each other. Schwartz [44] presents packing on the Ultracomputer and on perfect shuffle machines.

Besides its applications, packing is interesting from a theoretical point of view. The average amount of data to be moved is less than that for sorting (or routing), and the average distance for data items to move is inversely proportional to the number of data items. Unlike routing, there is no way to know in advance where the data items are to end up since we do not know how many holes exist in the system. But, unlike Sorting, we do not have to do any comparisons since the data is already in order. The serial complexity of packing is $\Omega_i(N)$ since each value must be examined to see if, indeed, it is a hole.

There are several types of parallel algorithms for packing. The first are simply extensions of the serial algorithms. For both the Star Pack and the BPM Pack algorithms the Bandwidth of the network only allows for the disposition of one hole (or one non-hole) at a time. There are also "data migration" algorithms, where the active processors move data into non-active neighbors, or non-active processors ask for data from their active neighbors. Running times for these algorithms are dominated by the Diameter of the interconnect, as illustrated by the Linear Pack and the Tree Pack algorithms. Our Full Pack and Shuffle Pack algorithms are "swapping" based algorithms. Values and holes are exchanged in a pre-specified pattern based on the shuffle pattern.

3.4.5 Sorting

After examining extrema finding and packing, it is natural to examine sorting on a multiprocessor. Baudet and Stevenson [4] present sorting on linearly connected processors in $O(N)$ time. Kung, Valiant, and Shiloach all extend their discussions of Extrema Finding to include sorting. Trees and one dimensional systolic arrays can sort in $O(N)$. Shiloach sorts in $O(\log(N)^2)$ with common memory. Dewitt [15] and Thompson [53] survey results in parallel sorting. The most recent and complete bibliography on parallel sorting is in Richards [42].

3.4.6 Minimum Spanning Tree

The minimum spanning tree problem is interesting because it has several different solution techniques both for serial and parallel computers. Deo [14] gives several algorithms for a common memory architecture, the best of which has a time complexity of $O(N * \log(N))$ for an N processor, N vertex system. Bentley [6] gives an $O(N * \log(N))$ algorithm for a tree machine of $N/(\log(N))$ processors. Savage [43] has a N^2 processor algorithm which runs in $O(\log(N)^2)$. Chang [10,11] and Yogen Dalal [13] give results for graph-structured parallel architectures, where the graph of the network is the graph for which the problem is to be solved.

There are several considerations which have a strong influence on how we can write parallel solutions for the MST problem. The first is a synchronization problem that occurs when we perform some of the operations in parallel. For instance, both vertices on an edge might have that edge as their minimum cost edge. Once one vertex claims that edge, the second vertex cannot use that edge too. Neither can it simply use its next best edge. The second best edge from that vertex might or might not be in the tree. Once two vertices share a common minimum edge, they must cooperate to find new edges. The general rule is that the edge that is picked must be the minimum of all edges of all vertices which have merged into a group, to a vertex which is not in that group.

A more subtle possible problem is that if we use a Sollin style "group merging" based algorithm, groups will merge in decreasing cost chains. Every vertex in all of the newly merged groups needs to know its "new" group identity so that it can evaluate its edges for the next merge. We call this propagation of group names across the network "contagion." The time to resolve the contagion is often the biggest time penalty for such parallel algorithms.

For our algorithms we also use the techniques of Prim, Dijkstra, Kruskal, and Sollin. The CAPP algorithm is the only one that uses the Kruskal style algorithm, and the Full with CAPP machine is the only one which uses the Sollin style algorithm. All the rest are variations of the Prim/Dijkstra algorithm. Our algorithms are for connected non-directed graphs. A summary of the running times of all the algorithms is presented in the next section.

4. Discussion

The details of the running of the algorithms, and their code is reported in [33]. The six problems: broadcast, report, max, pack, sort, and minimum spanning tree, on each of the machines: CAPP, Linear, Star, BPM, Tree, Shuffle, Full, and Full with CAPP, give a total of 48 parallel algorithms. These represent a fairly broad range of computational complexity which we have used for comparison purposes and as the basis of our analysis of the metrics.

The CAPP and Tree algorithms for broadcast, report, and max are based on standard algorithms for CAPPs and Tree Machines presented in Foster [19] and Bentley [5], respectively. The Linear Sort is based on Baudet and Stevenson [4]. The Shuffle Pack and Full Pack algorithms are based on Schwartz [44]. The

Shuffle Sort and Full Sort are based on Stone's mapping of the Bitonic Sort on the Shuffle-Exchange interconnect [48, 3]. The rest of the algorithms presented for these machines and all algorithms for the BPM, Star, and Full/CAPP machine are new.

We have also included a Serial machine and its times for reference. We define broadcasting and reporting for the serial machine as simply the act of performing a memory write or read respectively.

Some of the times in table 2 are given in terms of the value of the maximum value in the instance of the problem (Max). While this is accurate for the BPM, for the CAPP it is more correct to use the maximum possible value across all instances of the problem. Given that we are generally dealing with "almost N almost unique numbers," we will from now on take the value of N as the value of Max in our order relations. This follows the accepted practice of other researchers [53].

Table 2. Running Times of the Algorithms (order approx.)

Machines			Algorithms			
	Broadcast	Report	Extremum	Pack	Sort	MST
Serial	1	1	N	N	$N \log(N)$	$E \log(E), N^2$
Linear	N	N	N	N	N	N^2
Star	N	N	N	N	$N \log(N)$	N^2
Tree	$\log(N)$	$\log(N)$	$\log(N)$	$\log(N)$	N	$N \log(N)$
Shuffle	$\log(N)$	$\log(N)$	$\log(N)$	$\log(N)$	$\log(N)^2$	$N \log(N)$
BPM	1	1	$\log(\text{Max})$	N	N	$N \log(N), E$
CAPP	1	1	$\log(\text{Max})$	N	$N \log(\text{Max})$	E
Full	$\log(N)$	$\log(N)$	$\log(N)$	$\log(N)$	$\log(N)^2$	$N \log(N)$
Full/CAPP	1	1	1	1	1	$\log(N)^2$

4.1   Performance Summary of Diameter and Bandwidth

As we can see in table 2, some of the algorithm-machine performances are Bandwidth limited, some are Diameter limited, and some are simply not well predicted by either Bandwidth or Diameter. We first discuss how Diameter and Bandwidth can be used to explain the behavior of the machines on the simpler algorithms for broadcast, report, max, and pack. Then we proceed to an evaluation of the other metrics and other algorithms.

The CAPP is Bandwidth limited for any operation which involves decision making. The central control must either "orchestrate" the decisions, as it does in maximum finding algorithms, or it must do the decisions serially, as it does for the packing algorithm. The Star machine, also, is severely Bandwidth limited on all operations. Unless we more fully use the processing power of each Spoke processor, it is essentially a Serial processor with N words of memory. Our algorithms were all picked to minimize the processing that each processor would do, so it is no surprise that we get this result. More complex algorithms do somewhat better on a Star.

The BPM has a low Bandwidth, but it has the advantage of global communications. We have defined Bandwidth as the number of unique messages in the system so it has a Bandwidth of 1. However, for broadcast and report the network as a whole is really only transmitting (or receiving) one message. So the BPM does well with its one global message capability. The Star also has a Bandwidth of 1 but it is not a "global one." For packing the BPM and the Star perform the same since these algorithms require $O_i(N)$ different messages.

There are no real surprises in the Linear network performance. It is as diameter-limited as the Star was bandwidth-limited. It also does better on more complex algorithms. The Tree is also Diameter limited. The interesting facts come from the coefficients of the order results on broadcasting and reporting. For broadcasting binary trees are not very good compared to N-ary trees; however, for reporting they are nearly optimal. For these first algorithms the Shuffle Network behaves much like the Tree because it really is designed to be a set of interlocking trees where each node is a root. We see this is a real advantage on algorithms with more data movement. The Full interconnect, as expensive in hardware costs as it is, does not do much better than the Tree and Shuffle networks. This is because we have defined send and receive operations to take one unit of time. As we argued before this is not an unreasonable assumption. The Full interconnect also does better on harder algorithms.

The Full with CAPP machine, of course, does quite well as a lot more hardware is dedicated to communication tasks than in any of the other machines. If we were to call the CAPP an N "processor" SIMD machine, then there are N^2 equivalent processing units in this machine. There are also N broadcast buses and N receive buses.

4.2   Diameter, Bandwidth and Time

We can see a relationship between Diameter and Bandwidth for several of our algorithms:

$$(\text{Diameter} / \text{Bandwidth}) * N = \text{time}$$

This is true for the Star, the Linear, the Tree, the Shuffle, and the Full with CAPP Machines for broadcast, report, max and pack.

This equation reflects the nature of the problems we have chosen to examine and the way we have defined Diameter and Bandwidth. For broadcasting there is no constraint on data movement or timing. For reporting and maximum finding there are slight synchronization restrictions. However, all of these functions have the properties of commutativity, associativity, and replication-insensitivity or "stability." By stability we mean that adding redundant copies of the input will not change the computed value.

Packing does not have as many of the nice properties as the other functions. Its running time on the Linear and Star machines is caused by their Diameter and Bandwidth restrictions respectively. The Tree algorithm is not really fair since we pack towards the root, which is not strictly a solution to the problem. Packing on the Shuffle network is done by taking advantage of its routing properties [48], which allows it to move data between processors that are far apart in $\log(N)$ steps without collisions or congestion. Packing on the Full with CAPP is so fast for two reasons. First, only relative positions are involved, and the tagged architecture removes the data dependency that comes from absolute addressing. Also, the ability of the Full with CAPP to perform unique operations on each data item is used.

The equation is not true for the CAPP, the BPM, and the Full Interconnect. The issue for the Full Interconnect is that each processor cannot examine all N inputs in unit time. For example, in a report algorithm where everyone sends to processor 0, even though the messages get to the processor in time $O_i(1)$, it cannot operate on its inputs fast enough. As we showed, the best we can do is "tree it." The Full with CAPP machine solves the problem by using N broadcast and N receive buses.

For an explanation of the times for the BPM and the CAPP we are brought up against our definition of Bandwidth. We have defined it to mean the total number of different messages in the system. Both of these machines use single global broadcast buses for all communication. The BPM has one for both broadcast and report, the CAPP has one for broadcast and another for report. This means that they will do better than the equation predicts for problems which involve only one global datum. For maximum finding they both use a radix based solution which trades Bandwidth for Diameter. We use an SIMD algorithm which is feasible on networks with low Diameter. For packing both machines do no better than the Star, or Serial machines; this reinforces our intuition that their Bandwidth really is one.

This shows that a global bus is both more powerful and weaker than a full interconnect. For the case of non-unique messages that need to be sent, or formed by consensus, the bus is better. Some algorithms can take advantage of this multiple non-unique message capability. For cases where unique messages are needed, real point to point interconnect is clearly necessary.

For the BPM algorithms we can take advantage of the implicit synchronization imposed by the communication protocol [31, 32]. When there is an underlying ordering relation on the data, each processor can "make assumptions" on the values in other processors. These inferences can be based on what was *not* broadcast at a given time. We use this technique to sort with 100% efficiency; every value is mentioned exactly once in the process.

The BPM algorithms illustrate an important point about communication in multiprocessors. Communication and synchronization are really two aspects of the same coordination problem. In fact, synchronization is really just a special type of communication, where "control information" is sent rather than just "data."

Broadcast, report, max and pack are really not as communication intensive as other problems. These problems only require us to handle each data item once during the computation. Other problems, like sorting and the minimum spanning tree, often require that data items be moved and compared and moved again many times during the course of the algorithm.

4.3 Message Capacity Metrics

The rationale for our Message Capacity metrics is to capture the phenomena which lead to congestion and competition in network-based multiprocessors which cannot be described by Diameter or Bandwidth.

Diameter and Bandwidth are good metrics in the sense that they are easy to compute precisely and give hard bounds on performance. However, they do not always give tight lower bounds. Diameter gives bounds on the possible performance of algorithms which must move data from one "end" of the multiprocessor to the other. Bandwidth gives us a bound on how much data can be exchanged between processors during a computation. However, neither metric captures interference phenomena like the congestion near the root node in Tree networks. They do not predict well the running times for more sophisticated tasks like sorting.

This is not surprising. The information flow needed during sorting cannot be simply characterized. Sorting is more than sending a message down the length of the network or collecting a datum from each processor. There must be patterns of exchanges: in some cases fixed, in some cases data dependent. The data movement itself must not be "serialized" by the constraints of the network. This is really the crux of the matter. The weakest link of the chain (or network) is the rate-limiting factor. By weakest we do not mean non-homogenous processors. It is their position in the network which makes different demands on their equivalent resources. As we see later, the Worst-Case Path metric is the one which best characterizes the requirements of sorting and the minimum spanning tree.

The Message Capacity metrics do capture the actual ability of networks to support high data traffic. However, they are not "fine grained" enough to really bound any particular algorithm. This is as much a result of our lack of understanding of the data movement pattern of algorithms as it is a shortcoming of the metrics.

5.   Evaluating the Metrics

Evaluation of metrics is a subjective task. Different metrics are useful for abstracting different aspects of machine architecture, which in turn impose varying constraints for each algorithm. The values of the metrics were given in table 1.

The quality of the metrics as *predictors* of algorithm performance varies even more than their usefulness in establishing lower bounds. For instance, on the basis of Diameter alone we can set lower bounds for some algorithms on some machines; but Diameter tells very little about actual performance. Diameter is a very good predictor of algorithms on architectures which are "Diameter bound" like the Linear machine; however, it is a bad predictor of algorithms on other machines that have more severe constraints in their Bandwidth or Worst-Case Path Count.

We perform the evaluation of the metrics by defining a quantitative measure of how well the metrics can be used as predictors of algorithm performance. We rate the metrics by comparing their ability to rank our machines in the same order that the algorithms performed. We claim that this indicates how well they have captured the communication abilities of the machines to support real algorithms.

Our rating is done by a Rank Order technique [2]. First, we rank the machines by their performance on each algorithm. This is shown in table 3 where the machines are in "best-to-worst" order. We combine the broadcast and report rankings since they are the same. The horizontal lines indicate a change in running time, e.g., the Full-CAPP machine finds Extremum in $O(1)$ but the Full, Shuffle, Tree, CAPP, and BPM machines take $O(\log(N))$.

Next, we rank each machine by each metric as is shown in table 4. Again, horizontal lines indicate a change of value. We combine the Total and Average Path Counts and the Thickness and Narrowness rankings since these are the same. We use these two tables to generate our values of how well the metrics' ranking of the machines predicts the algorithm performance ranking of machines.

Table 3. Relative Ranking Machines by Algorithm Performance

Broadcast/Report	Extremum	Pack	Sort	MST
Full-CAPP	Full-CAPP	Full-CAPP	Full-CAPP	Full-CAPP
CAPP	Full	Full	Full	Full
BPM	Shuffle	Shuffle	Shuffle	Shuffle
Full	Tree	Tree	Tree	Tree
Shuffle	CAPP	CAPP	BPM	BPM
Tree	BPM	BPM	Linear	CAPP
Star	Star	Star	CAPP	Linear
Linear	Linear	Linear	Star	Star

Table 4. Relative Ranking of Machines by Metrics

Bandwidth	Diameter	Total/Average	Worst Case	Thick/Narrow
Full-CAPP	Full-CAPP	Full-CAPP	Full-CAPP	Full-CAPP
Full	Full	Full	Full	Full
Shuffle	Star	Star	Shuffle	Shuffle
Tree	CAPP	CAPP	Star	Star
Linear	BPM	BPM	CAPP	CAPP
Star	Shuffle	Shuffle	BPM	BPM
CAPP	Tree	Tree	Tree	Tree
BPM	Linear	Linear	Linear	Linear

We do this by giving points to each metric for correct predictions on the ordering of each machine by each algorithm. We give each metric a point for each pair-wise relationship among machines that it predicts. In the rank order of the machines one is either "better than," "equal to," or "worse than" every other machine. This gives us $N * (N - 1)/2$ possible relationships to get right; with eight machines a perfect score would be 28 for each algorithm.

As an example, the Thickness metric ranked the eight machines in the following order: Full-CAPP and Full, best at N^2; Shuffle and Star, N; then CAPP, BPM, Tree, and Linear, 1. To see how well Thickness predicts sorting we compare that ranking to the sorting algorithm rank of Full-CAPP, best at 1; Full and Shuffle, $\log(N)^2$; Tree, BPM, and Linear, N; CAPP and Star, $N * \log(N)$. We show this in table 5.

Table 5. Example Evaluation of a Metric

Thickness	Sorting	Points
Full-CAPP	Full-CAPP	6
Full	Full	5
Shuffle	Shuffle	4
Star	Tree	2
CAPP	BPM	1
BPM	Linear	0
Tree	CAPP	0
Linear	Star	0
Total Points =		18

The points are computed as follows. Sorting ranked the Full-CAPP machine as better than all others; Thickness ranked it better than six machines but equal to the Full machine. It got six (out of seven) relations right, so we give Thickness six points on that line. For the next line, sorting ranked the Full machine equal to the Shuffle machine and better than the rest. Thickness ranked the Full machine as better than all six of the other machines. It got five (out of six) relations right, so we give it five points for that line.

On each line, we only look down the list so that we do not count relations twice. We accounted for the relation between the Full-CAPP machine and the Full machine on the first line so we need not consider it again.

Moving down the table we give four points for the fact that Thickness predicted the Shuffle machine would be better than the Tree, BPM, Linear and CAPP machines. It missed the relationship of the Star machine to the Shuffle machine. It gets two points for predicting that the Tree machine would do as well as the BPM and the Linear machines, and we give it one point for grouping the BPM machine with the Linear machine. It gets all the other relationships wrong.

The total score for the Thickness metric to predict the ability of our parallel architectures to sort is 18 out of 28. Carrying out the same calculation for each of the metrics and each of the algorithms gives us the results in table 6. For comparison, a Monte Carlo estimate of 10,000 trials between pairs of random rankings of eight items (each divided into three groups) gives an average value of 9.8.

Table 6. Comparative Evaluation of Metrics

Metrics	Algorithms						
	Broadcast/Report	Extrema	Pack	Sort	MST	*Average*	*Rank*
Diameter	16	12	10	7	10	11.00	4
Bandwidth	8	10	18	16	13	13.00	3
Total/Avg Path	14	10	8	7	8	9.40	5
W-C Path	10	14	20	20	16	16.00	1
Thick/Narrow	8	12	16	18	13	13.40	2

This rank-order based evaluation method has two nice properties. First, it only uses the relative values of both the metrics and the running times. We do not have to normalize any values. Second, it gives credit for "clustering" of the machines in the predictions. It does not penalize a metric too much for getting one relation totally wrong and thereby ruining the absolute order of the lists. For instance, if three machines were predicted to perform the same and they did, the metric gets credit for the "equal to" relations even if the rank of all three machines was wrong.

Table 6 illustrates several important points about the metrics. Diameter is a reasonable predictor of broadcast and report, but not very good at predicting the other algorithms. Bandwidth, on the other hand, is better at predicting the pack, sort, and MST algorithms. Both Total and Average Path Counts do poorly.

However, the Worst-Case Path Count is the best metric for predicting all the algorithms except broadcast and report. Thickness and Narrowness are tied for second place among the metrics.

The next to last column of table 6 shows the average prediction ability of each of the metrics. As above, a 28 would be a perfect score. The last column gives the relative rank of the metrics as predictors.

6. Conclusions
6.1 Algorithms as Metrics

Since broadcast and report are primitive to very many parallel processing tasks, we are lead to consider using the machine running times for these algorithms as metrics themselves. We have extended our comparison of metrics to include all our algorithms, as metrics for predicting machine performance (table 7).

We can see that, without exception, every algorithm is a better average predictor than any of the metrics in table 6. The average values in table 7 do not include the ability of an algorithm to predict itself. The minimum spanning tree algorithm is the best predictor. We believe this is because most of the programs written for the MST were made up of applications of the other algorithms as subtasks. In the next section we build on these ideas and propose a "performance suite" for parallel architectures.

6.2 A Performance Suite

In another age of Computer Architecture, architects designed "Business Machines" and "Scientific Data Processors" and pretended that those machines needed to be different. People soon discovered that the basic operations of machines were really the same no matter what "high level" operation they were doing. They Moved data, Operated on data, Branched, Saved and Restored state.

Even though the "operate on data" aspect of computer design can be very complex, it is conceptually a single function of the machine. People are rediscovering this phenomena with such architectures as the Reduced Instruction Set Computer (RISC) architecture [16].

This is not to say that machines should not support high level operations. Rather that even when they are supporting high level functions, they are doing it by the same basic operations listed above, no matter if it is done in macro-code, micro-code, or even "nano-code."

In this respect, multiprocessors are not so different from uniprocessors. Not only are each of the processors that make up a multiprocessor performing the basic operations, but the parallel processing machine as a whole is also performing those same operations. Here we are not talking about a random collection of computers

Table 7. Evaluation of Algorithms and Metrics as Predictors

Algorithms	Algorithms						
	Broadcast/Report	Extrema	Pack	Sort	MST	*Average*	*Rank*
MST	17	22	22	20	–	20.25	1
Pack	16	18	–	20	22	19.00	2
Extrema	20	–	18	15	22	18.75	3
Sort	12	15	20	–	20	16.75	4
Broadcast/Report	–	20	16	12	17	16.25	5
W-C Path	10	14	20	20	16	16.00	6
Thick/Narrow	8	12	16	18	13	13.40	7
Bandwidth	8	10	18	16	13	13.00	8
Diameter	16	12	10	7	10	11.00	9
Total/Avg Path	14	10	8	7	8	9.40	10

connected together, although we could even make the argument for that case, but a homogenous multiprocessor which is performing a single task.

Looked at in this context our work has simply expanded on the basic machine operation "Move Data" to reflect the possible complexity involved.

Following this thought, a set of "Universal" tasks should be appropriate to test the relative merits of multiprocessors. This set should contain a representative mix of communication patterns in much the same way that the Gibson [23] mix of operations is useful as a universal test for the "operate" group of instructions. This idea of a set of test cases is not new [37]; rather, it is time to extend these ideas to parallel architectures.

It is clear from our work that the fundamental operations of broadcasting and reporting should be included in this mix. Additionally, we have seen that sorting, as the general extension of routing, is also a fundamental operation. Our experience with packing and the minimum spanning tree has shown that not only is data routing important, but also data-dependent decisions need to be considered. This is reflected in the serializations caused by the dependencies in packing and the contagion in the minimum spanning tree. Although these two tasks are not themselves universal, they do seem to encompass important aspects of interprocessor communication.

We propose the following tasks be used as a "Performance Suite" for evaluating the communication structures of parallel architectures:

- Broadcasting - - This is essential for any machine to support coordination of tasks. We need to broadcast simply to initiate most algorithms.

- Reporting - - Similarly, reporting is necessary for coordination and control.

- Selecting - - Maximum Finding or other "stable" functions which collect data from all nodes. Summing and other functions which are commutative and associative but not stable might be as good. The stable functions were easier to program, but their running times were not any faster than the corresponding non-stable functions would be.

- Sorting - - Routing and sorting reflect the machines abilities to support arbitrary communication patterns. This is the same as supporting all permutations.

- Propagating (Contagion) - - Packing or the Transitive Closure of a graph would all test the same abilities.

- Saturating (Many to Many) - - This is the task of each processor sending a message to every other processor. This measures bottlenecks or congestion in the machine.

There are many other possible tasks which could be included in such a mix besides these six. These are well documented in this paper and in the literature. We leave it to others to define the "ideal" set.

6.3   Future Work

As discussed in the Introduction, the ultimate goal of this research is to develop a theory of parallel complexity. We believe that a step towards that goal is a characterization of the communication structures of parallel architectures and the communication needs of parallel algorithms. Our goal has been to generate a set of metrics useful in this characterization. We have performed the exercise of solving a representative set of problems on a group of parallel machines and have discussed the techniques involved in writing those algorithms. We hope that this work will lay a foundation for describing the communication structures of parallel machines which is necessary for a usable theory of parallel complexity theory.

We propose that, until there is such a theory, an empirical approach is possible and desirable. This approach could be based on comparative studies of machine performance on a suite of algorithms. This would allow us to continue our exploration of parallel architecture communication structures.

Acknowledgments

This work was supported, in part by AFOSR/DARPA grant F49620-86-C-0041. We would like to thank Richard Weiss, and Beverly Woolf for their help with drafts of the paper. We would also like to thank Caxton C. Foster for his ideas, support and guidance.

References

[1]   Anderson, G. A., and Jensen, E. D., "Computer interconnection structures: taxonomy, characteristics, and examples," *Computing Surveys*, vol. 7, *4*, pp. 197-213, December, 1975.

[2]   Andrews, T. G., *Methods of Psychology*. New York: John Wiley and Sons, 1978.

[3]   Batcher, K. E., "Sorting networks and their applications," *AFIPS Spring Joint Computer Conference*, vol. 32, pp. 307-314, 1968.

[4]   Baudet, G., and Stevenson, D., "Optimal sorting algorithms for parallel computers," *IEEE Trans. on Computers*, vol. C-27, *1*, January, 1978.

[5]   Bentley, J. L., and Kung, H. T., "A tree machine for searching problems," CMU-CS-79-142, Dept. of Computer Science, Carnegie-Mellon University, Pittsburgh, Pennsylvania, August 30, 1979.

[6]   Bentley, J. L., "A parallel algorithm for constructing minimum spanning trees," CMU-CS-79-142, Dept. of Computer Science, Carnegie-Mellon University, Pittsburgh, Pennsylvania, August 30, 1979.

[7]   Bilardi, G., Pracchi, M., and Preparata, F. P., "A critique and an appraisal of VLSI models of computation," in H. T. Kung, B. Sproul, and G. Steele (Eds.), *VLSI Systems and Computations*. Rockville, Maryland: Computer Science Press, pp. 81-88, 1981.

[8]   Bokhari, S. H., "MAX: An algorithm for finding maximum," *Proc. of the 1981 Int. Conf. on Parallel Processing*, Bellaire, Michigan, IEEE Computer Society, pp. 302-303, August 25-28, 1981.

[9]   Borodin, A., and Hopcroft, J. E., "Routing, merging and sorting on parallel models of computation (extended abstract)," *Symposium on the Theory of Computing (STOC)*, pp. 338-344, May 1982.

[10] Chang, E. J.-H., "Decentralized algorithms in distributed systems," (Ph.D. Thesis), *Technical Report CSRG-103*, Computer Systems Research Group, University of Toronto, Toronto, Canada, October 1979.

[11] Chang, E. J.-H., "An introduction to echo algorithms," *Proc. First Int. Conf. on Distributed Computations*, Alabama, October 1-5, 1979.

[12] Chang, E., and Roberts, R., "An improved algorithm for decentralized extrema-finding in circular configurations of processors," *CACM, 22*, 5, pp. 281-283, May 1979.

[13] Dalal, Y. K., "Broadcast protocols in packet switched computer networks," *Technical Report No. 128*, Digital Systems Laboratory, Stanford Electronics Laboratory, Department of Electrical Engineering, Stanford University, Stanford, California, 94305, April 1977.

[14] Deo, N., and Yoo, Y. B., "Parallel algorithms for the minimum spanning tree problem," *Technical Report CS-81-072*, Computer Science Department, Washington State University, Pullman, WA, March 2, 1981.

[15] DeWitt, D. J., Friedland, D. B., Hsiao, D. K., and Menon, J., "A taxonomy of parallel sorting algorithms," *Computer Sciences Technical Report No. 482*, Computer Sciences Department, University of Wisconsin, Madison, Wisconsin, August 1982.

[16] Fitzpatrick, D. T., Foderaro, J. K., Katevenis, M. G. H., Landman, H. A., Patterson, D. A., Peek, J. B., Peshkess, Z., Sequin, C. H., Sherburne, R. W., and Van Dyke, K. S., "VLSI implementation of a reduced instruction set computer," in H. T. Kung, B. Sproul, and G. Steele (Eds.), *VLSI Systems and Computations*. Rockville, Maryland: Computer Science Press, pp. 327-337, 1981.

[17] Flynn, M. J., "Some computer organizations and their effectiveness," *IEEE Trans. on Computers*, vol. C-21, 9, September 1972.

[18] Foster, C. C., *Computer Architecture*. (Second Edition), New York: Van Nostrand Reinhold, 1976.

[19] Foster, C. C., *Content Addressable Parallel Processors*. New York: Van Nostrand Reinhold, 1976.

[20] Freeman, H. A., and Thurber, K. J., "Updated bibliography on local computer networks," *Comp. Arch. News*, vol. 8, 2, April 1980.

[21] Galil, Z., and Paul, W. J., "An efficient general purpose parallel computer,"

Symposium on the Theory of Computing (STOC), Milwaukee, Wisconsin., pp. 247-262, 1981.

[22] Gannon, D., "Notes on parallel algorithm taxonomy for numerical problems," Taxonomy of Parallel Algorithms Workshop, Los Alamos National Laboratory, Santa Fe, New Mexico, November 29 - December 2, 1983.

[23] Gibson, J. C., "The Gibson mix," TR00.2043, Systems Development Division, I.B.M. Corp., Poughkeepsie, NY, June 18, 1970.

[24] Gottlieb, A., and Kruskal, C., "Supersaturated ultracomputer algorithms," *Technical Report No. 024*, Department of Computer Science, Courant Institute of Mathematical Sciences, New York University, September 1980.

[25] Hirschberg, D. S., and Sinclair, J. B., "Decentralized extrema-finding in circular configurations of processors," *CACM*, vol. 23, *11*, pp. 627-628, November 1980.

[26] Horowitz, E., and Zorat, A., "The binary tree as an interconnection network: applications to multiprocessor systems and VLSI," *IEEE Trans. on Computers*, vol. C-30, *4*, April 1981.

[27] Jones, A. K., and Schwartz, P., "Experience using multiprocessor systems," *A.C.M. Computing Surveys*, vol.12, *2*, June 1980.

[28] Jordan, H. F., Scalabrin, M., and Calvert, W., "A comparison of three types of multiprocessor algorithms," *Proc. of the 1979 Int. Conf. on Parallel Processing*, Bellaire, Michigan, IEEE Computer Society, pp. 231-238, August 1979.

[29] Knuth, D. E., *The Art of Computer Programming Vol.3, Sorting and Searching*. Reading, Massachusetts: Addison - Wesley Publishing, 1973.

[30] Kung, H. T., "The structure of parallel algorithms," in M. C. Yovits (Ed.), *Advances in Computers*. Vol. 19, New York: Academic Press, pp. 65-112, 1980.

[31] Levitan, S. P., and Foster, C. C., "Finding an extremum in a network," *9th Annual Int. Symp. on Computer Architecture*, Austin, Texas, April 26-29, 1982.

[32] Levitan, S. P., "Algorithms for a broadcast protocol multiprocessor," *3rd Int. Conf. on Dis. Comp. Sys.*, Miami/Ft. Lauderdale, Florida, October 18-22, 1982.

[33] Levitan, S. P., *Parallel Algorithms and Architectures: A Programmer's Per-*

spective, Ph. D. Dissertation Computer and Information Sciences Department (COINS) Technical Report 84-11, University of Massachusetts at Amherst, May 1984.

[34] Levitan, S. P."Evaluation Criteria for Communication Structures in Parallel Architectures"; Steven P. Levitan; 1985 International Conference on Parallel Processing; St. Charles, Ill. August 20-23, 1985.

[35] Lint, B., "Communication issues in parallel algorithms and computers," Ph.D. Dissertation, University of Texas at Austin, TX. 1979.

[36] Lint, B., and Agerwala, T., "Communication issues in the design and analysis of parallel algorithms," *IEEE Trans. on Software Engineering,* SE-7, *2,* pp. 174-188, March 1981.

[37] Lucas, H. C. Jr., "Performance Evaluation and Monitoring", *Computing Surveys,* vol. 3, *3,* pp. 79-91, September 1971.

[38] Mead, C., and Conway, L., *An Introduction to VLSI Systems.* Reading, Massachusetts: Addison - Wesley Publishing, 1980.

[39] Metcalf, R. M., and Boggs, D. P., "Ethernet: distributed packet switching for local computer networks," *CACM,* vol. 19, *7,* p. 395-404, July 1976.

[40] Nassimi, D., and Sahni, S., "Data broadcasting in SIMD computers," *IEEE Trans. on Computers,* C-30, *2,* pp. 101-106, February, 1981.

[41] Preparata, F. P., "New parallel-sorting schemes," *IEEE Trans. on Computers,* vol. C-27, *7,* pp. 669-673, July 1978.

[42] Richards, Dana;"Parallel Sorting - a Bibliography"; SIGACT News, Vol. 18, No. 1; pp. 18-1.28, 18-1.48; Summer 1986.

[43] Savage, C., "Parallel algorithms for graph theoretic problems," Ph.D. Dissertation, Mathematics Dept., University of Illinois at Urbana-Champaign, Report ACT-4, Coordinated Science Lab., University of Illinois, August 1977.

[44] Schwartz, J. T., "Ultracomputers," *A.C.M. Trans. on Prog. Lang. and Sys.,* vol. 2, *4,* pp. 484-521, October 1980.

[45] Shiloach, Y., and Vishkin, U., "Finding the maximum, merging and sorting in a parallel computation model," *Technical Report 173,* Computer Science Department, Technion - Israel Institute of Technology, Haifa, Israel, March 1980.

[46] Siegel, H. J., "Analysis techniques for SIMD machine interconnection net-

works and the effects of processor address masks," *IEEE Trans. on Computers*, vol. C-26, *2*, pp. 153-161, February 1977.

[47] Siegel, H. J., "The universality of various types of SIMD machine interconnection networks," *Fourth Int. Symp. on Comp. Arch.*, March 1977.

[48] Stone, H. S., "Parallel processing with the perfect shuffle," *IEEE Trans. on Comp.*, vol. C-20, *2*, pp. 153-161, February 1971.

[49] Stone, H. S., "Dynamic memories with enhanced data access," *IEEE Trans. on Computers*, vol. C-21, *4*, pp. 359-366, April 1972.

[50] Sutherland, I. E., and Mead, C. A., "Microelectronics and computer science," *Scientific American*, vol. 237, *3*, pp. 210-229, September 1977.

[51] Thompson, C. D., "Area time complexity for VLSI," *Proc. 11th Annual ACM Symp. Theory Comput.*, pp. 81-88, April 1979.

[52] Thompson, C. D., "A complexity theory for VLSI," Ph.D. Dissertation, Computer Science Dept., Carnegie-Mellon University, Pittsburgh, Pennsylvania, August 1980.

[53] Thompson, C. D., "The VLSI complexity of sorting," *IEEE Trans. on Computers*, vol. C-32, *12*, pp. 1171-1184, December 1983.

[54] Valiant, L. G., "Parallism in comparison problems," *SIAM J. on Computing*, vol. 4, *3*, September 1975.

[55] Vuillemin, J., "A combinatorial limit to the computing power of VLSI circuits," *IEEE Trans. on Computers*, TC-32, *3*, pp. 294-300, March 1983.

[56] Weems, C. C., Levitan, S., and Foster, C., "Titanic: a VLSI based content addressable parallel array processor," *Int. Conf. on Computer Circuits*, New York, N.Y., September 29 - October 1, 1982.

[57] Weems, C. C., *Image processing with a content addressable array parallel processor*, Ph.D. Dissertation, Computer and Information Science Department, University of Massachusetts, Amherst, MA 01003, May 1984.

[58] Wittie, L. D., "Communication structures for large networks of microprocessors," *IEEE Trans. on Computers*, vol. C-30, *4*, pp. 264-273, April 1981.

Hierarchy in Sequential and Concurrent Systems

or

What's in a Reply?

Michael J. Manthey

Computing Research Laboratory
New Mexico State University
Las Cruces, NM 88003

The notion of hierarchy as a tool for controlling conceptual complexity is justifiably well entrenched in computing in general, but our collective experience is almost entirely in the realm of sequential programs. In this paper we focus on exactly what the hierarchy-defining relation should be to be useful in the realm of concurrent programming. We find traditional functional dependency hierarchies to be wanting in this context, and propose an alternative based on shared conserved resources. The new hierarchy is non-procedural, causally accurate in the face of non-determinism, and captures the emergent phenomena typical of concurrent systems.

1. Introduction

Anyone who has designed and built software systems cannot help having noticed that, unlike other kinds of programs which compute 'answers', operating systems have a rather different, existential quality. Operating systems are like the channel of a river which guides processes through itself without taking direct part in the flow. To coin a phrase, sequential processes *do*, but operating systems simply *are*.

We can make this intuitive notion precise by investigating the concept of hierarchy not just for operating systems, but for concurrent systems in general. In this paper, we examine the usual hierarchy of computer science, functional dependency, and a new hierarchy especially suited for concurrent systems. In a nutshell, it turns out that the criterion for what constitutes a *level* in a concurrent system is not based on the subroutine relationship, as it is in a sequential process, but rather on the conservation of shared resources among several processes. We also discuss some related historical and philosophical issues, namely reductionism, mechanism, and structuralism, which seem to have gone generally unmentioned in the computing literature.

2. The Concept of Hierarchy

In Computer Science and many other fields [1,7,17,19], hierarchy is a tool used to reduce the conceptual complexity of a system, i.e., as an aid to comprehension and understanding[16]. Hierarchies provide a controlled method of selectively hiding or exposing the detailed workings of a system's components. It appears in programming languages in many forms, e.g., block structure and procedure dependency, and has a great effect on the programmer's conceptualization of a program. Hierarchy is invoked when explaining a system's structure to someone, as well as in designing it in the first place. Given its crucial role in computing, it is therefore rather more than mere hair splitting to choose the *right* hierarchy for the particular tasks at hand.

In order to define the concept of hierarchy rigorously, we first choose a model of computation, the actor model [9,12], in which to embed it. An *actor* is an active entity, which in the present context corresponds to a processor-program pair. The program defines the actor's behavior, while the processor provides the 'action'. Examples of actors are network IMPs, disk controllers, and individual machine instructions and memory cells. In general, an actor may communicate with other actors, called acquaintances, via finite bit strings called *messages*. We restrict our discussion to actor networks where the set of acquaintances of any given actor does not change. Casting these definitions in terms of graphs, an actor network (or *system*) is a finite directed graph G whose nodes are actors and whose arcs connect acquaintances. The restriction to unchanging acquaintance sets means that the connectivity of the system graph is fixed; the restriction is made for the sake of of simplicity, and our results extend naturally to the dynamic case. Figure 1 shows an arbitrary system represented as a digraph; the nodes are actors and the arcs directed communication links over which messages flow.

A *process* is defined as a connected sequence of events, where an *event* is the receipt of a message by an actor. Thus in Figure 1, the sequence *1, 5, 3, 14* could be a process, as could *8, 12, 11, 8, 7*. We can state this more precisely by saying that a process is a *directed walk* in the given system graph G, which excludes as possible processes disconnected event sequences such as *1, 2, 8, 12*. A given system graph represents potentially *many* concurrent activities: the number of processes at any given instant is the same as the number of extant messages, which is a free parameter not expressed by the graph. Since to each event e in a given process we can associate a temporal instant, each process has its own local time frame, and processes have therefore a time-relativistic relationship to each other [11]. In the common case of a sequential program expressed in some language, we can say that the program is the *closed form* of the corresponding 'open form' process.

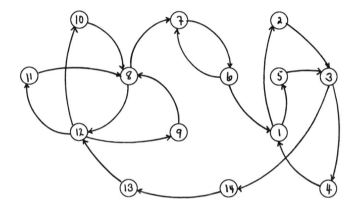

Figure 1. *A System of Actors Represented as a Directed Graph.*

As defined above, a process is a connected *sequence* of events. Since every event is associated with some message, it follows that message content coevolves with the 'movement' of the associated process, and that the instantaneous message content (together with its location in the graph relative to some given node) *is* the instantaneous state of the process associated with that message. This (standard actor model) definition of process therefore differs from the usual in that processes are 'node-to-node' rather than 'within a node'. The recursiveness of the process concept allows us to peek inside a node (presuming the node is not 'primitive'), but to do so requires a hierarchy concept, which is the subject of this paper.

Another way to understand this is to consider where the observer's 'eyeball' is located - if the eyeball is inside a node, then some message arrives and some 'new' message departs; messages are naturally immutable from this point of view. One also naturally focuses on the 'process' inside the node. On the other hand, if the eyeball is in the message, then processes are naturally node to node, and messages naturally mutable. An analogy: a flower's point of view versus a bee's. From the flower's viewpoint, bees arrive and depart, and the focus is on what they do there. From a bee's viewpoint, life is a succession of flowers. The definition of a process as a connected sequence of events is very robust, and can be applied in virtually any setting. See also Wall[18].

Notice that the notion of a process is an abstraction over the activities of individual actors: processes are composed of sequences of actor activations. Because of this, it is a *fact* of this model of computation that *while actors communicate via messages, processes can only communicate via shared memory,* i.e., by visiting (writing and reading) common memory-containing actors. (See [13] for a surprising consequence of this.)

Finally, we note that the dual of G gives the other common graphical model of computer systems, in which the nodes are states, and the arcs are relations between states.

In this paper, we are concerned with the concept of *hierarchy*, with particular reference to hierarchy in computational systems. Generally speaking, a hierarchy is an ordering on a set; in our case, the set is a series of condensations of the system graph G which leaves the properties of the constituent processes intact. The key properties we are interested in here are the identity of the processes themselves, which includes the modularity implicit in their existence in the system or design; and the causal relationships both within and between processes, which subsumes all issues of variable values and states, and therefore functionality.

Definition. *(hierarchy)* Let $G_0 = G$ be a finite digraph. G_0 need not be connected. A hierarchy h on a digraph G_0 is a totally ordered sequence of condensations $G_{i+1} = X_i(G_i)$ induced on G_i by X_i, X_i a partition of G_i, and induced initially by X_0 on G_0. Each G_i is called a *level* of h.

Thus

G_0 the system as given, at the lowest level of description

$G_1 = X_0(G_0)$ the next 'higher' level of description of G, obtained by condensing G_0 with X_0

$$\vdots$$

$G_k = X_{k-1}(G_{k-1})$ the highest level of description.

Higher levels do not exist when for some k, $\forall j > k$, $G_j = G_k$.

In the next two sections, we will discuss two hierarchy relations, h_{fd} (functional dependency) and h_{cyc} (cycle-based). Their definitions are:

Definition. h_{fd} is the hierarchy of graphs $G_{i+1}=F_i(G_i)$, $F_i = \{S_1, S_2, \cdots, S_{k_i}\}$, where each S_i is the set of leaf nodes of G_i having the same parent together with that parent. F_i together with any nodes of G_i not in F_i, taken as singletons, constitute a partition of G_i which can induce the next condensation in the hierarchy h_{fd}.

Note that this definition assumes that G is a tree, which is a basic assumption of the functional dependency viewpoint. If a node has more than one parent as in the case of a shared subroutine, then the node must be replicated, at least conceptually, to satisfy the implicit single-parent rule. See Figure 2 for an example of h_{fd}.

In order to define h_{cyc}, we first must define the concept of a cyclic subgraph.

Definition (*cyclic subgraph*). Given any finite digraph G:

(1) if $S_0, S_1, \cdots, S_{n-1}$ are mutually disjoint strong subgraphs of G, *and*

(2) if $\exists$ arcs of G not in $\bigcup_i V(S_i)$, these arcs at most connect the pairs S_i and S_j acyclically, where $i \neq j$, *and*

(3) if each S_i is a non-trivial, maximum, minimal strong subgraph with respect to (1) and (2), i.e., each S_i is non-trivial and contains no proper subgraphs which satisfy (1) and (2) within S_i

Then each S_i is a cyclic subgraph of G.

Definition. h_{cyc} is the hierarchy of graphs $G_{i+1}=M_i(G_i)$ obtained by condensing G_i by M_i, where $G_1=M_0(G_0)$, $M_i = \{S_1, S_2, \cdots, S_{k_i}\}$ together with any nodes of G_i not in M_i, taken as singletons, is a partition of G_i, and M_i is the *maximum* set of cyclic subgraphs of G_i. Non-conserving S_i's (see §4) are not condensed.

h_{cyc} is discussed more fully in §4. Figure 4 shows an example of the hierarchy generated by h_{cyc}.

An adequate hierarchy relation h must satisfy the following constraints:

    1. It must yield a unique hierarchy on any given graph;

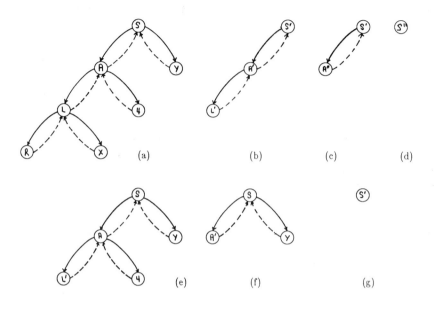

Figure 2. *Two Functional Hierarchy Condensation Rules.* (a) Tree structure of a sequential computation expressed using function composition. The dotted arcs represent *reply* messages containing the values computed by the subtrees; they are usually suppressed. (b,c,d): Condensations of (a) with a simple *reduce leaves* importance rule, $h_{f\,d}$ as defined in the text. (e,f,g): Reduction of (a) with a different importance rule, $S_j \in F_i$ including only those parents and leaves farthest from the root (see §3).

2. It must reflect the connectivity of the graph;

3. It should build itself up slowly, so all intermediate levels are seen.

The first constraint guarantees objectivity: different observers of a system, having the same notions of what constitutes hierarchy, will arrive at the same hierarchy without the need for any decisions based on subjective or personal criteria. We subscribe to the notion that even in non-deterministic systems, there is "at the bottom" a unique causal conceptualization of a system. The second constraint comes from the need to capture causal relations in the form of message transmissions and receipts, and provides an objective basis for the definition of a binary ordering relation between events. Thus, h will preserve causal relationships across levels. The third constraint means that the hierarchy relation should not ignore essential aspects of the system's structure. This reflects the fact that typically, the largest equivalence class, rather than the smallest, has been the one condensed, burying several potential levels of abstraction in one step.

In this paper, we are interested in the question of erecting a hierarchy on a given system, rather than generating a system according to some hierarchy criterion. We choose this approach because it illuminates the prejudices one always carries to the task of figuring out the structure of an existing system, e.g., one inherited upon a job change. It is however also worth noting that the task of inferring structure from a wealth of event sequences is essentially that of understanding the natural world in a rigorous way, that is, the task of science. It is therefore of interest that our proposal for a hierarchy relation for concurrent systems makes central use of the principle of conservation, in the form of resource invariants. Howsoever, once the hierarchy relation is given, systems can be designed and built top-down using the technique of successive refinement, which operates independently of the particular hierarchy relation.

In order to loosen up our thinking, suppose that instead of an operating system one were posed the following gedanken experiment: There exists a large table in a room. On the table are thousands of intricately inter-connected blinking lightbulbs. A lightbulb is on from the time it receives a coded pulse on a line until it has sent all of its reaction pulses. At the base of each lightbulb is some arbitrarily complex digital circuitry, generally different for different bulbs, which determines which of the connected bulbs will be sent a pulse. We assume that information flow between bulbs is discrete and all operations and flows are observable at the bit level. The problem is: given the interconnection pattern and a precise specification of the circuitry at the bases of the bulbs, explain the operation of this system to someone who has never seen it.

The verb 'explain' in the previous sentence is the key word, for it embodies two principles: abstraction and causality. With regard to abstraction, merely showing someone the connection graph and circuit specifications in all their gory detail does *not* constitute an 'explanation' of the system's operation. The construction of abstractions is equivalent to placing a hierarchy on the operation of the system, a hierarchy which says which aspects of the system's operation are in some sense subservient to others. Likewise, an explanation must include a component which says 'how the system works', which is to say causality. Any abstraction or description which, e.g., leaves out some of the arcs, cannot in general yield a *complete* explanation.

The situation is thus the following. We are given a directed graph representing a concurrent system's message traffic and local state transitions. We know exactly what the message contents are and what the nodes do to them. How does one then erect a hierarchy on such a graph?

3. Functional Dependency Hierarchies

In this section, we examine the hierarchy built on the concept of functional dependency, or 'who calls whom', in detail. This entails a discussion of the differences between the mathematical and computational concepts of what a function is. Because of our mathematical training, the computing community has accepted much about functions, which is honored mostly in the breach in real computer systems. For example, subroutines and functions are considered to be identical; and as will be shown, data flow is in spite of appearances fundamentally sequential in concept. We seek therefore to raise, by various considerations, a skepticism about the universality of the notion of functional dependency when concurrency and non-determinism are present.

The concept of a functional dependency hierarchy is a child of the functional view of computation. Included is the mathematical concept of function composition as a means of expressing the very concept of a computation. For example, the straight line assembly language program *LOAD R,X; ADD R,4; STORE R,Y* becomes *STORE(ADD(LOAD(R,X), 4), Y)*. From Figure 2a we can see that a tree structure is implicitly imported as the underlying abstraction, although as we will see below, many acyclic structures are acceptable as well. Indeed, the computation can be viewed as a depth-first walk of the tree structure. In the functional view, cyclic structures are treated as iterative loops and are replaced by subtrees which correspond to tail recursive function calls. If the computation is to terminate, these subtrees must be bounded. In any event, STORE is functionally dependent on ADD, which is functionally dependent on LOAD; either of the latter could have been more complex routines, with corresponding sub-trees.

Figure 2b-d shows the hierarchical condensation of a typical functional dependency tree by our definition of h_{fd}. Here we have used one of several possible rules for capturing the relative 'importance' of nodes, namely that all leaves are less important than their parents. Figure 2b might be read "first do a Load, then an Add, then a Store, and don't bother with the details". Figure 2e-g shows condensation via a different importance rule, based on distance from the root. Figure 2f might be read "store the result of the Add into Y". Depending on one's purposes, either importance rule may be reasonable. The crucial point is that each condensation yields an *accurate*, though increasingly less detailed, description of the computation.

The second importance rule (distance from root) is the one typically used in discussing the levels of a computer system. Notice that the *sequence* of condensations shown in Figure 2e-g does not correspond to our usual use of a functional dependency hierarchy. This is because we usually draw a functional dependency hierarchy at the lowest level of detail (Figure 2a), and directly (and visually) infer the hierarchical relationships which the series of condensations makes explicit. What the series of graphs resulting from the condensing steps is telling us is that there are several *levels* of detail at which we can view the structure of the computation. Common speech reduces this to references to the nodes at a given level, e.g., the ADD is at level 1 (counting from 0).

Shared routines are a common occurrence in real systems, so a basic question to be resolved is how to condense shared sub-functions (Figure 3a). If we persist in a functional view, then the single parent rule can be saved by *replicating* the sub-node, to get Figure 3b, which condenses nicely (Figure 3c). Notice however that there exist computational situations where one *may not logically* replicate, e.g., where the node represents a semaphore or other resource counter. Consider for example a binary semaphore. Being binary, it stands for a *single* allocatable resource. If one replicates it, one implies the availability of *two* resources. Since we are *given* the system (graph) on which the hierarchy is to be erected, replicating resources out-of-hand is simply indefensible, especially in the sense of "leaving certain properties of processes intact", one of our goals. Insofar as semaphores are a creature of concurrent systems but do not occur in sequential ones, we see for the first time how functional dependency takes an implicitly sequential world view.

A second way in which sequentiality implicitly enters is that the sequence of function compositions must generally be pre-specified. This tends to remove any explicit consideration of non-determinism, which is a paramount characteristic of concurrent systems. We have hedged the previous two sentences because of the distinction between determinate and indeterminate computations, and between deterministic and non-deterministic. We define these now:

determinate - the outcome is fixed given the initial state;

indeterminate - the outcome is unpredictable from the initial state;

deterministic - the sequence of elementary events is fixed given the initial state;

non-deterministic - the sequence of elementary events is unpredictable from

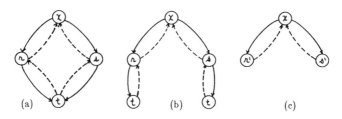

Figure 3. *Condensing Shared Routines.* (a) Routine t is shared by r and s . Allowing replication (b), condensation follows (c). A common structure in real systems, where routines at different levels share t , requires further embellishment of $h_{f\,d}$'s importance rule.

the initial state.

where by 'outcome' we mean some selected 'final state' of the computation in question.

Traditional sequential programs, such as algorithms specified in Pascal or Fortran, are both determinate and deterministic. Data flow programs are determinate, but non-deterministic, i.e., every time a given data flow algorithm is run, the same 'answer' will be produced, but the sequence of elementary computational events along the way may not be the same in each instance. In contrast, the value of a resource counter is in general indeterminate because it is produced by a non-deterministic event sequence, the actual asynchronously arriving resource allocation and deallocation requests.

Data flow, which is clearly based on function composition, is a clever idea exactly because it produces determinate results in the face of the non-determinism. This obtains because the various concurrently executing subfunctions synchronize implicitly through facilities hidden in the parameter-passing mechanism. What in effect is happening in data flow is that the individual function *bodies* specify determinate and deterministic computations, but the functions themselves execute concurrently. In other words, the steps of the tree walk may occur in roughly any order, but no upper level node may 'complete' until its entire subtree has been traversed (ignoring the issue of garbage collecting obsolete activations). From a hierarchical point of view, the interior of any non-leaf node may contain non-determinism, but this is invisible to other nodes at the same or higher functional levels. It is for this reason that one should distinguish between parallelism and concurrency, because the former implies determinacy, while the latter does not.

The synchronization hidden in the parameter mechanism of data flow yields an implicitly specified parallelism, and removes from the programmer the responsibility of explicitly synchronizing on the access of global variables. In fact, globals are anathema to data flow precisely because they require explicit synchronization. Referring to our earlier definition of the process concept, and the means by which processes may communicate, we see that the determinate outcome of data flow programs rests on the fact that the various processes cannot communicate through the shared memory which globals supply. The only communication allowed is through the (hidden but synchronized) local shared memory of the formal parameter slots. In principle, and we are arguing principles here, data flow simply *cannot* produce an indeterminate value.

The incompatibility of shared memory with data flow is not just a quirk of implementation. Rather, it reveals what we think is a fundamental incompatibility between the mathematical concept of a function and the computational concept of memory. Strictly speaking, a function should always return the same value given the same input parameters. That we may have bastardized the original mathematical concept of a function with our use of it in computing should not obscure this fact. If a function retains, via its use of memory, some knowledge of its activation history, and uses this knowledge in computing subsequent values, this is a violation of the spirit of the function concept. Such considerations have not bothered theoreticians of course, since they either just consider each such memory-modified function to be a different function, or move all local memory into the parameter list (which is not allowed for semaphores, as noted above). Nevertheless, a basic issue is being swept under the rug here, and that is the issue of *time*.

Ideally, a function is timeless - it computes the same value given the same input parameters regardless of 'when' it is invoked. If one reflects on the tools which mathematics offers, the vast bulk of them have this same timeless quality. A theorem proven true is true, not just true at certain 'times'. An equation (with its equal sign) states a time-wise unconditional equality; notice that program invariants are equalities. It is truly no accident that we so carefully distinguish equality from replacement (:=). Nor is it any accident that language theoreticians have discovered that the replacement statement is the root of much semantic evil in programming language design. As soon as the concept of memory is introduced, as it is with the replacement statement, one gets immediately the concept of state, and thence of state sequences, and thence of *time*. The place where the tools of mathematics have most successfully dealt with time is with differential equations. The formalism of the calculus and

differential equations assumes either one-dimensional time or globally synchronized time. But the multiplicity and relativity of the time frames of the
individual concurrent processes [cf. §2] are incompatible with these assumptions.

We note finally that without the concept of memory, processes cannot communicate. This is a consequence of our model of computation, but since this
model is universal, this same conclusion must be found in any model, even if
unobvious on the surface.

The critique to this point includes 1) functional dependency hierarchies cannot
deal with functions which represent resource counters, 2) the functional concept of a computation contains implicitly the concept of a determinate 'value'
for the computation's outcome, 3) implicit in the production of a value is the
termination of the associated computation, 4) the concept of a function is not
compatible with the concepts of memory and state, 5) the concept of time
(and hence such things as asyn*chron*y) inheres in the concepts of state and
memory, 6) processes cannot communicate without memory.

To these we add the comments (re 1) that a resource counter is memory-
bound, and (re 2&3) that concurrent systems, such as operating systems, do
not really compute a value as such, nor are they intended to terminate. In addition, synchronization, although very real, is intentionally transparent to the
state of the synchronizing process, and hence is *invisible* to the functional
viewpoint. Consider also a request to print a line on a printer - replies (as opposed to acks) really don't make any sense, because the relationship between
caller and callee is process-process, not function-subfunction, where the former
is the more general relationship. For these reasons alone, it seems reasonable
to be skeptical of the concept of functional dependency as a basis for constructing a hierarchy and defining the concept of level for concurrent systems,
especially real ones.

There is yet another reason, not related to the issues of memory and time, for
being dissatisfied with functional dependency as the basis for an abstraction
hierarchy on a concurrent system. One reason we chose an event-oriented
model like the actor model as the basis for our considerations is its intuitive
closeness to the way real computer systems work. The message passing paradigm makes explicit the informational and control relationships between actors. More specifically, the *implementation* of a function requires two messages,
a *request* from the caller, and a *reply* containing the computed 'value'. But
when the functional dependency hierarchy is erected, replies play no role, only
requests (i.e., who *calls* whom). However irrelevant replies may be to the functional view, they nevertheless represent distinct and very real causal events.

Any hierarchy mechanism which systematically eradicates roughly one half of the causal events in the system it purports to describe cannot make a very strong claim to being complete in its description.

4. Resource Hierarchies

In searching for an alternative to functional dependency as a hierarchy generating relation, we considered numerous possibilities. A principle criterion for eliminating candidate relations was that of preserving the connectivity of the graph, i.e., the causal relations implied by the given event sequences. We have already seen one application of that principle in the preceding paragraph. A second elimination criterion was that of objectivity - that questions of personal taste or opinion should not play any role.

As an example of the application of this second criterion, consider the fact that any straight-line piece of code can be clumped arbitrarily, which corresponds to splitting or combining nodes in the graph. Reasonable people can differ on what is the 'best' such partitioning. The same applies to the choice of a 'modular set of primitives'. The result of such considerations was that we should seek a *topological* invariant, so that the particular partitioning would not play any role. This led to the idea of looking at cycles in the graph. [It is interesting here to note that a different way to formulate the problem is that of finding some island of objective stability in a vast sea of generally non-deterministic event sequences - something which does not change even while everything is changing.] The cycle idea connected immediately to the concepts of resources and their invariants, which we will describe momentarily.

The next and harder problem was to find a way to order the cycles by relative 'importance', in order to define the hierarchy relation h_{cyc}. Wanting again to avoid the partitioning question, we again sought a topological means of constructing this ordering. After much experimentation, we settled on the criterion of the coupling of the cycles with each other, which is reflected in our definition of the concept of a cyclic subgraph.

The use of the directed cycles of the system graph as the basic element in defining a hierarchy relation has several very attractive properties. First, as already mentioned, they are a topologically invariant, objective property of the graph. Second, they allow naturally for non-terminating event sequences, which are typical of any system which is not supposed to ever terminate, e.g., a network IMP. Third, and most important, cycles represent the *only* path by which non-consumable resources can be conserved in a system; conserved resources such as mutex tokens are typical of any concurrent system. Points of dynamic stability are also found in cyclic structures, just as a vortex is a point of dynamic equilibrium in a river.

We define a conserved resource to be any resource for which the relationship $c = c_0 + \Sigma deallocations - \Sigma allocations$, where c_0 is the initial amount and c the current amount of the resource, holds for the life of the system. The relationship between cyclicity and conservation is most easily seen in the case of a simple binary semaphore, where a wait-operation results ultimately (presuming no deadlock or starvation) in allocation of the resource in the form of permission to use it. The subsequent signal operation on that same semaphore deallocates the resource by relinquishing the permission. There is clearly a directed cycle from wait to signal.

A less abstract example of the cyclicity of resource conservation is a company which rents cars. If they insist that customers always return the car to the particular location from which they originally rented it, then an observer of the company's parking lot could eventually conclude that the number of cars (barring accidents) was conserved. If customers are however allowed to drop off a car at a different company location, then observation of any one of the lots would not give any reason to infer conservation. Cars would appear to come from apparently inexhaustible sources, and disappear unaccountably into sinks. If one connected these sources and sinks together, i.e., enlarged the bounds of the system under consideration, then the conservation would be observed at a higher level, where 'level' is understood in the sense of h_{cyc}.

The h_{cyc} hierarchy is built out of the shared conserved resources in a system, and is based on the observation that a (re-usable) resource is one whose amount is conserved over time. Conservation of resource cannot occur without a cyclic message-passing relationship among the system components, although conservation is not thereby guaranteed. The cycle represents the return of the "access right" to the issuing "resource manager". Clarke's algorithm [4] for calculating the conservation-determining resource invariants given the allocate/deallocate (P/V) sequences of the constituent processes can be applied. Although this algorithm is defined for the restricted class of concurrent computations specifiable by some number of individually cyclic processes, it appears that a large number of practical systems nevertheless are covered, especially if one treats a sequential process as one which loops zero times, just as an aperiodic function is considered to be periodic for the purposes of Fourier analysis.

With this background, the method for constructing h_{cyc} on an arbitrary system graph G, which at each level is the same as calculating the maximum set of cyclic subgraphs of G_i, is as follows: Denote by $C_0 = (c_0, c_1, ... c_n)$ the set of all cycles of G_0. C_0 empty implies that h_{cyc} doesn't exist. Notice that each c_i is an equivalence class of potentially resource-conserving processes. Construct from C_0 the *maximal* subsets of mutually disjoint cycles to obtain the set M_0. The maximality criterion establishes "least importance", since it favors local cycles over global ones; it also embodies the third hierarchy criterion, that of building up the hierarchy as slowly as possible.

Consider an element $m_j \in M_0$. Then $m_j = \{c_k, c_l, ..., c_n\}$. The mutual disjointness criterion means that each of the $c_l \in m_j$ shares no nodes with any of the other c 's in m_j. Since they are only indirectly coupled via arcs external to them, each cycle can be individually considered. If the processes in the cycle equivalence class conserve any resources, then replacing the cycle by a single node will simply bury a behavior which cannot be seen from outside the cycle anyway. As long as no *other* behavior is buried, but rather used to characterize the *new* node, the description of a system whose graph has been so condensed will remain accurate.

If M_0 contains more than one maximal set, this can only mean that some cycles are shared by different m_j's. Said differently, there exist *cyclic subgraphs*. We could not find any way to pick apart such closely coupled processes, and so we simply conjoin the associated equivalence classes. If a cyclic subgraph so formed conserves resources, then it is condensed as if it were a simple cycle. The algorithm for producing the cyclic subgraphs of G_0 given M_0 is given in the Appendix.

When all the conservative cyclic subgraphs in M_0 have been condensed to single nodes, we will have calculated $G_1 = M_0(G_0)$, which is the next most detailed level of abstraction of the system. See Figure 4. We note that highly interconnected graphs, such as complete graphs, typical n-cube structures, and chains where each node both sends and receives from its neighbors, generally condense in one step. This is a consequence of restricting h_{cyc} to the purely topological properties of the system. Whether this phenomenon is ultimately an advantage or a disadvantage depends on one's purposes.

M_1 is now calculated, ignoring any non-conservative cycles from M_0 graphically, but *not* while calculating the invariants, and the condensation repeated, etc. In this way, a sequence of abstractions $G_1, G_2, ..., G_k$ of G_0 will be produced. The sequence terminates when an acyclic graph, or a graph all of whose cycles are non-conservative, is reached. G_k is the simplest description of the original system that can be obtained by h_{cyc}. At this highest level, the system may be describable functionally, which does not contradict our earlier critique of functional dependency hierarchies. We observe that a proof that this highest level satisfies a formal system specification implies that the system has been

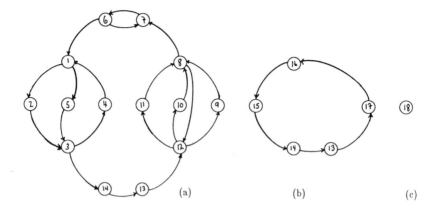

Figure 4. *Condensation Using* h_{cyc} . This graph is Figure 1 redrawn, showing the necessity of having an algorithm to pick out the substructures, rather than relying on visual cues. The text gives the details of the condensation calculation.

implemented correctly.

In detail, the calculation of the condensations shown in Figure 4 is as follows; we assume all cycles are conservative. The cycles of G_0 are

$$a = \{3, 14, 13, 12, 9, 8, 7, 6, 1, 2\}$$
$$b = \{3, 14, 13, 12, 10, 8, 7, 6, 1, 2\}$$
$$c = \{3, 14, 13, 12, 11, 8, 7, 6, 1, 2\}$$
$$d = \{3, 14, 13, 12, 9, 8, 7, 6, 1, 5\}$$
$$e = \{3, 14, 13, 12, 10, 8, 7, 6, 1, 5\}$$
$$f = \{3, 14, 13, 12, 11, 8, 7, 6, 1, 5\}$$
$$m = \{6, 7\}$$
$$n = \{1, 2, 3, 4\}$$
$$o = \{1, 5, 3, 4\}$$
$$p = \{8, 12, 9\}$$
$$q = \{8, 12, 10\}$$
$$r = \{8, 12, 11\}$$

So $C_0 = \{a,b,c,d,e,f,m,n,o,p,q,r\}$, and the maximal mutually disjoint subsets of C_0 are $M_0 = \{ \{m,n,p\}, \{m,o,p\}, \{m,n,q\}, \{m,o,q\} \{m,n,r\}, \{m,o,r\} \}$.

Since M_0 contains more than one element, we must calculate their conjunction. Using the algorithm in the Appendix, we find that the maximum set of mutually disjoint cyclic subgraphs of G_0 is $M_0 = \{ \{n,o\}, \{m\}, \{p,q,r\} \}$. Label the condensations of these cyclic subgraphs *15*, *16*, and *17* respectively; the result of condensing these is shown in Figure 4b. Iterating the algorithm on Figure 4b, we get $C_1 = \{17, 16, 15, 14, 13\}$, and $M_1 = \{ C_1 \}$ since there are no other cycles. Label the condensation of this single element of M_1 node *18*, and we have Figure 4c. Iterating the algorithm on Figure 4c we find there are no cycles, so the hierarchy calculation halts.

The sequence of graphs shown in Figure 4, which is a hierarchy on G_0, is unique on the given graph[15] and captures that which is most characteristic of concurrent systems - shared conserved resources. It has the quality of focusing on invariant rather than arbitrary or evanescent qualities of the system, and there is a sine qua non relationship between those processes exhibiting such invariant behavior and the hierarchy which they form. Finally, we note that the cycle hierarchy is a purely structural concept, and hence is divorced from the actual details of resource management, e.g., deadlock avoidance schemes or nested monitors.

5. Examples

In this section we give several examples of the applicability of h_{cyc} to concurrent systems. Our goal is to demonstrate that h_{cyc} creates those abstractions of concurrent systems we *expect* it should, and hence is a valid tool for conceptualizing and designing such systems.

Our first example is a typical von Neumann sequential computer. One normally regards the memory of such a computer as a reasonable unit entity consisting of address decomposition logic and a large number of flip-flops. But what justifies regarding this large collection of circuitry as a 'reasonable unit entity'? It is a fact of the actual implementation of a memory module that of the (say) 4K cells it contains, only one may be accessed at a time. This is of course a result of certain economically driven design decisions, but nevertheless this is how memory units work. From our point of view, some process accesses a memory cell by first obtaining an exclusive *access permission* to the module, visits the desired memory cell, returns the access permission for use by the next accessor, and goes on its way. There are clearly 4K touching cycles here in each of which the number of access permissions ($=1$) is conserved. h_{cyc} specifies that such a cyclic subgraph should be condensed, and hence, and in accordance with our expectations, the collection of circuitry and cells is correctly abstracted to a single unit entity, a memory module.

Let us now examine the control unit of a von Neuman computer. Again we ask, whence comes the presumption that a 'cpu' is a 'reasonable unit entity'? Whether or not it is in fact implemented as such, we can correctly view the operation of the control unit as a micro-program in the form of an Ifetch loop. The form of the loop *implicitly* expects that there is only one process executing the micro-code and manipulating the machine's control registers. This does not preclude that there *could* be more. Rather, if there were more than one process executing the micro-code, either there would be multiple register sets and re-entrant code, which doesn't change the issue; or the sharing of the single register set would imply extra synchronization, which would condense in its own way according to h_{cyc}; or serially reusable micro-code, which is the case most relevant in the present context. In the archetypical case of a single process executing the Ifetch loop, the implicit sequentiality postulates an equally implicit access permission, which permission is conserved over the Ifetch cycle. h_{cyc} thus properly condenses the Ifetch loop and produces the unit entity we call a cpu.

Let us now widen our scope to a computer network, which one usually conceptualizes as a collection of nodes and communication links. Is this conceptualization reasonable, considering that such a network is in reality made up of large numbers of communicating routines, complex data structures, and lesser black boxes? To answer this question, we first imagine the network at a much lower level as a vast directed graph, considerably larger than the network itself. The nodes of this graph are the memories and cpu modules constituting the network. The arcs are the paths which the protocol and other control processes follow in carrying our their duties. In this analysis we will not distinguish between short and long haul communication arcs over which the messages of whatever kind flow. For the sake of simplicity, and without loss of generality, we will assume that each memory module consists of a single cell, holding a single datum or machine instruction.

Considering first that local collection of nodes we would identify with a single computer, a given (say) protocol process will, on a multiprogrammed interleaved basis, visit the micro-code engine, acquire the Ifetch access, execute an instruction, including the implied trips to various memory cells, and return the Ifetch access. The state of this process is encoded in some message in accordance with our basic message-passing model of computation. Among other items in this state will be access permissions to various data structures in the memories. Just as the Ifetch access permission is acquired and released, so are the access permissions to these data structures. These acquisitions and releases must, as argued previously, necessarily imply cyclic graphical structures. h_{cyc} will therefore specify a sequence of condensations, taking in the first step the individual Ifetch loops and memory cells.

The next step will condense the appropriate permission-conserving software-level cycles. Assuming no access permissions are passed over long-haul arcs, these software-level condensations will eventually encompass everything we normally associate with a 'network node'. Using h_{cyc} we have once again reached our usual conceptualization of a computer network.

If we continue our condensations from the original (computer and comm-link) level upwards, we might find cyclic traffic flows representing (probably approximate) conservation of messages. We note that from the point of view of the network, messages are non-consumable resources. h_{cyc} would condense such a cycle, which might for example represent the New York - Boston communications corridor, or perhaps a collection of LANs characterized by high intra-mural traffic. Such flows are the computational analog of the river vortices mentioned earlier.

The vortex analogy is but one example of what should be obvious: the natural world is a concurrent system. We mention therefore one more example from this domain. It is well known that atoms are made up of protons, neutrons, electrons and other such particles. Viewing an atom as a concurrent system, the characterization of (say) a helium atom is an entity with two protons, two neutrons, and two electrons. The fact that a helium atom normally endures from one time instant to the next, and that its constituent particles (really, their quantum numbers) are in fact what *define* its helium-ness, is but another way of stating that the numbers of these various particles are conserved. The cycle hierarchy once again produces the abstraction we expect (see also [13]).

6. Discussion

Our objections to functional dependency as a general purpose hierarchy relation are three, in order of increasing seriousness: 1) it ignores very real causal events (replies); 2) it is incompatible with the concept of memory, a necessary pre-requisite for inter-process communication; and 3) data flow notwithstanding, it reflects a deterministic model of computation, while concurrent systems are generally non-deterministic. In constrast, the hyper-cyclic shared resource hierarchy which we have proposed suffers from none of these maladies. Instead, it focuses on that which is most characteristic of concurrent systems: conserved shared resources.

It is interesting to note that there is an historical parallel, in quantum physics, to this conflict between the sequential and the concurrent; see for example [10,14]. The non-determinism of quantum mechanical systems was very difficult for the physicists of the early part of this century to accept, and for decades it was argued by some that the non-determinism was perhaps due to "hidden variables", i.e., as yet unknown variables at a yet deeper level of the (function composition) hierarchy. The statistical approach, in contrast, accepted the non-determinism at face value, and has proven to be the more successful approach by far. From our concurrent and h_{cyc} viewpoint, it is not hard to see that this outcome was the only one possible.

One final point we would like to make concerns the scientific philosophy called reductionism; see Crowley's excellent essay[6], to which this article can be seen as a continuation. A strict reductionist will claim that a detailed understanding of the operation of the lowest level of a system is sufficient to understand all higher-level system behavior. As an example, a reductionist would claim that if we understood completely the neural level of the brain, then we would directly understand our intelligence as well; thus, at least one neuron *must* possess some function which *is* intelligence.

An integral part of the reductionist position is determinism: event sequences at the lowest level determine exactly all higher level phenomena; the whole is exactly the sum of its parts. This is h_{fd} in a nutshell. In addition, the reductionist position is typically (and incorrectly) equated to that of the mechanist, which is that all higher-level phenomena are grounded in mechanism at a lower level. Both h_{fd} and h_{cyc} are mechanistic hierarchies, but as we explain next, h_{cyc} allows one to be a mechanist without being a reductionist.

In opposition to the reductionists stand the structuralists, whose position is that the whole can be greater than the sum of its parts, i.e. that new phenomena can *emerge* as a result of level closure; such phenomena are called *emergent*. Vitalism, the concept that there is a "life force" peculiar to living systems, is an example of a (now discredited) emergent phenomenon; on the other hand, morphogenesis is rife with examples of emergence. An example of an emergent phenomenon in computing is deadlock: it arises out of the *collective* behavior of the individual processes, and cannot be found anywhere in the scripts specifying the individual behavior of any of these [5]. General Systems Theory [2,3] is basically a structuralist approach to systems, although most treatments abuse the notion of emergence. Quantum interference [8] is clearly an emergent phenomenon, since *two* entities are required for its appearance. It should be apparent that h_{cyc} captures the emergence associated with conservation laws, as well as other types of emergence which depend on cyclic interaction.

Real-life examples of emergent phenomena are, e.g., the chemical properties, as distinct from the nuclear properties, of atoms and molecules. Traffic based phenomena in computer networks also yield many examples. We stress that we are operating here with a very precise notion of what constitutes an emergent phenomenon, and that nothing magical is being claimed. Nothing happens in an emergent phenomenon which is not grounded in mechanism at a lower level, ad infinitum. Rather, due to the non-determinism of interactions and the suppression of more detailed mechanism through hierarchical condensation, it makes sense to shift perspectives to concentrate on newly dominant behaviors. What the cycle hierarchy does is to point out which sets of process interactions should not be considered further, with the picture becoming simpler and less detailed at each successive level upwards. It should be apparent that one cannot expect emergent phenomena within a sequential paradigm, i.e., a functional dependency hierarchy.

On the one hand, no computer scientist would accept that anything happens in a computer system, no matter how complex, that does not ultimately result from the execution of primitive machine instructions. Hence, computer scientists are fundamentally mechanists. On the other hand, only a fool would claim that it is possible to explain the dynamic behavior of an airline reservation system by studying, however closely, the micro-code and micro-program dynamics of the networked computers in question.

[Although one can argue that this is an unfair example, the non-determinism being due to the entry into the reservation system of 'external' processes, the same is ultimately true even of a closed system. The process of arbitrating competing requests for resources inevitably introduces non-determinism, since completion times for the processes will vary as a result. Memory bus arbitration is a simple example. Practical considerations such as slight variations in component performance (e.g. crystal oscillators and communication links) also dictate the presence of non-determinism. Finally, replacing any component with another (e.g., a faster cpu) destroys any design decisions based on predicted completion times. It is thus hopeless to design concurrent systems assuming determinism, and this is the reason that in concurrent system design, one always assumes non-determinism.]

We believe that our analysis of the two types of hierarchy clarifies the debate between reductionism and structuralism. As long as one considers only sequential systems, a strict reductionist position is entirely defensible. In the case of concurrent systems, however, the determinism which sees higher level processes as simple concatenations of lower-level causal sequences fails entirely, for at each level of a concurrent system, as defined by h_{cyc}, the processes of that level interact non-deterministically to produce behavior characteristic of that level. To study a concurrent system, a computer scientist must therefore become a structuralist, and abandon reductionism. The crux of the matter is what is meant by the phrase "an explanation". As stated in the introduction, an explanation must include both abstraction and causality. A corollary is that a system may legitimately, and often considerably more clearly, be described in terms of interactions among these abstractions, rather than among their constituents.

Relative to statements of causality in the two types of hierarchy, we have seen that functional dependency equates causality to determinism and hereby attempts to jam concurrent systems into a mold that they in principle cannot fit. In a phrase, functional dependency hierarchies express global determinism. The cycle hierarchy, on the other hand, accepts the non-determinism and indeterminacy and extracts from it whatever invariants exist. Causality is by no means abandoned, but determinism is restricted to the individual sequential process segments which together compose the system at hand. It is unnecessary to assume or establish either global determinism or global determinacy.

Thus a reductionist description of a concurrent system has little explanatory power, and the reductionist position makes little sense when concurrent systems are at issue.

In summary, computer scientists are fundamentally mechanists, since they accept that no higher-level phenomenon can occur which is not grounded in lower-level mechanism. The split comes when one considers sequential processes versus concurrent systems. The former cannot give rise to emergent phenomena, and hence the reductionistic hierarchy relation h_{fd} is entirely appropriate. Concurrent systems do give rise to emergent phenomena, and hence require a hierarchy relation, h_{cyc}, which accepts this and the other unique properties of such systems. Non-determinism is *the* basic emergent phenomenon, a phenomenon which we speculate forms the basic intuitions underlying the notion of randomness.

7. Conclusion

We have posed the question of "explaining" an existing system (e.g., the light-bulb network) in order to expose some important issues of hierarchy in concurrent systems. Functional dependency as a hierarchy relation was found to be wanting in that 1) very real causal events - reply messages - are not represented; 2) functions are ideally memoryless, yet memory is the sine qua non of process communication and resource management, hallmarks of true concurrent computation; 3) functional dependency is determinate by nature, yet concurrent systems are indeterminate; and 4) functional dependency captures the reductionistic notion that the whole equals the sum of the parts, a notion which is demonstrably inapplicable to the concurrent context.

We conclude that the concept of functional dependency, though undoubtedly useful, is a creature of the sequential world of computation. The hierarchy defined by h_{cyc}, on the other hand, does not ignore any causal events, and accurately reflects the characteristic properties of a concurrent system, including emergence. In addition, a much richer set of graphs (i.e., system structures) is naturally encompassed. We stress that we are not postulating that functions cannot compute anything we might wish, but only that functional dependency hierarchies should not be used to describe concurrent systems.

On the basis of this analysis, we perceive the current trend toward applicative (function-based) languages, especially when prejudiced in the direction of high-speed computation, also known as 'parallelism', with some misgiving. Their strength (no memory implies no side effects) is also their weakness (no communication), so general insight into or specification of concurrency will be hard to come by via this route. Object-based and data-abstraction languages in turn we find more attractive; these languages embrace concurrency with much less reservation. The construction of large systems using these and their descendants will require a suitable concept of hierarchy. For such languages, built-in facilities to maintain hyper-cyclic relationships appear to be a valid area for extension.

The concept of emergent phenomena is one which we think computer scientists should adopt and apply. Ignoring emergent phenomena will generally result in an unreliable or inefficient design. Similarly, the art of inspired design is the exploitation of such phenomena for the purposes of the design as a whole. Examples of both can be found in bridge building. The Tacoma Narrows bridge was destroyed by emergent oscillations. The keystone arch, which creates a space-time deadlock, laid the foundation for centuries of architecture.

Acknowledgement

I am grateful to Bernard Moret, Don Morrison, Anita Jones, and particularly to an anonymous reviewer of an earlier version of this paper for their careful readings.

References

1. Atkin, R.H. *Mathematical Structure in Human Affairs*. Crane, Russak, and Co. 1974. (See also M. Snyder in Computing Reviews, 1978, #33514.)

2. Bahm, A.J. "Five Types of Systems Philosophy". Int. J. General Systems, 1981, Vol. 6, pp. 233-237.

3. Bertalanffy, L. *General Systems Theory*. George Braziller, New York, 1968.

4. Clarke, E.M. Jr. "Synthesis of Resource Invariants for Concurrent Programs". ACM TOPLAS 2,3. 1980.

5. Coffman, E.G. Jr., and Denning, P.J. *Operating Systems Theory*. Prentice-Hall. Englewood Cliffs, NJ. 1973. p.35.

6. Crowley, C.S. "Structured Programming is Reductionistic!". SigPlan Notices 15,5. May 1980.

7. Eigen, M. and Schuster, P. *The Hyper-Cycle - A Principle of Natural Self-Organization*. Springer Verlag, N.Y. 1979.

8. Feynman, R.P. et al. *The Feynman Lectures on Physics*, vol. III, §1. Addison-Wesley, 1965.

9. Hewitt, C. et al. "Behavioral Semantics of Non-Recursive Control Structures". Lecture Notes in Computer Science. Vol 19, p. 385. Springer Verlag, N.Y. 1973. or Hewitt, C. Bishop, P., and Steiger, R. "A Universal Modular ACTOR Formalism". Proc. IJCAI 1973.

10. Jauch, J.M. *Foundations oj antum Mechanics*. Addison-Wesley. 1968.

11. Lamport, L. "Time, Clocks, and the Ordering of Events in a Distributed System". CACM 21,7. July, 1978.

12. MacQueen, D.B. "Models for Distributed Computing". IRIA Technical Report 351, Rocquencourt (France). 1979.

13. Manthey, M.J. and Moret, B.M.E. "The Computational Metaphor and Quantum Physics". Communications of the ACM February 1983.

14. Manthey, M.J. "Non-Determinism Can Be Causal". International Journal of Theoretical Physics 23,10, October 1984.

15. Manthey, M.J. "A Unique Hierarchy on the Successive Condensations of a Digraph." CRL Technical Memo MCCS-86-49. Computing Research Labaratory, NMSU, 88003.

16. Simon, H. "The Architecture of Complexity". In *The Sciences of the Artificial*, 2nd Ed. MIT Press. Cambridge, MA. 1981.

17. Thissen, W. "Investigations into the World3 Model: Lessons for Understanding Complicated Models". IEEE Trans. on Systems, Man, and Cybernetics. SMC-8,3. March 1978, pp183-193.

18. Wall, D. "Messages as Active Agents." ACM POPL 1982.

19. Warfield, J.N. *Societal Systems - Planning, Policy, and Complexity.* John Wiley and Sons. 1976.

Appendix

Algorithm for resolving multiple maximal sets of cycles into cyclic subgraphs

{The algorithm below shows in detail how the maximal set of cyclic subgraphs of a graph is calculated from the maximal sets of mutually disjoint cycles. Not shown are the calculation of the original cycle set itself and the consequent maximal set(s) (C and M in the text). The basic strategy is an iterative union of cycle sets which share nodes. This code has been extracted from a program which calculates and interactively displays the cycle hierarchy. }

```
type
    node  =     1..maxnode;
    cycle =     set of node;
    cyclst =    ^cycrec;
    cycrec =    record
                cyc : cycle;
                next : cyclst
                end;

procedure joinoverlaps (var L : cyclst);
```

{Merge any cycles (elements of the list L) whose intersection is non-nil.
The result in L is a set of cyclic subgraphs}

```
var
    R,      {local result -- assigned to L at end}
    T,      {temporary}
    prev, cur   {used for working our way down L and possibly
                deleting an element of L as we go}
```

: cyclst;

begin
    R := nil;
    while L <> nil **do**
        begin
        {Merge with the first element of L any elements in the remainder of L that
        have a nonempty intersection. Note: elements that are merged with the
        first element of L are removed from L.}
        prev := L
        cur := L^.next
        while cur <> nil **do**
            if L^.cyc $\cap$ cur^.cyc <> ϕ **then**
                begin
                {Merge and remove the current cycle.}
                L^.cyc := L^.cyc $\cup$ cur^.cyc;
                prev^.next := cur^.next;
                dispose (cur);
                cur := prev^.next
                end
            else
                begin
                {Get the next element of L.}
                prev := cur;
                cur := cur^.next
                end

        {Put the first element of L onto the front of R and remove this element from L.}
        T := L^.next;
        L^.next := R;
        R := L;
        L := T
        end

    L := R
end

Section II. Application Domain Characterizations of Parallelism

The driving force behind parallel architecture design has been the need to execute time critical or large scale applications as fast as possible. This might mean real-time response for speech recognition or image understanding, or reasonably rapid turnaround for aircraft designs based on numerical simulation. In some cases entire families of computers have been designed around the organization of a few algorithms. In fact, the area of algorithmically specialized computer design is now a well established component of the computer industry.

In this section we consider the characteristics of concurrency from the point of view of the applications specialists. There are six chapters that cover four basic areas: image processing, speech recognition, partial differential equations and numerical linear algebra. In each case the authors have focused on those characteristics of the application that have a direct influence on machine characteristics needed to exploit the natural algorithm parallelism.

In the case of image processing, Stefano Levialdi points out that image processing is far more than simple pixel level array operations. Rather, there are a host of higher level processes that put different demands on the architecture. He shows that some algorithms are well suited to cellular processors, others are natural for pipelines, but many can best exploit the organization of a "pyramid" design. In one view of pyramids, at the bottom level cellular pixel based computations can be performed, while at the top, higher level decisions are computed. In an alternative scheme for the use of pyramids, the bottom level mesh supports neighborhood operations, while the two-dimensional tree structure of the upper levels is used to provide fast global communications.

When considering the case of speech recognition, Anantharaman and Bisiani examine the class of algorithms that can be characterized as searching a Markov network. Systolic algorithms are shown to be well suited to exhaustive search strategies, but are not well suited to non-exhaustive searches in which data-dependent decisions are used to customize the search to the data at hand. They therefore focus on the architectural requirements for non-exhaustive search. In particular, they show that a decomposition of the process yields a data-flow style pipeline and a system with up to 128 processors. They conclude with a simulation study of this design.

In the third chapter, John Rice considers the problems associated with the parallel solution to partial differential equations. This is a topic that has been at the heart of parallel computation research for 20 years. Traditionally work in this area tended to focus on the fact that PDEs are solved by building a discrete mesh or grid structure that approximated the differential operator. The evaluation of the approximating numerical operator is a process that can be cast in terms of numerical linear algebra and, as such, it is highly parallel. Rice argues that this view is both unnatural and, ultimately, limited in application. Rather than extract the concurrency from the abstraction of numerical linear algebra, one should turn to the natural physical geometry of the problem. Viewed from this perspective we give up none of the fine grained parallelism provided by the traditional approach, but we gain a new dimension of concurrency provided by a decomposition of the geometry into superelements. Another important point he makes is that special application domains like solving PDEs are best programmed in very high level languages. Because these special languages allow the programmer to express the computation more in terms of the geometry and natural mathematics of the problem, it is possible to convey more information to the compiler. Consequently, there is a greater potential to exploit automatically some of these higher forms of concurrency.

The last three chapters in this section consider the issues associated with parallel numerical linear algebra. This area is the one in which the greatest amount of both theory and experimentation have been done. Geist, Heath and Ng study the general issues of parallelism for matrix computations. This group of researchers has been one of the most active in the area of designing mathematical software for hypercube parallel systems. Consequently, their research has focused on the tradeoffs between task granularity and communication complexity for non-shared memory MIMD processors. In particular, they observe that the differences between a shared memory and non-shared memory machine have important consequences in the selection of algorithm features for which parallelism can be applied. Factors that influence algorithm design include the need to do dynamic load balancing, data structure partitioning, and the organization of the interface between the calling application and the matrix algebra software.

The last two papers in this section deal with issues that are critical to the design of software for the class of machines known as "vector multiprocessors." The systems in this family have shared memory with a relatively small number of processors where each processor has a powerful vector instruction set. On one hand, the shared memory environment makes it very easy to move programs from a standard serial "vax world" machine and achieve relatively good performance. On the other hand, the system has two dimensions of parallelism: multiple processors and vector operators. This implies that there are a large number of tradeoffs in the design of concurrent algorithms for these machines. Because there are a large number of different machines in this class, one critical issue in algorithm design is portability. In fact, one can argue that a reasonable way to classify parallel processors into related families is to see how much algorithm restructuring is needed to move a program from one machine to another.

One approach to building portable numerical programs is to write them in terms of high level mathematical functions such as the BLAS (Basic Linear Algebra Subroutines). If the hardware vendor provides an optimal implementation of the BLAS for his system, the issue of portability is solved provided that the optimal algorithm for that machine can be implemented in terms of these primatives. This is the issue addressed by Bill Harrod. He shows that this may not always be the case. In fact, he demonstrates that both the BLAS and the vector oriented extension BLAS-2 are both too low level. In order to fully exploit the power of a vector multiprocessor, it is necessary to work with higher level operators based on matrix algebra.

The last chapter in this section deals with another important factor in parallel algorithm design that is often overlooked in the more theoretical literature. Jalby and Gannon argue that the design of the memory hierarchy in a shared memory system plays a critical role in the structure of parallel algorithms. They consider the design of fast Fourier transformation software in a system with a shared parallel cache. It is shown that there is an analytical model that relates the number of processors, the length of vector operands and the cache size. This model provides the algorithm designer with critical parameters needed to build fast software.

Computer Architectures for Speech Recognition

T.S. Anantharaman and R. Bisiani

Computer Science Department
Carnegie Mellon University
Pittsburgh, PA 15213

This paper describes the computational characteristics of the most representative algorithms for speech recognition and addresses the problem of designing computer architectures that are suitable for such algorithms. The paper also presents the design and performance of two custom architectures tailored to a beam search algorithm. Both architectures have been simulated using real data and the results of the simulation are presented.

1. Introduction

The object of speech recognition is to recover a text string **w** from a preprocessed speech signal **y**. Most speech recognition algorithms share the same strategy: Search for the best feature vector sequence (a feature vector is a transformation of the input speech signal) given some knowledge of the task (for example, a description of legal words). In order to illustrate this we will model the speech process as shown in Figure 1-1.

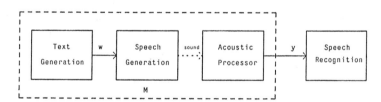

Figure 1-1: A model of the speech process

Depending on the kind of speech recognition task, a text string **w** may consist of a complete sentence or just a word. The speech signal is transformed into a sequence of feature vectors **y** by an acoustic processor. There is one set of feature vectors per time sample, and there exists a suitable distance measure for comparing two feature vectors. The final accuracy of a recognition system depends on the kind of feature vectors and distance measures used. For example, LPC feature vectors together with the *Itakura-Saito distortion measure* [14], are often used in speech precessing. Although much of the research in speech recognition has concentrated on the acoustic preprocessor, there is no agreement as to which is the best preprocessing technique. On the other hand, since all preprocessing algorithms (with the possible exception of auditory modeling [16]) pose no problems as far as computational requirements are concerned, we will not tackle this aspect in this paper (see Brodersen [7] and Lyon [16] for an example of the technology that can be used to do speech preprocessing).

We will consider the first three sections in Figure 1-1 as a 'model' of the process that generates **y**. The problem of speech recognition can now be formalized as follows : Given a model **M** of the combined text, speech generation and acoustic preprocessing, find the text string **w'** that is most likely to have produced the sequence of feature vectors **y**. If $P_M(\mathbf{w}|\mathbf{y})$ denotes the probability of string **w** given the model M and the feature vectors **y**, our aim is to find **w'** which maximizes $P_M(\mathbf{w}|\mathbf{y})$. Since one is only interested in finding the string **w'** with the best probability, the exact probability is not required. Any matching score $S_M(\mathbf{w}|\mathbf{y})$ that is a monotonically increasing function of this probability will do just as well. The model M encodes the system's 'knowledge' and directly influences the accuracy of the system, therefore it should model the actual speech generation and acoustic processing as closely as possible. Realistic models should reflect the possibility of insertions, deletion and substitutions in the phonetic stream due to variation in duration, noise in the acoustic channel, and errors in the acoustic processor. Variations in pronunciation of individual words including the variation due to word junctures should also be modelled. An accurate model of the text generation (*i.e.* the possible **w**'s) is also desirable, in order to be able to bias the recognition process in favor of grammatically more likely strings. Once a good model is chosen, the problem of speech recognition reduces to a problem of efficiently finding the string that maximizes $P_M(\mathbf{w}|\mathbf{y})$. Direct maximization of the match score may not be computationally feasible, and it may be necessary to use heuristics to reduce the amount of computation.

A number of well-known speech recognition algorithms including Dynamic Time Warping algorithms [10], Level Building [15], DRAGON [9], HARPY [13], IBM's Stack Decoding [17], Bridle et al's connected speech recognition algorithm [4] and Two Level DP matching [19] can be characterized as special cases of a general class of algorithms based on Markov Modelling. Each of the algorithms in this class can be characterized by an implicit model M which is a special case of a general finite state Markov model, and a decoding strategy which may or may not be optimal. This makes it possible to directly compare the computational characteristics of these algorithms, including the way these algorithms scale with vocabulary size and complexity of the word grammar.

2. The Markov Model

The Markov model used here is a slight modification of that used by the Stack Decoding Algorithm described in [17], [8]. All possible legal utterances as well as their various pronunciations are modelled by a single Probabilistic Finite State Machine (PFSM). At each time frame the Machine transitions from the current state to one of a set of successor states (which may include the current state), with probability determined by transition probabilities $p(S_i \rightarrow S_j)$. All transition probabilities to a given state add up to one. In this model, outputs occur on transitions. Associated with every transition is a conditional output probability distribution $q(y_k|S_i \rightarrow S_j)$, where the y_k's are the set of possible feature vectors. q() can be viewed as a matching function between any arbitrary feature vector y_k and the feature vector corresponding to the transition from state S_i to state S_j. An example of a PFSM is shown in Figure 2-1.

In this model every path **t** from the initial state to a final state maps to some legal utterance **w** = W(**t**). (This mapping can be encoded directly in the graph, by labelling states at word

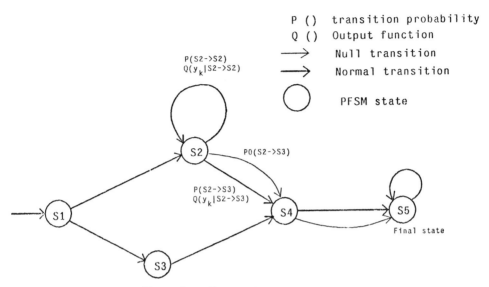

Figure 2-1: Format of the Markov Model

boundaries, with word labels). Thus, each path **t** corresponds to an utterance **w** and a number of possible feature vector sequences **y** produced by it. The problem of speech recognition is to find the best **w** given **y**.

The two most common decoding strategies that can be used are :

1. Best Match Decoding : Find the path **t** that maximizes the joint probability P(**y**,**t**). The corresponding **w** = W(**t**) is taken to be the recognized utterance. The probability of a path **t** (unnormalized) is simply the cumulative product of the transition probabilities $p(S_i \rightarrow S_j)$, and the matches $q(y_k|S_i \rightarrow S_j)$ along the path. Here the y_k's are matched up in sequence against the transitions. (If such a match is not possible, the joint probability is zero).

2. Maximum Likelihood Decoding : Find the **w** that maximizes $P_M(\mathbf{w}|\mathbf{y})$ or, equivalently, the joint probability P(**y**,**w**). This can be shown [17] to be equivalent to finding **w** that maximizes the sum of the path probabilities P(**y**,**t**) over all paths **t** such that **w** = W(**t**).

The Viterbi algorithm [21] is a dynamic programming algorithm which can be applied if the Best Match decoding method is used. Dynamic programming can also be used for Maximum Likelihood decoding by finding the best score at every state for every possible word sequence at that state. Independently of which decoding strategy is used, it is often not possible to exhaustively examine all possibilities, and hence some kind of heuristics is required to reduce the amount of computation. The possible heuristics may be divided on the basis of the search method used :

1. Exhaustive Search : The degenerate case, which is only feasible by using the Best Match decoding strategy, and for small vocabularies. Examples in this class include template matching algorithms [10], Level Building [15] and DRAGON [9].

2. Best First/Best Few Search : The set of possible candidates (utterances **w** or paths **t**) is successively subdivided into smaller and smaller sets, by expanding partial candidates, (e.g., utterances in a left-to-right manner). A scoring function is applied to these candidates. The scoring function should be an estimate of the best possible score possible for any extension of the candidate. At all times the best few candidates are selected for further expansion, and the scoring function is applied to these new candidates. Backtracking will occur if the new candidates have low scores, so that the best few candidates at the next stage include some that were rejected for expansion previously. The first complete candidate is taken as the best candidate possible. IBM's Stack Decoding algorithm [17] is an example.

3. Beam Search : This is similar to Best First Search, but eliminates back tracking. Only the best few partial candidates selected for further expansion are retained at any stage, and the rejected ones are eliminated from all further consideration. This usually results in implementations that are more efficient. For comparable performance the number of 'best' candidates retained needs to be larger than in the Best First Search. The HARPY algorithm [13] is an example.

The heuristics used can also be divided into the following classes:

1. Admissible Heuristics : Heuristics that guarantee the best solution.

2. Opportunistic Heuristics : Heuristics that do not always find the best solution, but tend to find good solutions faster.

Exhaustive search is obviously admissible. Best First search can fall into either category depending on the kind of function used to estimate the score of partial candidates. If the scoring function gives an upper bound on the best possible score for possible extentions, an admissible heuristic is obtained. IBM's stack decoding algorithm for example uses an opportunistic scoring function. Beam Search is always Opportunistic unless the Beam width is so large that candidates are never discarded, in which case Beam Search reduces to Exhaustive search.

3. The Algorithms

Single word TM: Single word template matching [10] attempts to match the input feature vectors with a template (the stored feature vectors of a reference word). This algorithm attempts to locally expand/compress the two feature vector streams in order to get the best match (time warping). In Figure 3-1 the time-warp path shown indicates which part of the input is matched with which part of the reference template. The matching score is the product of the scores at each point in the path. Dynamic programming (the Viterbi algorithm [21]) is used to efficiently compute the score of the best possible path by retaining only the best-scoring path at each grid point. The best score is then the score of the end grid point of the path. Paths are usually restricted to lie near the main diagonal to reduce computation. The local restrictions placed on the shape of the path determine how the best score S_{ij} of a grid point is obtained from the best score of neighboring grid points, and the match D_{ij} between input and reference at the grid point itself. This is illustrated in Figure 3-1.

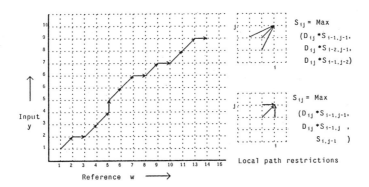

Figure 3-1: Single word template matching

Single word TM can handle deletions, duplications and insertions in the phonetic stream, and can directly deal with variations in duration. It does not directly deal with substitutions and variations in pronunciation, though these cases can be handled by having multiple templates per reference word. Since Single Word TM only matches single words, it does not have to deal with word junctures.

Equivalent Markov model:
> Shown in Figure 3-2 for the case of a simple local path shape. The Markov model for most other path shapes follows similarly. (The main restriction is that with horizontal path segments, a distance score D_{ij} may only be computed for the first grid point of the segment.)

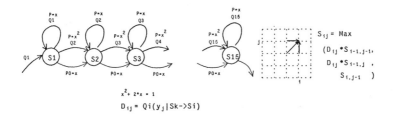

Figure 3-2: Markov model for single word TM

Equivalent decoding algorithm:
> Best Match.

Equivalent search technique:
> Exhaustive search.

Other Heuristics: Restrictions on grid point translate directly into restrictions on the state/input combinations whose scores need to be considered when using the general Viterbi algorithm.

Single word TM can be adapted to perform maximum likelihood decoding by adding the scores of neighboring grid points instead of taking their maximum.

Level Building : This is an example of a multiple word TM algorithm using Dynamic Time Warping. See [15] for an example. The algorithm uses ordinary dynamic time warping for each single word. First, every possible one-word hypothesis starting with the first input is considered and the best possible word ending at each different time frame is found. Then every feasible two-word hypothesis is obtained by extending the best one-word hypothesis with every possible word. This is repeated until some limit L of the number of words in an utterance is exceeded, at which time the best scoring multiword hypothesis is retrieved. This is illustrated in Figure 3-3. Note that in contrast to normal Dynamic Time Warping diagrams, input **y** is shown along the X-axis, while the references are along the Y-axis.

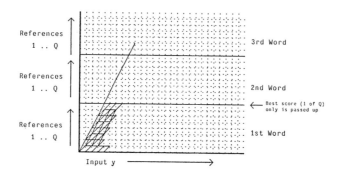

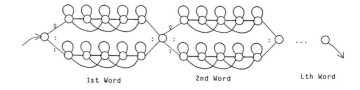

Figure 3-3: Level Building

Equivalent Markov model:
>   See Figure 3-3. The output functions $q(y_k|S_i \rightarrow S_j)$ are the same for all transitions from a given state, hence, they can be considered to be part of the state.

Equivalent decoding strategy:
>   Best Match.

Equivalent search technique:
>   Exhaustive. In using the Viterbi dynamic programming algorithm, the scores for state/input combinations are filled in state by state *i.e.* all scores for the first state are filled before any scores for the next state are considered. This is equivalent to filling in the scores on grids points in Figure 3-1 column by column.

Both Sakoe's Two Level DP Matching [19] and Bridle's Connected Word Recognition Algorithm are very similar to Level Building.

DRAGON : The DRAGON speech recognition algorithm [9] models speech as a Markov process with heuristic state transition probabilities. The possible paths through the Markov graph correspond to all possible phonetic streams produced by the possible text strings. Self looping arcs on nodes model the variation in duration of any segment of the path. See Figure 3-4 for an example.

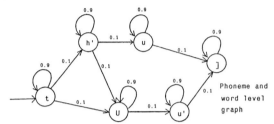

Figure 3-4: A DRAGON Markov graph

Equivalent Markov model:
Already uses a Markov model. The output functions $q(y_k|S_i \rightarrow S_j)$ are the same for all transitions from a given state and hence can be considered a part of the state.

Equivalent decoding strategy:
Best Match.

Equivalent search technique:
Exhaustive. Scores are evaluated one input at a time since this allows evaluation to start before all inputs arrive. This corresponds to row wise evaluation in fig 3-1.

Markov Model with Level Building : This is combination of Level Building and Markov Modelling. Markov modelling as applied to words is the same as presented in Figure 3-2, and it does not change the Level Building algorithm except for the use of Markov modelling techniques to obtain the model parameters (transition probabilities, output functions) from real speech data. The algorithm differs from the normal Level Building algorithm in that a word level Markov Graph is used to restrict the possible word sequences that are being considered. The 'Levels' now correspond to states in the word level Markov Graph. After each 'level', recognition results from predecessor 'levels' are combined by taking the best result for each input instant.

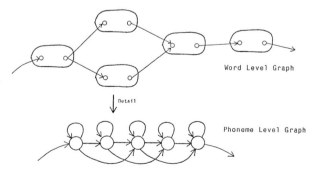

Figure 3-5: HMM and Level Building: The equivalent Markov model

Equivalent Markov model:

See Figure 3-5. The output function is restricted in the same way as in Level Building and DRAGON.

Equivalent decoding strategy:

Best Match.

Equivalent search technique:

Exhaustive. Scores evaluated one state at a time (as in Level Building).

Other heuristics : A special heuristic is used to penalize excessively long or short words. This is equivalent to multiplying the scores at the last state of each word by a vector that is a function only of the word (with penalties at lengths corresponding to too long and too short), shifted by an amount equal to the length of the previous words.

Apart from the heuristics used to penalize excessively short or long words, this algorithm can be viewed as a special case of the DRAGON algorithm.

Beam Search. The first example of Beam Search, the HARPY speech recognition system [13], evolved as an improvement and optimization of DRAGON. HARPY penalizes excessively long or short phonemes by making the transition probabilities depend on the duration at a state (due to self looping). This corresponds to locally constraining the paths along the diagonal in single word TM. HARPY further constrains the paths by using Beam Search, which restricts the number of paths at any (input) time instant to those within a threshold of the best scoring path.

Equivalent Markov model:

Already uses a Markov model. The output functions $q(y_k|S_i \rightarrow S_j)$ are the same for all transitions from a given state, and hence can be considered a part of the state.

Equivalent decoding strategy:

Best Match.

Equivalent search technique:

Beam Search. Scores are evaluated one input at a time since this allows evaluation to start before all inputs arrive. This corresponds to row wise evaluation in Figure 3-1 (with only the best elements in each row being considered).

Other heuristics: Excessively long or short phonemes (word segments) are penalized. The penalty function happens to be such that this heuristics can be exactly modelled by a more extensive Markov model, however it is more convenient to think of the heuristics as a dynamic modification of the scoring procedure.

Stack Decoding : The Stack Decoding algorithm as applied to speech recognition [8], [17] models **M** as a Markov grammar. Unlike in the case of DRAGON or HARPY, transition probabilities are derived from real data. The scoring function is also defined for partial text strings (prefixes of **w**), and stack decoding expands legal strings in a left to right manner in best-first order (the term 'stack' comes from the ordered stack used to store information about partial strings). The scoring function used can be viewed as a generalization of the case of single word TM. The score assigned by the Viterbi algorithm to each grid point in Figure 3-1 can be interpreted as the score of a partial text string $\mathbf{w_i}$ matching a partial input $\mathbf{y_j}$. The first change is to modify the Viterbi algorithm so that scores of all paths converging on a grid point are added, instead of the maximum being taken. Next the representation of the reference **w** is changed from a linear sequence to a Markov graph. Fig 3-6 illustrates this case, with **w** again a single word.

The points along the horizontal axis are now states of the Markov graph (*i.e.* not necessarily in chronological order), and the possible paths through the grid may be more complex, being

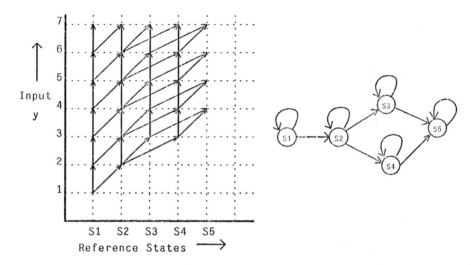

Figure 3-6: Maximum Likelihood Decoding

determined by the possible state sequences of the Markov graph. Path scores can still be computed incrementally, combining paths at the same grid if they correspond to the same text string **w**. However, for multiword Markov graphs more than one score may have been kept per grid point. The score for a partial reference w_i is obtained by taking the scalar product of the vector of scores of the column of grid points corresponding to w_i with some heuristic constant vector (which may be a function of the input but not of **w**). Simply summing up the column vector will result in an admissible heuristics since the score is now an upper bound of the final score that is to be maximized. However, better performance is obtained in [17] by using a heuristic vector that is a function of the input (precomputed before the search is started) and reduces the amount of backtracking.

The basic model used in stack decoding is similar to that of HARPY or DRAGON but more accurate as it is based on actual data. The use of a true maximum likelihood scoring function, and a heuristic best-first search, improve recognition accuracy at the cost of additional computation and memory.

Equivalent Markov model:
> Already uses a Markov model. The output functions $q(y_k|S_i \rightarrow S_j)$ are precomputed for each of a finite number of feature vectors.

Equivalent decoding strategy:
          Maximum Likelihood.
Equivalent search technique:
          Best First.
Other heuristics: An opportunistic heuristics is used to bias the search is favor of an already
                      large hypothesis. Other heuristics are used to avoid computation of scores
                      with very low values, in a manner similar to restricting attention to scores
                      near the main diagonal in Figure 3-1.

4. Algorithms vs. Computer Architectures

4.1. Computation Power

Before launching into the design of specialized architectures one should be really sure that no general purpose processor can solve the problem in a cost effective way. Recently, the gap between the cost effectiveness of general purpose and special purpose machines, when executing AI applications, has been narrowing down. A good example is given by the competitive performance of general purpose machines when executing Lisp. The algorithms that we described in the previous sections have characteristics that make them inefficient on general purpose machines: they require more state than can easily be stored in the registers of a general purpose processor. Therefore, they become very memory intensive since part of the state has to be stored in main memory. This is apparent from the structure of the many specialized processors that have been proposed or built for Dynamic Time Warping [10, 22, 5, 11]: either many registers are provided on chip or multiple input/output buses are provided. Typically, general purpose processors become too slow when the vocabulary is larger than about a thousand words (this number is based on a system demonstrated by Dragon Systems Inc. based on an Intel 80286 processor).

The analysis of the computation power required by speech recognition algorithms is complicated by the fact that their requirements are influenced in a complex way by the characteristics of the task domain. For example, in the case of the Harpy system, the branching factor of the grammar and the number of words were the main factors controlling the number of instructions per second of speech required. Table 4-1 shows how the performance of the Harpy system (expressed in Millions of Instructions per Second of Speech, MIPSS) is influenced by changes in the number of vocabulary words or in the branching factor.

The table shows a factor of three increase in the amount of computation when the number of words increases by a factor of four (from approximately 1 MIPSS and 250 words to 2.6 MIPSS and 1000 words). An increase of the static branching factor with the number of words (250) held constant caused an equal increase in the amount of computation required (.7 MIPSS to 6 MIPSS).

In order to extrapolate the results of these experiments to more complex tasks we can use the data in Table 4-1 to compute the effect of having larger branching factors and more words

Grammar	MIPSS	Static branching factor	Vocabulary size
AIS10	1	8.2	250
AIM12	1.7	10.5	500
AIX05	2.6	9.5	1000
AIS06	.73	4.6	250
AIS10	1	8.2	250
AIS15	1.4	11.9	250
AIS30	4.5	33.3	250
AIS40	6	39.5	250

Table 4-1: Effect of the vocabulary size and grammar branching factor on the performance of the Harpy system. The performance is measured in millions of instructions per second of speech (MIPSS).

on the Harpy system. For a 1000-word vocabulary *where any word can follow any other word* (i.e. with branching factor 1000), we have to multiply the amount of computation (2.6 MIPSS) by the branching factor ratio of the two tasks (1000 versus 9.5 for the AIX05 grammar). This brings us to about 270 million instructions per second of speech for a Harpy-like system. This should be considered an upper bound on the computational requirements of such a system. In practice, experiments performed by Zue [18] and others seem to indicate that the use of filtering based on coarse phonetic features and prosodic patterns can reduce the effective vocabulary size by one to two orders of magnitude with a corresponding decrease in the amount of computation required. Better parameter extraction and phonetic transcription modules will also help in limiting the amount of computation required. In general, a number of factors will contribute to reduce the amount of computation required and will counterbalance the growth of computational requirements caused by the increase in the vocabulary size and in the branching factor.

4.2. Architectural Alternatives

In the previous sections we have characterized speech recognition algorithms along the following dimensions:

- the function they use to compute the path score;
- how much of the search space they examine.

These characteristics have a very different influence on selecting the best architecture: the first characteristic influences the "data paths" of the processor(s) used while the other two characteristics dictate how an algorithm can be decomposed into parallel computations.

Exhaustive search algorithms are the ones that are most used because they are the simplest. For these algorithms recognition time on a uniprocessor is proportional to the vocabulary size. When the vocabulary size is not too large (say less than a few thousand words), very effective single processor solutions have been demonstrated. The strength of these architectures lies in the tailoring of the data paths, the large number of registers, the use of specialized functions and the use of more than one memory. See, for example [5].

If the vocabulary is very large, this type of algorithm can be executed on a systolic multiprocessor architecture which can be scaled with increasing vocabulary size. Architectures of this kind are surveyed by Wood [23]. Even though the amount of hardware required can grow very large for large vocabularies, the hardware is very regular and can be easily designed. Although full 2-dimensional arrays can be used, processor utilization can be very low since the computation is limited to a few diagonals of the array. Therefore, either more computations can be pipelined through the array (e.g. the computation of the score of different words) or the structure limited to a few diagonals that are cyclically fed by suitable peripheral circuitry. Although prototypes of some of these architectures have been fabricated, we do not know of any speech recognition product that uses one of these systolic architectures.

Non-exhaustive search algorithms pose very different problems. In the case of HARPY, the average number of active states is generally independent of the size of the grammar or vocabulary and is limited by a heuristic function that is continuously recalculated during the search. The IBM's stack decoding algorithm can be even more data-dependent because, besides having the same data dependent search pattern, it can generate more than one hypothesis per state and input time combination, depending on the input data. This is because different hypotheses can only be merged if they correspond to the same word string.

This type of algorithms have not been successfully implemented on a systolic multiprocessor. The reason is that in the case of dynamically restricted search, the exact computation performed depends on the input data, which is hard to predict in advance, and requires dynamic reallocation of resources, which is hard to do with systolic processors. This kind of algorithm can be handled much better by custom architectures based on event queues in which the active states of the computation are queued for further expansion. The kind of hardware required tends to be highly irregular, but in general less total hardware is required than in the case of systolic architectures. Examples of this kind of architecture are described in [6, 3, 2].

Non-exhaustive search algorithms contain:
1. A heuristic evaluation function which is executed each time an alternative path in the graph is extended.

2. A network access function that fetches the task specific information that is stored in the nodes of the network representing the task knowledge.

3. A "collision-detection" function that detects and chooses between different sub-paths that terminate at the same node.

4. A pruning function that discards non-promising paths.

For example, in beam search, the collision function has to check if there is already an active transition into a node. This requires a full search in the list of nodes already expanded unless some associative or hashing mechanism is used. In stack decoding the same function is more complicated because it must recognize if two colliding paths correspond to the same sequence of words. Data-dependent computations arise in 1 and 3 (evaluation and collision-detection functions). Function 2 (graph access) is not data-dependent but is memory intensive. Function 4 (pruning) is very simple and inexpensive. In the next sections we will describe a class of custom architectures that are appropriate for non-exhaustive search algorithms.

5. Non-exhaustive Search Architectures

Non-exhaustive search algorithms are data dependent and some of their dependencies (both data and control flow dependencies) cannot be resolved at compile time. One very powerful technique for resolving dependencies at run time is to use a data-flow architecture. There are currently several drawbacks of this technique :

1. There is a basic execution overhead due to the more complex execution engine, which slows down sequential regions of the program.

2. data-flow has difficulty in handling large and complex data structures efficiently. This problem, as well as proposed solutions are discussed by Gaudiot [12]. All of them suffer from the drawback of increased storage (multiple copies of data structures), suboptimal speedup, or partial abandonment of the data-flow principle to obtain solutions known from vectorizing compiler techniques.

3. Some of the parallelism that is extracted at run-time in data-flow architectures, such as multiple instances of a loop body, can be extracted at compile time with no run-time overhead.

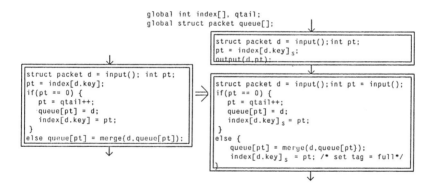

Figure 5-1: Example of synchronized reads/writes

We use a full/empty tag bit attached to each word in shared memory to resolve data dependencies dynamically. This is similar to the synchronization primitve used by the HEP multiprocessor except that the synchronization takes place within a single bus cycle. Reads/Writes to shared memory may be normal or synchronized. The former proceed as usual and have no effect on the full/empty bit. If a processor performs a synchronized read, the read blocks if the tag is set to empty, otherwise the read proceeds and the tag bit is toggled. If a processor performs a synchronized write, any other blocked processor whose address matches that on the bus, grabs the data and proceeds without having to repeat the blocked read. The tag bit gets set to empty/full according to whether a blocked processor was unblocked or not.

This hardware primitive is relatively easy to implement. The following example, taken from the custom speech recognition machine design, illustrates the power of this simple technique, and is shown as a transformation on a piece of code in Figure 5-1.

In this example the program (actually the body inside a deeply nested loop) stores all data packets destined to a Queue in global memory, while merging all data packets that have the same 'key', using an unspecified merge function denoted by merge(). The call to input() just denotes a copy from the input link of the processor. Memory references that are synchronized are denoted by subscripts 'S'. Reads marked with an 'S' are synchronized and block if the word is 'empty' and writes marked with an 'S' are likewise synchronized as described above. A potential data dependency is created between successive instances of the code because the d.key value may be the same. If two or more consecutive d.key values are the same, the corresponding code instances must be executed sequentially. However if a number of consecutive d.key values are all different the corresponding code instances can be executed in parallel or pipelined. The dependency cannot be resolved at compile time and it can be shown that to resolve this kind of dependency (within a nested loop) on a general purpose processor $O(N^2)$ hardware is insufficient [20]. Here, by making the references to the index[] array synchronized, and by adding one additional statement (to set the tag to the right value), the hardware automatically resolves the dependencies at run time. It is now possible to apply the **Pipeline** transformation as illustrated. For cases where the d.key values are different, the resulting hardware executes as fast as if no dependency had existed at all.

The transformed program allows two instances of the code to execute simultaneously. While the first instance is executing the second code section, the second instance could already be executing the first code section. By introducing synchronized memory references the second instance is forced to wait for the first instance to complete execution *iff* the array element d.key value is indeed the same for both instances : in that case the read of the array element in the first part of the code (by the second instance) hangs, until the write in the second part of the code (by the first instance) completes. Work on minimizing synchronized memory references in the context of general purpose MIMD architecture has been reported in [1], and the results apply directly to this technique as well.

Note that further transformations can now be applied, since the statement added can be factored out of the conditional, so that the additional statement actually simplifies the hardware. In general the additional statements required by this technique may require additional pipeline stages to implement, but rarely reduce the potential speedup.

6. Example of a Beam Search Customization

In this section we describe the partitioning of a speech recognition beam search algorithm in a number of pipeline stages by applying transformations that partition straight code, loops and conditionals into pipeline stages. The custom architecture for which the algorithm is partitioned consist of a set of simple processors communicating through both dedicated links and a number of shared memory modules, as well as a specification of the code executed by each processor (see Figure 6-1). Dedicated links act like read-only or write-only registers whose references are synchronized (in hardware) with those of the corresponding other processors. Each processor contains: a simple hardwired controller whose "program" can be generated automatically at the end of the decomposition, the functional units required by the program and register storage. Instructions can read data from registers , operate on them and store the result in one or more clock cycles (depending on the speed of the functional unit), as well as access shared memory by read/write instructions. Several such processors can reside on a single chip. The hardware communicates with the host computer through the shared memory, which can be accessed from the host bus.

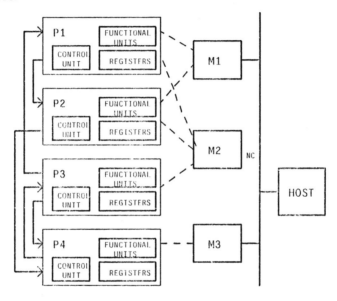

Figure 6-1: Scheme of the custom hardware

The algorithm was divided into two parts: The part doing the graph search, which takes most of the time, was selected for compilation into custom hardware, and the part dealing with initialization, speech input and output of the result was assigned to the 'host' processor. All function calls were eliminated by inline expansion of functions (there were no recursive functions) and the entire code was assigned to a single processor.

Communication with the host was done via synchronized memory references to a 1 bit wide circular buffer of size 16. The host successively inputs (into one of the shared memory modules) a segment of speech data (corresponding to 1 to 60 msec of speech) that has been preprocessed into likelihood estimates, and writes a 0 into the next position of the buffer. When the custom processor finds a valid 0 in the buffer it starts recognition on the utterance using the data available. The processor blocks as long as no valid data are present. If a valid 1 is found, this is taken as an end-of-utterance signal and the processor empties its pipeline and waits for data of the next utterance to be input by the host computer. In addition to the circular buffer there were 7 large data arrays in the code resulting in 7 shared memory blocks. In the following simulations these memory blocks are implemented as 2,4 or 7 physical memory blocks.

The parallel algorithm is illustrated in Figure 6-2 where boxes represent processors and circles indicate shared memories. The number within each circle indicates the shared memory size (in bytes) required for a 1,000 word speech recognition task. The memory requirement will grow less than linearly with vocabulary size.

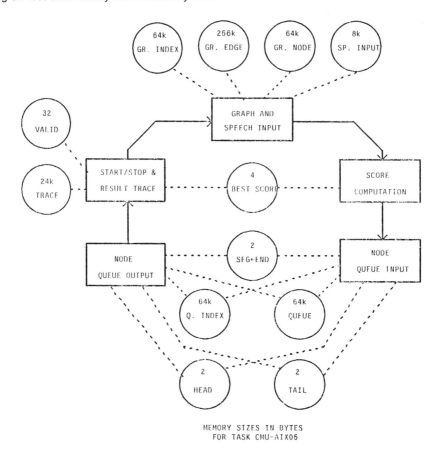

MEMORY SIZES IN BYTES
FOR TASK CMU-AIX05

Figure 6-2: The structure of the pipelined search algorithm

The diagram in Figure 6-2 can be used to understand the behavior of the transformed algorithm. Partial paths are stored in the memory labeled FIFO together with enough information to expand the path to which they belong. Each node is sent around the pipe where it is expanded into its successors (by accessing the graph stored in the memories labeled GRAPH). The newly expanded paths are then evaluated (in the block called SCORE COMPUTATION) and then pruned before being stored in the FIFO. Partial results of the search are stored in the

SEARCH TRACE memory during each iteration. The LEXICON, SPEECH INPUT, and SEARCH RESULT memories can be read or written by the host to set up the task, input the speech signal, and retrieve the result. This five stage decomposition corresponds to the five-processor simulation of Figure 6-3.

Each of the five processors can be further subdivided until the 'code' of most processors consists only of a single instruction, or of a few independent instructions that could be scheduled in parallel. Only two new shared variables are necessary and the final design used 28 processors.

The simulated performance of the 1, 5 and 28 processor designs obtained are shown in Figure 6-3. The three curves correspond to using 7, 4 and 2 physical memory blocks (fewer memory blocks must be used for a VLSI implementation, since seperate sets of pins are required for each physical memory). Each processor and memory is simulated clock cycle by clock cycle. The simulation is implemented in C++ using an event driven simulation package and assumes 400nsec for a memory read or write, 200 nsec fro a register to register 32 bit add or 16 bit multiply and 100 nsec for a register to register move (or and on-chip memory read or write), and a 100 nsec branch condition delay. These figures are typical for a CMOS VLSI design that has not been heavily optimized (and therefore might be generated semi-automatically using standard cell design techniques). Real (preprocessed) speech input data was used during the simulation.

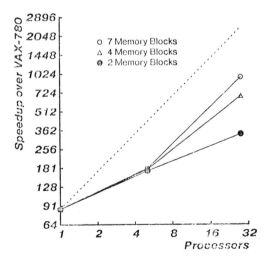

Figure 6-3: Speedup vs. processors

It should be noted that with a single processor the custom hardware (running in parallel with the host computer) is running about 85 times faster than the VAX-780. Partly, this is because there are two processors (the host and the custom hardware), and the two times faster clock cycle, but largely this is because of the RISC type custom processor used in the design. This processor uses a custom ALU capable of executing all required arithmetic operations (including unconventional ones like adds which saturate at the maximum value instead of overflowing, which normally takes 3 instructions on the VAX-780) in 100-200 nsec and contains a 128 by 16bit dual ported register file. Similar favorable results for custom uniprocessors with hardwired code, have been reported by Brodersen [6].

The speed up for 5 processors is low, compared to the speed up for 28 processors, indicating that the transformation sequence used (which was derived manually) is less optimal in the 5 processor case than in the 28 processor case. The total speed up for 28 processors over a single processor is about 15. This does not appear impressive until it is realized that each of the 28 processors is much smaller (in terms of hardware) than the original single processor. This fact is illustrated graphically in Figure 6-4 which plots the speedup against the transistor count of the design. The transistor count was obtained by adding the number of transistors in actual layouts of the various functional units and registers, and is a crude estimate of the amount of silicon area required in a VLSI design.

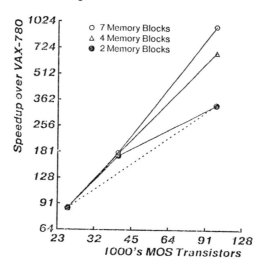

Figure 6-4: Speedup vs. transistors

The speedup vs. number of transistors is actually super-linear; this can be explained by the fact that that the number of instructions as well as data registers increases only very slowly as the number of processors is increased. It should be noted that the final design is not well balanced in terms of processor utilization, and a subsequent simulation revealed that a 20 processor design would give essentially the same speed-up.

7. An alternate multiprocessor architecture

To evaluate the effectiveness of the compilation technique an alternate multiprocessor implementation was considered, in which the algorithm was divided amongst a number of identical processors running the same program. The speech graph was statically subdivided amongst a number of processors so that the original program could be executed on each sub-graph separately. Whenever a path (i.e. an alternative) is expanded and is found to connect to a node stored at another processor, all information about it (56 bits) is sent to that other processor, using a regular network between the processors. Each processor has local copies of shared variables. Careful analysis of the algorithm revealed that only one of them (the 16 bit score of the best path) had to be kept consistent across all processors for correct operation. We assumed that a serial broadcast could be performed (using 1 extra pin and round robin fix slot scheduling) in 16*N clock cycles, where N is the number of processors. In addition, we assumed that at each iteration of the algorithm, all processors must flush their pipes and synchronize (so that they always all work on paths of the same length). This is done by using a global open collector signal, which is released at the end of a segment by each processor (and the network) when it is done processing. When all processors have synchronized, the signal is pulled down again, and processing on the next segment starts.

Since each processor is identical and executes identical code, we can apply the hardware compilation technique to each of these processors to obtain the best of both techniques. In the simulation of this architecture presented below, each of the high-level processors is actually implemented as the processor-memory cluster shown in Figure 6-2 whose performance is shown in Figure 6-3 for 5 processors and 1 memory. The network between the processors was a simple ladder network with a bandwidth of about 45 MBytes/sec. The network was chosen mainly for ease of simulation. Since the network bandwidth turned out not to be the limiting factor no effort was made to replace it by a more appropriate network. The architecture is illustrated in Figure 7-1. The simulation results are shown in the lower curve in Figure 7-2.

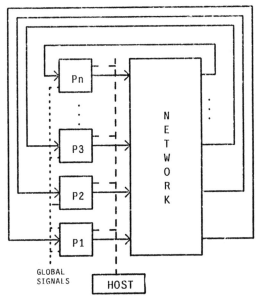

Figure 7-1: Multiprocessor implementation

Speed up saturates rapidly after about 16 processors. Since the network was found not to saturate we postulated that there was not enough parallelism in the task. To check this interpretation of the results the simulation was rerun with a three times larger task. Moreover the network bandwidth was increased proportionately. The result is shown in the upper curve of Figure 7-2. The speed up levels out at a level about 3 times higher than before, confirming our belief that we are approaching the intrinsic parallelism of the problem. With 128 processors (each built with 5 smaller processors) the custom hardware is about 8500 times faster than a VAX-780.

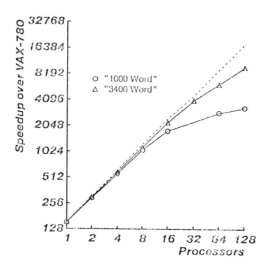

Figure 7-2: Multiprocessor Performance

8. Conclusions

Speech recognition algorithms can all be modeled as a search of a Markov network. We have used this characteristic to present the most used algorithms within the same framework and then compare their computer architecture requirements. It turns out that most current, practical solutions involve single processors with a highly customized data path. Extremely high performance, but not necessarily high cost-effectiveness, can be obtained with systolic architectures if the speech algorithm uses an exhaustive search. If a non-exhaustive search algorithm is used, systolic solutions are not possible and some form of exploiting data-dependent, fine-grain parallelism is necessary. We have presented a specific custom architecture for a non-exhaustive speech recognition algorithm and its evaluation when applied to real data. The architecture has a good cost-effectiveness and allows a three order of magnitude speed-up over general purpose processors. Different versions of the architectures can be easily derived to obtain a number of different hardware-size vs. speed trade-offs.

Acknowledgements

This research is sponsored by the Defense Advanced Research Projects Agency, DoD, through ARPA Order 5167, and monitored by the Space and Naval Warfare Systems Command under contract N00039-85-C-0163. Views and conclusions contained in this document are those of the authors and should not be interpreted as representing official policies, either expressed or implied, of the Defense Advanced Research Projects Agency or of the United States Government.

References

[1]   Ahiyuan Li and Walid Abu-sufah.
        A technique for reducing synchronization overhead in large scale multiprocessors.
        In *12th Computer Architecture Conference Proceedings*, pages 284-291. Computer
            Society Press, June, 1985.

[2]   Ananthamaran, T. and Bisiani,R.
        Hardware Accelerators for Speech Recognition Algorithms.
        In *Proceeedings of the 13th International Symposium on Computer Architecture*. IEEE,
            June, 1986.

[3]   Bisiani, R., Mauersberg, H. and Reddy, R.
        Task-Oriented Architectures.
        Proceedings of the IEEE , July, 1983.

[4]   John S. Bridle, Michael D. Brown and Richard M. Chamberlain.
        An Algorithm for connected word Recognition.
        IEEE Icassp :899-902, 1982.

[5]   Kavaler, R.A. and Noll, T.G. and Lowy, M. and Murveit, H. and Brodersen, R.W.
        A Dynamic Time Warp IC for a One Thousan Recognition System.
        In *ICASSP '84*. IEEE, San Diego, March, 1984.

[6]   Rober Kavaler, R.W.Broderson.
        A Dynamic Time Warp IC for a one thousand word recognition system.
        In *IEEE ICASSP*, pages 25B.6.1-25B.6.4. IEEE, 1984.

[7]   Fellman, R.D., Brodersen, R.W.
        A Switched Capacitor Adaptive Lattice Filter.
        Journal of Solid State Circuits , February, 1983.

[8]   Lalit R. Bahl and Frederick Jelinek.
        Decoding for Channels with Insertions, Deletions, and Substituions with Applications to
            Speech Recognition.
        IEEE transactions on information theory IT-21(4):404-411, July, 1975.

[9]   J.K.Baker.
        The DRAGON system - An overview.
        IEEE Trans. Acoust., Speech, Signal Processing ASSP-23:24-29, Feb, 1975.

[10]   Nel Weste, David J. Burr, Bryan D. Ackland. .
        Dynamic Time Warp Pattern Matching Using an Integrated Multiprocessing Array.
        IEEE Transactions on Computers C-32(8):731-744, August, 1983.

[11]   Frison, P. and Quinton, P.
        A VLSI Parallel Machine for Speech Recognition.
        In *ICASSP '84*. IEEE, San Diego, March, 1984.

[12]   J.L.Gaudiot.
        Methods for handling structures in Data-Flow systems.
        In *12th Computer Architecture Conference Proceedings*, pages 352-358. Computer
            Society Press, June, 1985.

[13]   Bruce. T. Lowerre.
        The HARPY Speech Recognition System.
        PhD thesis, Carnegie-Mellon University, April, 1976.

[14]   F. Itakura and S. Saito.
        Analysis synthesis telephone based upon the maximum likelihood method.
        Conf. Rec. 6th Int. Congr. Acoust. , 1968.

[15]   C.S.Myers and L.R.Rabiner.
        A level building dynamic time warping algorithm for connected word recognition.
        IEEE Transactions on Acoustics, Speech, and Signal Processing ASSP-29(3):284-296,
            June, 1981.

[16]   Lyon, R.F.
        Processing Speech with the Multi-Serial Signal Processor.
        In *ICASSP*. IEEE, 1985.

[17]   Lalit R. Bahl, Frederick Jeliknek, Robert L. Mercer.
        A Maximum Likelihood Approach to Continuous Speech Recognition.
        IEEE Transaction on Pattern Analysis and Machine Intelligence PAMI-5(2):179-190,
            March, 1983.

[18]   D.W. Shipman and V.W. Zue.
        Properties of Large Lexicons: Implications for Advanced Isolated Word Recognition
            Systems.
        In *ICASSP '82*, pages 546-549. IEEE ASSP, 1982.

[19]   Hiroaki sakoe.
        Two-Level DP-Matching-A Dynamic Programming Algrithm for Connected Word
            Recognition.
        IEEE Transactions on acoustics speech and signal processing ASSP-27(6):588-595,
            December, 1979.

[20]   Augustus K. Uht.
        Exploitation of Low-Level Concurrency : An Implementation and Architecture.
        PhD thesis, Carnegie-Mellon University, December, 1985.

[21]   G.D.Forney, Jr.
        The Viterbi algorithm.
        Proc. IEEE 61:268-278, March, 1973.

[22]   Feldman, J.A. and Gaverick, F.M. and Rhodes, F.M. and Mann, J.R.
        A Wafer Scale Integration Systolic Processor for Connected Word Recognition.
        In *ICASSP '84*. IEEE, San Diego, March, 1984.

[23]   Wood, D.
        A survey of algorithms & architectures for connected speech recognition.
        in New Systems and Architectures for Automatic Speech Recognition, Springer-Verlag ,
            1984.

Issues on Parallel Algorithms for Image Processing

Stefano Levialdi

Dip. di Matematica
Università degli Studi
Ple. Aldo Moro 2
00185 Rome, Italy

Image processing demands heavy computational throughput essentially because of the vast amounts of image data that must be handled in short times (sometimes in real time). Since the first cellular computers became available a number of different architectures (using a multiprocessor approach) have been suggested, built and evaluated on many typical algorithms. There is no optimal solution to the image task since there are many, completely distinct tasks, but an insight may be gained by discussing the different factors influencing performance of given architectures for some classes of algorithms. Some examples are given for the cellular and pipeline systems and, as a hint on a promising future, the pyramid architecture is briefly considered as a good candidate for solving both low-level and high-level vision tasks.

1. Background

The requirement of higher computational power has always been in the mind of computer architects, algorithm designers and computer users since automatic data processing started invading most of the fields of practical application and mental speculation. The "more" the "better" has generally been the driving force behind the developments of new computer systems, this in turn, has led to find the "correct" problems to be solved by such new computers since, in general, there was no strong feedback between the people designing machines and those designing algorithms. In fact recent efforts were devoted to build a taxonomy for algorithms that could embed inherent parallelism (to be possibly matched to specific new computer architectures), to introduce new definitions of parallel "levels" that could be implemented on different multicomputer systems. In general, paralellism has proved to be an evasive concept, difficult to grasp in a global way (conceptually), difficult to define (formally) and difficult to apply (constructively).

Within the realm of the image processing community, which sprang from the evolution of other disciplines like pattern recognition, human visual perception and, more recently, computer vision, the need for higher throughput (of processed images) was clear from the beginning (the 60s) although neither the technology nor the algorithms were available at that time. The amount of data to be processed began to grow as soon as new input devices became operational like satellites (carrying remote sensing equipment), medical imaging (in X-ray, tomography, thermography, ultrasound, etc.) were screening of an entire population could be foreseen or in robotics aimed at production automation (from assembly lines to the restructuring of the factory like in the data-driven approach). Color, high definition, three-dimensionality and movement still added other dimensions to the two-dimensional, binary images which were originally considered as the input data to be processed. Nevertheless the work done on those elementary binary images was instrumental in the development of new strategies to tackle the more sophisticated tasks that were to come.

Typically an image processing operation is the computation of a function of non-negative, generally integer values (corresponding to the grey levels, or intensity levels) of all the image elements (pixels) where the arguments are the values of the neighbors of the pixel on which the computation is performed. If the value of a pixel is indicated by $p_{i,j}$ and its new value (after computation) is $p'_{i,j}$ then

$$p'_{i,j} = f(N(p_{i,j}))$$

where $N(p_{i,j})$ are the neighbors of pixel $p_{i,j}$ (which may also include the same pixel $p_{i,j}$). For the most common local neighborhoods they are the 4-(or 8-) connected neighbors placed vertically and horizontally (or also including the diagonal neighbors) but may become a much larger number if the neighborhoods are not local. In fact, a well-known taxonomy of image processing algorithms is based on neighborhood size and considers three classes: local, global and statistical [3]. The first one corresponds to those algorithms only concerned with the immediate neighbors, the second one with those using all the pixels in the image to compute the image transformations (like in Fourier analysis for example) and the third one is a class where the position of the considered pixels is unimportant like in histogram calculations where only the frequency (i.e. number of occurrences) of a certain pixel value is relevant.

The algorithms, used in image processing and understanding, may also be classified, in terms of their task solving aim as belonging to three groups: those for low level vision, for intermediate level and for high level vision.

The functions are computed on all elements (pixels) of the image, if possible, simultaneously or in parallel. Typically, low level algorithms may extract a border of an object (provided the border detection technique is based only on the evaluation of particular neighborhood configurations). High level algorithms may be those required for object detection in a scene (where image partitions must be performed based on an object model, on values coming from the whole scene/image) whilst the intermediate level algorithms are more difficult to define since there are no well specified borders in this classification. Some tasks require a combination of the different levels, like counting objects where the objects must be firstly located and then counted, when distances between objects must be computed (from baricenters, defining a special metric, etc.) or in region growing methods where a seed pixel must be defined and then a propagation signal (to contact neigboring elements) is released.

This last classfication was originated by an analogy with our visual system which uses the retinal vision for low level tasks and the cortical vision for the high level ones. Its high performance is due to: the use of motivation and cultural context in relation to the specific visual task, the (micro and macro) sacadic movements of the eye, and to the flexibility in establishing varying interconnections between different cortex areas in the brain; all of which act in a cooperative way. The algorithms, to be properly defined, require the discovery of an adequate (finite) set of primitive operations, (at the lowest "unbreakable" level), the recognition of the most suitable data structures to represent the image information and, later, the possible layers of concurrent execution of the operations: this is more easily said then done! Many years ago, a basic work on the equivalence of sequential and parallel operations in image processing paved the way to the design of parallel algorithms using parallel local operations since these where independent [20]. On this account the new value of each pixel could be found just by computing the function f (linear or non-linear) written above and if a sequential simulation was to be performed, the neighborhood $N(p_{i,j})$ values were to be stored before f was computed.

2. Models & Algorithms

The studies on parallel computing originated a number of computational models [12,28] which may be seen as a way to describe and interpret both the information flow and control taking place in a given class of machines (Von Neuman model, cellular automata, etc.). A resemblance may be found between the syntactic and semantic aspects in programming languages and the control structure and interpretative capacity of a parallel computational model. The control structure is defined in terms of a state space, a transition function, an alphabet of operation symbols, a start state and a set of final states (which, in some cases, may also correspond to one single state). When a computation takes place, it may be represented by a sequence of elementary states such that the first one is the start state and then the intermediate states succeed this start state and precede the final state which is the last state of the sequence.

Models corresponding to the cellular automata (two-dimensional, finite version of the standard automata) for control-driven computation and computation graphs for data-driven computation have been suggested in order to deepen the understanding of the situation and to have a predictive potential for different instances of the same situation. In order to fully exploit the computational system all processors must be active most of the time performing useful operations, i.e. calculations on image data and not instruction fetch, housekeeping, etc. Unfortunately algorithms vary in their structure from requiring to perform simple operations on large amounts of data to just the opposite, a long complex set of operations on a narrow stream of data which must necessarily flow in a sequential manner; moreover, if moving images are considered (or any sequence of images) the input-output of the images is in itself a heavy task requiring special resources for positioning all the data on the right processing elements. These two extreme conditions, typical of image processing, make a universal solution to the "best" parallel architecture (based on a flexible computational model) an impossible one.

In conclusion, although parallel computing systems are a valid solution to enhance computational power (and throughput) in many scientific applications and particularly so in image processing (where typically the data involved is of a high order, $10^6/10^9$ bits) a convenient architecture with its software environment and a set of well-matched algorithms is still far to come [2]. The amount of work done both from a theoretical standpoint (in the analysis of compact graph structures, representing multiprocessor systems, leading to small diameter and high degree graphs) and experimentally (choosing benchmarks, applying "best" algorithms for specific architectures, evaluating performance by simulation, etc) has shown that an optimal solution (for maximizing efficiency) is the least flexible one and that the general solution is for some algorithms a little better than the sequential one but that high gains are rarely obtained on all phases of the computation.

3. Architectures & Algorithms

Many authors have stressed the importance of having the "correct" architecture to implement classes of algorithms and a number of solutions have been suggested to enable this match between the operations to be performed [7,19] and the computing resources implementing such operations. In the wide field of multiprocessor architectures (sometimes called multicomputers) the number of processors is dependent on the machine design. An array of processors, where each processor corresponds to a pixel in the image will have, up to 10^{4-5} processors (vertically and horizontally interconnected) whilst a pipeline of processors may have up to 10^2 processors as stages where the pixels enter sequentially and, by means of suitable delays, may perform the required f local computations. Combinations of both systems give rise to other architectures where special forms of pipelining take place based on the regularity of data (and the simultaneous input of streams of data) like in systolic systems [14].

The possiblity exists of building systems which reconfigure themselves [29] according to the algorithm structure: the number of simultaneous processors available for given computation steps, the interconnections between processors allowing data exchange, the memory readout from local processor memory or from a global one, etc. The reconfiguration (which accounts for the flexibility of the system) may be fixed (for a class of algorithms) or variable (and therefore algorithm-driven) so that in different parts of the computation the system modifies its connections to maximize efficiency. A number of workshops dealing with the general point of finding a suitable architecture for the most typical image processing tasks, have shown the difficulty of giving a definitive answer to this problem [11,8,19,17,9], both from the standpoint of evaluating the performance (by finding appropriate benchmarks) and from that of the generality of the solution (final architecture) in an application-independent way.

4. Image Algorithms

4.1. Basic Operations

As mentioned before it is difficult to group all image processing algorithms into one single class because images are transformed for a great variety of reasons [22]: for restoration, enhancement, coding, storing, transmitting, displaying, monitoring, measuring, understanding. Each of these trasnformations is performed on data of a particular nature: images from remotely sensed origin, from television, from medicine and biology, from industry, from natural scenes (three-dimensional, in color, with moving objects, etc). Moreover, the economical importance of the task will also determine the image definition, the pixel value accuracy, the instrumentation to be used (for displaying, plotting, etc.) and the computer system that will process the images. Another important factor is whether the images are processed for research purposes in an academic or laboratory environment or if they must undergo regular, massive tests (like in a population screening, a weather forecast station, etc.): this difference will have a strong consequence on the chosen algorithms, coded programs, selected computer system.

The basic operations (see for instance [15]) used in image processing (identical to those used in numerical processing) are arithmetical and boolean and the data on which these operations must be performed are typically non-negative integers having a range of $0-10^3$. In some cases real numbers are also used and the accuracy of the computation must be considered but rarely do floating point operations become significant within the most common image processing tasks. It has been often said that the computer representation of images corresponds to matrices of integer numbers; this description is misleading in the sense that properties of matrices have no relevance to those of images nor do matrix operations (like multiplication or inversion) correspond to image operations. In fact, by the very nature of the digital computer and the required discretization of the input image information (both in space and intensity) we obtain a set of integer numbers which are spatially organized but, as will be even clearer in multiresolution computations, the interpretation of these values depends on their overall distribution, on their intensity ratio within given environments, on their time variation (for a constant location) and not on their specific row/column position. For this reason, many new high-level languages that are oriented towards parallel image processing, consider the image as one single piece of information and not as a collection of values (matrix numbers, pixels) since all the operations must be performed on all values regardless of their position.

The basic operations may sometimes be performed on a single pixel (like in thresholding), on corresponding pixels from different images (like in map registration) being therefore unary or binary operations; in turn each one may have single bit or multiple bit operands for binary or grey level images respectively. A stack of binary bit planes generally contains the grey level information of all pixels in an image and there are many schemes for coding (in binary form) the intensity level of every pixel, this has a strong impact on the way arithmetic is performed for instance on cellular machines (based on the cellular automata scheme) to conveniently use all the processors of the machine. These basic operations are easily implemented with very simple processors and, as soon as technology allowed to assemble large quantities of tightly coupled processing elements in a reduced space, some prototypes of cellular machines began to appear having from 32x32 to 128x128 processing elements (PICAP, DAP, CLIP, MPP, etc., for a discussion on the principles of these architectures see [6]) and a series of experiments were conducted to see how did these parallel machines (for image processing) outperform the conventional sequential computers.

4.2. Cellular algorithms

Two extreme algorithms may be quoted which show the best and worse cases to be run on binary static images having a size equal to that of the array (a very favorable situation).

Best: thinning algorithms, ([22] contains a brief introduction to these transformations and more specifically see [24]), generally used in image preprocessing whenever a stylized version of an elongated shape must be obtained (like for instance in OCR applications). The resulting image, of stick-like appearance, is one pixel wide and preserves the connectivity structure of the original image. A thinning algorithm [1] (there are many around in the literature and still new ones keep appearing...) is based on a set of 3x3 non linear masks which must be applied iteratively in sequence on all the pixels of the image. When the 3x3 configuration of the considered central pixel matches the mask, the corresponding central pixel is deletable (it can be erased without changing the connectivity of the object). The algorithm terminates when no more pixels become deletable after all the masks of the set have been successively applied to the image. Clearly each mask of the set may be applied in parallel on the whole image and the number of iterations of the set will depend on the maximum width (half of the width to be more precise since each iteration of all the masks of the set will erase the object from both sides of its breadth and width). In brief, the mask-based thinning algorithm is $O(w_m)$ for w_m the maximum width of any object in the image. The computation steps of this algorithm are independent from the number of objects, from their connectivity degree and from their position in the image.

Worst: a combination of operations requiring location of an object made of connected (according to a connectivity definition given a-priori) 1-elements in a background of 0-elements by labeling one of its pixels (and therefore making one processing element different from all the others), some numerical computation on it (like obtaining the value of the object area) and the output of this value in numerical form. We are forcing the cellular computer outside its typical range of operations: processing elements are typically non-addressable, the addition is a well-known sequential process, unless special hardware is available, and the output of numbers instead of images is hardly adequate for an array of processors (even if nested row adders may produce, in a few steps, the result of an addition).

Nevertheless there are propagation techniques (see [10]) which allow to locate an object (like the top leftmost one) by generating a row of 1-elements (a rectangled shaped object) which is AND-ed with the binary image row by row from the top downwards until one 1-element of this row overlaps one 1-element (or more) of the object(s) present on the array. When this situation is met a single 1-element on this found row is shifted from the leftmost column towards the right until, again, it overlaps with one 1-element of an object: at this moment we will have found the top leftmost 1-element corresponding to an object which, itself, is the top leftmost object of the array. If n is the linear size of the array this algorithm will operate in $O(2n)$ for a worst case object located in the bottom rightmost corner of the array. Once the object is located it will have to be labeled so that the area measurement may be performed on it. Labeling takes place by propagating a signal from the top leftmost 1-element of the object in steps, each of which will contact the immediate neighbors until no elements of the object remain untouched (this propagation may include 4-neighbors or 8-neighbors depending on the chosen connectivity for the objects of the image). The order of this algorithm is dependent on the maximum side of the circumscribing rectangle of the object L, and is therefore $O(L)$. Finally, in order to have the area of the object, a sum of all its 1-elements must be performed and unless the sum may be obtained simultaneously by rows (or columns) with special hardware, the order will be dependent on the object area, $O(L^2)$.

In conclusion, the global order of this algorithm for measuring the area of an object in the image will be

Area Algorithm = $O(2n) + O(L) + O(L^2)$ so that the second term may be ignored and the relative weight of the other two terms depends on the n/L ratio, i.e. on the size of the object with respect to the size of the array.

We see that, typically, a sequential computer would take $O(n^2)$ for both algorithms in the best case (only one full raster scanning on the image without any computation) and that in practice, specially in the first algorithm (the thinning one), a number of local operations on the border of the original object must be performed so that the computation time will certainly exceed the single raster time.

More complex tasks, like histogram construction for instance, where all the pixels having a given value must be counted so as to obtain the total grey value distribution of the image has also been analyzed if performed on a cellular machine. The results show that even if a number of arithmetic operations must be performed and therefore integer numbers (representing grey values) coded in binary form had to be manipulated, the cellular array (for large arrays) is slightly better than a sequential machine. This is particularly so for the detection of the minimum of the histogram (valley) when only one minimum is present as in typical bimodal histograms [5].

These analyses are only a very rough approximation of reality: in general it is very difficult to obtain a significant gain (or speed-up) by means of a cellular machine with an algorithm unless its basic operations are local, binary and require small memory (under 1k bits). Fortunately this situation is frequent in image processing were a great number of transformations are local, independent and may be performed on binary planes (also called bit-planes). A counter example of this is rotation of an image where the new values of the pixels must be recalculated (according to the rotation centre and angle) and therefore some geometrical computation must be performed.

In short the class of local computations (for small neighborhoods) well matches cellular machines (for a survey of these operations see [18]) and many image transforms may be decomposed as a series of parallel operations which are essentially local. Nevertheless other transformations exist, like scaling, rotation, rectification, mosaicing, gridding (within the geometrical field) and feature extraction, automatic classification and interpretation (within pattern recognition) or storage management, program control, input-output, test pattern generators, etc (as utilities) that are essentially non local, imply numerical calculations, use memory heavily, in other words have both some intrinsically sequential requirements and number crunching requirements.

It has also been said, and advocated, that algorithm reformulation [16] is the key to fully exploit non-conventional architectures or, even better, to design new architectures for obtaining good speed-ups and this is certainly true for narrow classes of algorithms not for the majority of the ones used in image processing which have a mixture of basic operations some of which have conflicting requirements as shown above.

There are many other solutions to the general problem of designing the algorithms that will run efficiently on a special image processing architecture, one of these is to consider the task as broken down in data movements (as when the image is scanned or when a series of images is loaded to the system) and data operations: the goal is always to have all the data required by each processor at all instances of time so that no time is lost by the processing element waiting for data. If different processing elements perform different operations (on independent data) and the image data is running through them we have a pipeline structure which is very adequate for performing a fixed set of local operations on an image. The locality is fixed (typically three by three), the computational burden is balanced between all the processing elements (so that they are all active most of the time) and pixel data keep entering the system until no more images are left to process. This arrangement overcomes the input-output delay inherent to cellular processors which must wait until the image is loaded before processing may start. Since the processing stages of the pipeline architectures are generally more powerful than those of the cellular computers, number crunching may be performed more easily.

4.3. Pyramid Algorithms

An interesting combination of the advantages introduced by the cellular array architecture with those belonging to the pipeline architecture is represented by the pyramid architecture. Typically this structure is made of a series of square planes; the bottom plane (the base) having nxn processors followed by a tapered sequence of planes each of which has one quarter of the number of processors of the plane underneath: n^2 (the base), $n^2/4$ (or $(n/2)^2$), $n^2/16$,...until the topmost plane is reached which will only has one processor and is the apex of the pyramid. Each processor has four co-planar neighbors (its brothers), one father (on the plane above) and four children (on the plane below). There are many research groups round the world that believe this architecture (with slight variations) may solve most of the problems that are causing bottlenecks in other existing multiprocessing systems but there is also another reason for the growing interest in the pyramid approach and is its analogy with the human visual system. The high definition of the base (pyramid base equivalent to a cellular array) models the retinal vision and may perform pre-processing tasks like smoothing, sharpening, etc; whilst the top part models the cortical vision where high level decisions are taken on the basis of some computations which have been performed in lower levels. Moreover, the possibility of having many resolution levels and of allowing the intercommunication between these levels (from father to children or in the reverse direction) may be conveniently used in a fast ($O(\log_2 n)$) strategy both for low level vision (object detection, feature extraction, etc) and high level vision (object from background, perimeter and area computation, etc.). A number of books are appearing [26,23,27,4] containing algorithms and performance evaluations of pyramidal processing, (also called multiresolution approach since it uses a collection of arrays having different numbers of processors).

There are different, tentative, classification schemes for pyramid algorithms [25] according to the data flow patterns along the pyramid, the data structures that are manipulated, the paradigm employed. As mentioned above, another possibility is to classify the pyramid algorithms according to the task (edge detection, region growing, etc.).

In the first scheme, (suggested by Hansen and Riseman) the upward flow (termed reduction) indicates the movement of data from coarser to finer levels of the pyramid, the reduction applies to image definition. The downward flow (projection) is the opposite and therefore indicates data movement from the coarse level to the finer level and, finally, lateral flow implies movement of data on the same plane along rows or columns; i.e. between brothers.

In the second scheme if the images are purely binary we speak of bit pyramids, if the images are coded with 8-bit approximations of grey values we have byte pyramids and if three-dimensional objects are located in space their 3-D coordinates may be stored in the single memory of a processing element.

In the third scheme, the paradigm stands for the relationship between the control structure and the data-flow seen together. For instance building pyramids, like in bit pyramids which are produced by some boolean function of the bits at the lower plane (pyramid base): an AND pyramid propagates bits toward the apex provided that the four children are all 1-elements; the OR pyramid requires that only on child be a 1-element. Real pyramids are those using real numbers obtained by averaging, finding the median, maximum, etc. between subsets of pixels on a given plane.

For instance the location of the first 1-element in an image may be obtained by firstly having a bottom-up pass (in an OR pyramid) followed by a top-down pass (in an AND pyramid) perhaps with a propagation along only one of the possible children (e.g. top leftmost) so as to detect the 1-element in $O(\log 2n)$.

Pyramids may also be seen as trees where a search may be performed so as to rapidly locate interesting points (maximum intensity or maximum contrast, etc.) bottom-up and then, in order to obtain the pixel coordinates, a top-down process is triggered. Propagation may also ennance computation (even compared to cellular arrays) since, again, by propagating upwards and then downwards, less steps are required than by propagating sideways (within the plane); typically we are gaining from $O(n)$ to $O(\log n)$ whre n, as usual, is the linear size of the base array (number of processors in a row or column).

From a perceptual point of view and considering high-level vision tasks, a research proposal has been made [24] to use pyramidal computation for rapidly locating structures (straight lines even if interrupted, clustered dots in a random field, smooth curves, etc.). The visual system is very good at achieving this task using of the order of 100 neuronal computational steps. The goal of the research is to detect and measure bimodality of the image histogram (assuming an object and background to account for the modes) by a recursive divide-and-conquer procedure on two dimensional data. To compute the histogram of L grey levels we would require, on the pyramid, if d is the object diameter (length of the maximum straight line contained in the object) O(logd) to add all the pixels having that level and LxO(logd) for all grey levels. In order to avoid this we infer the presence of histogram modes by forcing each father to find the partition (of its children) so as to minimize the variance of each subset about the mean (this partition is called bimean). A comparison of the found means will tell us if the means are far apart (population strongly bimodal) or near (unimodal); this process is repeated bottom-up until a cell appears that signals bimodality, this event triggers a propagation downwards (top-down) until the base of the pyramid so that all pixels belonging to the same population will be labeled. The visual analogy is that bimodality is perceptually conspicuous if a cell on a sufficiently high level of the pyramid has a bimodal population. Here again the reducing/projecting mechanism is exploited to convey local information (from the base upwards) and condense it in a more global way so as to detect features that require a coarser digitization, moreover vertical communication on a pyramid is an extremely fast process and therefore will guarantee good computational efficiency.

5. Conclusions

Hishashi Horikoshi from Hitachi's Central Research Laboratory in Tokyo has declared [13] that the new supercomputers to be manufactured in the 1990s will reach processing speeds of 1 TFlops (three orders of magnitude higher than today's supercomputers: Cray-2 at Lawrence Livermore Labs operates at 1 GFlops) but the mere increase in clock frequency is not the deciding factor in the real speedup that may be obtained. The reformulation of algorithms in the light of the new computer architectures is one of the many factors affecting performance but many others are equally important: choice of homogeneous (having the same atomic level) primitive operations, of convenient control structures, of optimal data paths and movements, of a good high level language able to express and communicate the coded algorithm. As hardware becomes automatically designed and produced, algorithm-driven architectures seem to become practical provided their programmability should not be hindered by too much overhead due to management of computing resources. In a near future we will be able to see specific architectures for performing many different classes of image algorithms. In a production mode, images will enter (perhaps directly from sensors either locally or remotely) and be pre-processed, coded, displayed, enhanced, interpreted and re-displayed by different hardware subsystems, some of which are likely to be of pyramidal nature.

References

[1] C. Arcelli, L. P. Cordella, S. Levialdi, Parallel thining of binary pictures, *Elctronics Letters*, vol 11, n° 7, 1975.

[2] V. Cantoni, C. Guerra, S. Levialdi, Towards the evaluation of an image processing system, in *Computing Structures for Image Processing*, M. J. B. Duff edit., Academic Press, London, pp. 43-56, 1983.

[3] V. Cantoni and S. Levialdi, Matching the task to an Image Processing Architecture, *Computer Vision, Graphics and Image Processing* 23, pp. 301-309, 1983.

[4] V. Cantoni and S. Levialdi, edits., *Pyramidal Systems for Image Processing and Computer Vision*, Springer, Berlin, (to appear) 1987.

[5] L. P. Cordella, M. J. B. Duff, S. Levialdi, Thresholding: A Challenge for Parallel Processing, *Computer Graphics and Image Processing*, vol 6, pp. 207-219, 1977.

[6] P.E. Danielsson and S. Levialdi, Computer Architectures for Pictorial Information Systems, *IEEE Computer*, pp. 53-67, November 1981.

[7] M. J. B. Duff, The Elements of Digital Picture Processing, in *Real-Time/Parallel Computing, Image Analysis*, M. Onoe,

K. Preston, Jr., A. Rosenfeld, edits., Plenum Press, New York, pp. 1-10, 1981.

[8] M. J. B. Duff, edit., *Computing Structures for Image Processing,* Academic Press, London, 1983.

[9] M. J. B. Duff, edit., *Intermediate-level Image Processing,* Academic Press, London, 1986.

[10] M. J. B. Duff and T. J. Fountain, *Cellular Logic Image Processing* , Academic Press, London, 1986.

[11] M. J. B. Duff and S. Levialdi, edits., *Languages and Architectures for Image Processing*, Academic Press, 1981.

[12] C. Guerra and S. Levialdi, Computational models for image understanding, in *Progress in Pattern Recognition* 2, L. N. Kanal, A. Rosenfeld, edits., Elsevier Pub., pp. 39-56, 1985.

[13] Peg Killmon, Computers Tackle Challenges of the 90s, *Computer Design,* December, pp. 47-56, 1985.

[14] H. T. Kung, Why Systolic Architectures?, *IEEE Computer,* pp. 41-47, 1979.

[15] S. Levialdi, Neighborhood Operators: An Outlook, *Pictorial Data Analysis,* R. M. Haralick, edit., NATO ASI Series, Vol. F4, pp. 1-14, Springer, Berlin, 1983.

[16] S. Levialdi, One, two...many processors for image processing, in *Computer Architectures or Spatially Distributed Data,* H. Freeman and G. Pieroni, edits., Springer, Berlin, NATO ASI Series F, vol 18, pp. 159-186, 1985.

[17] S. Levialdi, edit., *Integrated Technology for Parallel Image Processing,* Academic Press, London, 1985.

[18] K. Preston, Jr., M. J. B. Duff, S. Levialdi, P. E. Norgren, J. I. Toriwaki, Basics of Cellular Logic with Some Applications in Medical Image

Processing, *Proc. IEEE,* Vol. 67, pp. 826-856, 1979.

[19] K. Preston, Jr. and L. Uhr, edits.,*Multicomputers and Image Processing: Algorithms and Programs,* Academic Press, New York,1982.

[20] A. Rosenfeld, J. L. Pfaltz, Sequential operations in digital picture processing, *Journ. ACM,* 13, pp. 471-494, 1966.

[21] A. Rosenfeld, A characterization of parallel thinning algorithms, *Inform. and Control* 29, pp. 286-291, 1975.

[22] A. Rosenfeld, A. Kak, *Digital Picture Processing,* Academic Press, 2nd Edition, 1982.

[23] A. Rosenfeld, edit., *Multiresolution Image Processing and Analysis,* Springer, Berlin, 1984.

[24] A. Rosenfeld, Pyramid Algorithms for Perceptual Organization, NATO ARW on *Pyramidal Systems for Image Processing and Computer Vision,* Maratea, Italy, May 1986.

[25] S. L. Tanimoto, Lecture Notes on Pyramid Algorithm Design, Draft for the NATO ARW on *Pyramidal Systems for Image Processing and Computer Vision,* Maratea, Italy, May 1986.

[26] S. L. Tanimoto, A. Klinger, edits., *Structured Computer Vision: Machine Perception Through Hierarchical Computation Structures,* Academic Press, New York, 1980.

[27] L. Uhr, edit., *Pyramid Multi-Computers,* (tentative title) Academic Press, New York, to appear,1987.

[28] S. Yalamanchili, J. K. Aggarwal, Analysis of a Model for Parallel Image Processing, *Pattern Recognition,* Vol.18, N° 1, pp. 1-16, 1985.

[29] S. Yalamanchili, J. K. Aggarwal, Reconfiguration Strategies for Parallel Architectures, *IEEE Computer,* December, pp. 44-61, 1985.

Parallel Methods For Partial Differential Equations

John R. Rice

Department of Computer Science
Purdue University
West Lafayetee, IN 47906

This paper examines the potential of parallel computation methods for partial differential equations (PDEs). We start by observing that linear algebra is not the right model for PDE methods, that data structures should be based on the physical geometry. We observe that there is a naturally high level of parallelism in the physical world to be exploited. An analysis is made showing there is a natural level of granularity or degree of parallelism which depends on the accuracy needed and the complexity of the PDE problem. It is noted that the granularity leads to the use of superelements and that computational efficiency suggests that these should be of higher accuracy. We discuss the inherent complexity of parallel methods and parallel machines and conclude that dramatically increased software support is needed for the general scientific and engineering community to exploit the power of highly parallel machines. The paper ends with a brief taxonomy of methods for PDEs, the classification is based on the method's use of three basic procedures: Partitioning, Discretization and Iteration.

1. Introduction And Summary

This paper examines the potential for the use of parallelism in the solution of partial differential equations (PDEs). There are eight principal points made as follows:

1. Linear algebra is not the right model for developing methods for PDEs and it is particularly inappropriate for parallel methods.

2. The best data structures for PDE methods are based on the physical geometry of the problem.

3. Physical phenomena have large components that inherently parallel, local and asynchronous. Parallel methods can be found to reflect and exploit this fact.

4. There is a natural granularity associated with parallel methods for PDEs. The best number of "pieces" and processors depends on the complexity of the physical problem, the accuracy desired and properties of the iteration used.

5. Partitioning the computation into pieces is equivalent to using superelements in the discretization.

6. More accuracy in the superelements can enormously reduce the computational task.

7. Parallel machines are very messy and it is essential for most users that one have very high level PDE systems to hide this mess.

8. There is much to be gained to using regularity in parallel methods, but one should not carry this to extremes.

The final section presents a taxonomy of PDE methods. The classification is based on how the methods use three procedures: Partitioning, Discretization and Iteration. It is conjectured that the most promising methods for parallelism are those created in the order: iterate, partition, then discretize.

2. Parallel Methods For PDEs Is Not About Linear Algebra

In recent years there have been numerous papers written about linear algebra on parallel/vector machines (see [5], [2], [11], [12] and [9] for surveys and further references). Many machines have been designed to provide very high performance for linear algebra computations (see [7] and [6] for surveys and further references). Most of this work is motivated or justified in some part by applications to solving PDEs. Everyone sees that solving large linear problems is an inherent step in solving PDEs and it is usually the most expensive step. Yet the thesis of this section is that most linear algebra approaches are only tangentially relevant to solving PDEs and, in fact, they are often misleading.

A case in point is *nested dissection*. This was a breakthrough in solving PDEs, one that many people (including myself) had searched for over a period of decades. The original presentation [3] of nested dissection was inscrutable. If one starts (as everyone did) with the linear algebra problem $Ax = b$, then to discover nested dissection, one had to see that the matrix rearrangement such as shown in Figure 1 was the ''right'' way to eliminate the unknowns. However, if one expresses the reordering in terms of the underlying geometry of the PDE, one sees that nested discretion is a natural divide and conquer algorithm. It is then easy to understand why the method works so well, to see how to extend it to nonrectangular domains or to 3 dimensions or to finite element methods.

If one starts with a conventional matrix/vector representation of a PDE computation one is almost sure not to find efficient methods to solve the PDE. This is because the inherent structure of the PDE problem is so distorted by conventional matrix/vector representations that it is infeasible to uncover the natural problem structure. This is further illustrated in Figure 3 which shows the conventional matrix structure obtained by discretizing a second order PDE using a 9 point star on the domain shown in Figure 2. It is a computational tour-de-force to recover from Figure 3 the information that is superficially apparent in Figure 2.

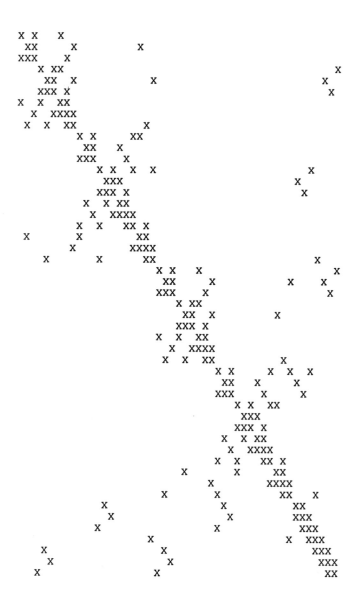

Figure 1. The pattern of non-zeros that occurs in solving Laplace's equation using the nested dissection ordering of the conventional matrix formulation with finite differences.

Figure 4 gives a geometric visualization of nested dissection. First the domain is divided in four parts (by the open bars), each of these parts is then, in turn, divided into four parts (by the solid bars). Then each of these is divided into four parts (by the open circles). This leaves the solid circles completely isolated from one another. The discretization equation (using a 5-point or 9-point star) at one of the solid circles does not involve any variable at the other points. Thus the solid circle unknowns can be eliminated independently (and simultaneously). After that the unknowns at the small connected groups of open circles can be eliminated independently. The order within the groups of five is immaterial. Then the solid bar unknowns can be eliminated followed by the open bar unknowns. This is the geometric pattern behind the matrix structure seen in Figure 1.

The shortcomming of the conventional linear algebra approach is that the right data structure is not used, one should base the data structure on the underlying physical geometry. Figure 2 shows a domain which has been "exploded" to group "like-kinds" of elements together in a PDE problem. A method that is really successful in exploiting parallelism in this problem must "know" this structure, the most practical way to know it is to have it given explicitly in the data structure. More complex problems have other structure (interfaces, singular points, etc.) that could be incorporated in a similar way.

It is not just parallelism that needs information such as seen in Figure 2, the control of numerical methods also need it. Numerical models need to be more accurate (e.g., grids need refining) near special locations. The partitioning of the computations for rapid convergence in iterative methods is strongly influenced by this information. To underscore this view of parallel methods for PDEs, we claim (conjecture?) that:

The really good parallel methods for PDEs do not require a global numbering of the unknowns in the computation.

A global indexing is often useful to create a specific, efficient computational representation, but it should not be an essential ingredient in the method. We note that the work to obtain a global numbering can become the dominant component of the computation if it is done carelessly. A good parallel method for a PDE problem with K unknowns and N processors $(K >> N)$ asymptotically should use time which is $O(K/N + \log N)$ or so, a numbering that requires $O(K)$ time must be avoided.

3. Parallelism Is (Almost) Unlimited In Solving PDEs

We claim that the physical phenomena that PDEs model are inherently local and asynchronous. Locality means that they are naturally amenable to parallel methods, the computation done at point A does not depend on anything being done at the physically distant point B. There are logical limits to the potential parallelism, we do not foresee much parallelism in time (as opposed to space) except for very special situations. There is also some sequentiality in local computations, one must compute values of coefficient functions in an equation before one can use the equation. For specific applications one can often reduce the sequential work dramatically by preprocessing computations (i.e., computing everything possible as soon as possible).

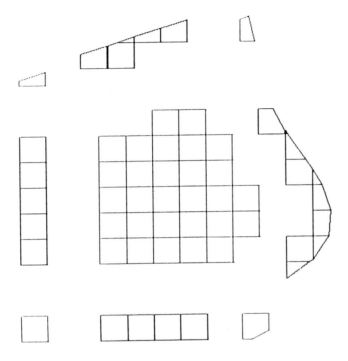

Figure 2. An exploded view of a physical domain which shows the elements of a "like" nature grouped together. The groupings are the first step in determining an appropriate structure in the problem of an efficient parallel method.

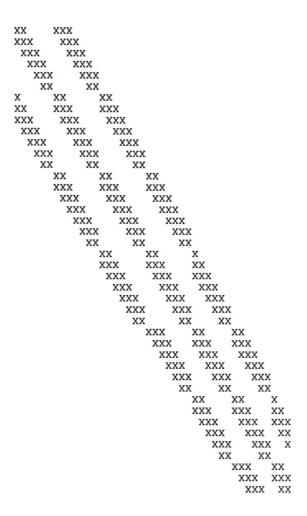

Figure 3. The conventional matrix structure obtained from a 9-point finite difference discretization on the domain seen in Figure 2.

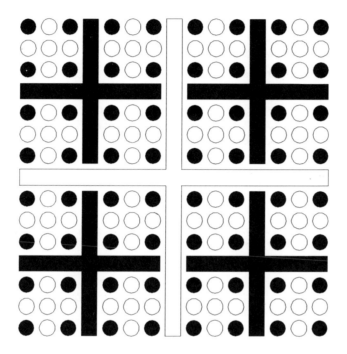

Figure 4. A visualization of the nested dissection ordering shown on a two dimension grid. The solid circles unknowns are eliminated first, followed by the open circles, then the solid bars and finally the open bars. The order of elimination within groups is immaterial.

The preceding observations are based on asymptotic considerations, i.e., if the physical domain is big enough and the accuracy required is high enough then any fixed number N of processors can be used profitably. We argue, however, that there is natural optimal or appropriate granularity and number N of processors associated with any particular PDE computations. We measure granularity in terms the number N of *elements* of the computation or model of the physical object. For simplicity we ignore any cases where computational elements do not correspond naturally with physical elements. The two extremes are:

(i) $N = 1$ processor gives 1 element which gives a sequential computation which gives very limited speed.

(ii) N very large (one Cray 2 per atom in a river?) gives a huge number of elements which gives very high parallelism which gives almost unlimited speed.

There are four considerations (at least) besides cost which lead to the existence of an optimal granularity, they are

1.   Every problem has an *acceptable solve time* beyond which solving it faster does not matter.

2.   Every problem has an *acceptable accuracy* beyond which more accuracy does not matter.

3.   For a fixed physical size, the number of interfaces between elements grows with the number of elements, thereby increasing the complexity and communication requirements of the computation. This growth might be very slow.

4.   For a fixed problem and method, the total work might eventually grow faster with N than parallelism reduces it because of slower convergence of iterative methods, etc.

Having identified granularity with N, we see that the independent variables in an application design are N, the desired elapsed time T and the required accuracy ε. We assume for now that ε behaves in a known way, that it is fixed and we only consider choosing N to achieve a specified T value. Figure 5 shows an idealized plot of cost versus time to solve a particular problem using a fixed number N of processors. The key points are that there is a lower limit on time (because processors can go only so fast) and that cost quickly reaches a plateau as the time increases. Figure 6 shows a different view of the situation, cost versus N for a fixed time to solve a particular problem. Again there is a lower limit because processors can go only so fast, but there is also an optimum. As N increases the cost starts to increase because of idle processors and/or increased communication (overhead) costs.

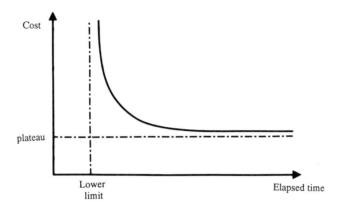

Figure 5. Cost versus elapsed time to solve a particular problem using N processors.

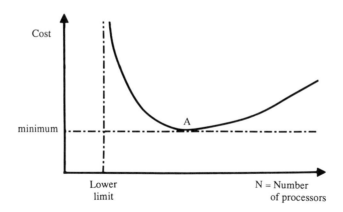

Figure 6. Cost versus the number of processors N used to solve a particular problem in a fixed elapsed time. The point A gives the minimum cost using an optimal number of processors.

We can replot the information of Figures 5 and 6 in the (N, T) plane and show two curves: the limiting curve of what is possible and the curve of optimal combinations of T and N. This is shown in Figure 7, the shapes are purely conjectural, one does not know what they are. It is true that cost decreases monotonically from point C to D.

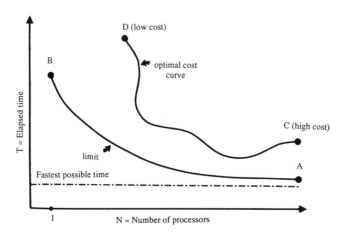

Figure 7. The (N, T) plane showing the limiting curve (A to B) of what is possible and the locus (C to D) of optimal cost combination of N and T.

Thus we see that while in principle there is no limit on the amount of parallelism that can be used in solving PDEs, there is definitely such a limit for any fixed application. Very little is known about actual values for real applications. I believe we are very far from the methods that give optimal time or cost in solving PDEs. On the other hand, I find it very convincing to argue that many real problems are quite complex and that to achieve "engineering" accuracy and "reasonable" elapsed time with even a low cost method (never mind optimal cost) will use thousands of processors.

4. Parallel Methods Create Superelements

We have argued that parallel methods will tend toward partitioning the computations into elements or pieces of "moderate" size. There will usually correspond to dividing physical space into elements and, to give a concrete example, if the numerical model requires 10,000 grid point in a 100 by 100 space, you can expect N to be perhaps 25 to 500 so that each processor "handles" something like a 4 by 4 grid (16 points) to a 20 by 20 grid (400 points).

Given that this argument is correct, let us recall that we are striving for a computationally optimal method. So, within the smallish subgrid handled by each processor, we want the numerical model that achieves the accuracy requirements with minimal computations. We might rephrase this to say that each processor is to handle 16 to 400 degrees of freedom and this is to be optimal (or at least reasonably good) in efficiency. That is to say, each processor will handle a *superelement*, one with a fairly large number of degrees of freedom. We will, of course, never know the best numerical model. However, there is abundant evidence that the standard approach of using a large number of simple finite difference formulas or a large number of simple finite elements is *very far* from optimal. The really good parallel methods will use complex superelements with high accuracy.

The question of what kind of elements to use is not one intrinsic to parallelism because the same is true for sequential methods. Since the essence of the question is outside the topic of this paper, it is not pursued further here. See [10] for some of the evidence of the value of higher accuracy elements. We do believe that the inherent use of superelements in parallel methods will make it more natural for one to introduce a little more complexity into the elements in order to gain considerably in accuracy and enormously in efficiency.

5. Parallel Methods Require New Software Systems

Parallel machines are already rather complex, much more so than previous operations of computers. They will become even more complex as it is discovered that a mixed set of capabilities provides more efficient computing. There will be variety is everything: processors (integer, floating point, graphics, vector, FFT, ...), memory (local, global, cache, archival, read only, ...), I/O (keyed, text, graphics, movies, acoustical, analog, ...), communication (message passing, packets, buses, synchronous/asynchronous, hypercubes, high/low speed, long haul, ...). The difficulty in managing (programming) this complexity is easily an order of magnitude higher than for present machines. The difficulty is compounded by the fact that changes in the capabilities available will become much more frequent.

The current programming methodology for solving PDEs is that of Fortran. One has a fairly intelligible language where one can exert fairly direct control of the machines resources. Each Fortran statement is typically implemental by 5-10 machines instructions. There must, I believe, always be such a language and I believe that Fortran will be expanded to handle the greater complexity of the machines. It might also be replaced by another moderate level language with such capabilities, e.g. Ada or C suitably enhanced. However, it will no longer be reasonable to expect the end-user scientists and engineers, the people who solve PDEs, to learn how to manage this complex computational environment. They will generally not do a very good job of it and, even if they did a good job, it would be a great waste of talent and duplication of effort. The potential benefits of parallel computation will not be achieved if every user has to master (even partially) how to manage such complex machines.

The solution to this problem is to substantially raise the level of the user's "programming" language. He must be able to say in a natural and succinct way what is to be done. In the PDE context they should be able to say things like:

1.   Solve $(1 + x^2)u_{xx} + u_{yy} - \sin(\alpha y)u = $ Force $2(x, y)$
     on the Domain #12
     with $u = 1$ on the boundary.

2.   Use finite differences with a 40 by 40 grid
     plus SOR iteration

3.   Show me plots of u, u_x and u_y

In fact, we should aim for the situation where statement 2. is replaced by

2a.   Obtain an accuracy of about 0.5 percent

Then, between such a program and the Fortran level is a layer of software which has two components. The first is a set of *problem solving modules* written by people who are relatively expert in solving the problems at hand and experienced in how parallelism (or other special capabilities available) can be exploited. There will be different methods (or, at least, different implementations) in the modules suitable for important subclasses of machines.

The second component of this layer is a set of *computation management facilities* written by people who are relatively expert in memory management, network scheduling, program transformations, etc. They have spent the time to learn how to provide such facilities well and have embedded much of their expertise into their software. These two components are then integrated to provide a bridge between the high level user input and a Fortran-like program targeted for the particular machine (or machines) to be used to solve the problem.

The obvious advantage of this methodology is that, if it works, there is a dramatic reduction in programming effort. This is, of course, the goal of introducing the methodology. Note that this not being done just to reduce software costs, the "mass-market" viability of parallel computation depends on introducing a methodology which hides the underlying complexity from most users.

The obvious disadvantage of this methodology is that the intermediate layer might introduce so much in efficiency that the power of parallelism is seriously weakened or even lost. It is clear that no foreseeable software for managing a computation can be as clever, resourceful and effective as clever, experienced people. This fact is a smokescreen that obscures a much more relevant "fact": people, even clever and experienced ones, almost never get close to "optimal" computations because they do not take the time to do it, it is inordinately expensive to do so. The result is that a good software system, one with many flaws which does many obviously stupid things, consistently can produce moderately good implementations which are significantly better than the ones people consistently produce. Scientific evidence to support this fact is scarce, but there is one solid data point.

Figure 8 shows a program written in DEQSOL, a high level PDE problem solving language under development at Hitachi [13]. No attempt is made here to explain DEQSOL. Hitachi has two PDE application programs that were written in FORTRAN prior to their vector supercomputer and DEQSOL efforts. There programs were brought into their vectorizing Fortran compiler environment and hand tuned to run well on their machines. The problems being solved were later reprogrammed in DEQSOL which produces a Fortran program which then use the vectorizing Fortran compiler but no hand tuning. The results of this experiment are shown below.

	A	B
FORTRAN:		
lines of code	1361	1567
execution time (sec.)	2.3	5.8
DEQSOL:		
lines of code	127	132
execution time	0.6	1.8
speed up factor	3.8	3.2

We see that not only was the programming effort reduced by the least an order of magnitude, but there was also a very worthwhile *gain* in execution speed. Keep in mind that a speed up of 3 or 4 is the typical total benefit achieved from using vector hardware on Cray and Cyber 205 machines.

We illustrate the power that can be achieved using such high level languages by considering the Plateau problem:

$$(1 + u_x{}^2)u_{xx} - 2u_x u_y u_{xy} + (1 + u_y{}^2)u_{yy} = 0$$

$$u(x, y) \text{ given on the boundary of a region } R \tag{1}$$

This is classical difficult PDE problem, its solution is the surface that a soap film takes on for a wire frame bent according to the value specified on the boundary of R. We solve this problem for the domain R and wire frame shape seen in the later figures (and explicitly defined in Figure 9). The high level language used is that of ELLPACK [10], one that provides modules and facilities for solving linear PDEs.

Newton's method is a natural candidate to try to solve a nonlinear problem by iterating on a sequence of linear problems. In this case one differentiates (1) with respect to u and rewrites the standard iteration (F represents the PDE operator in (1))

$$u^{(N+1)} = u^{(N)} + (dF/du)^{-1}F(u^{(N)})$$

```
dom      x = [0:1] ,          /* 3D DIFFUSION PROBLEM */
         y = [0:1] ,
         z = [0:2] ;
tdom     t = [0:5] ;
mesh     x = [0:1:0.1] ,
         y = [0:1:0.1] ,
         z = [0:2:0.1] ,
         t = [0:5:0.001] ;
var      T ;                  /* Temperature */
const
    rho = 1 ,        /* Density */
    c = 1 ,          /* Constant */
    k = 1 ,          /* Diffusion Constant */
    u = 0 ,          /* x-axis Velocity */
    v = 0 ,          /* y-axis Velocity */
    w = 5*(1.0-x**2)*(1.0-y**2) ,   /* z-axis Velocity */
    S = exp(-x**2-x**2-(1.0-z)**2) ;
                              /* Source Distribution */

cvect    V = (u, v, w) ;      /* Velocity Vector */
region
    In = (*, *, 0) ,          /* In */
    0 = (*, *, 2) ,           /* Out */
    X0 = (0, *, *) ,          /* Left */
    X1 = (1, *, *) ,          /* Right */
    Y0 = (*, 0, *) ,          /* Bottom */
    Y1 = (*, 1, *) ,          /* Top */
    R = ([0:1], [0:1], [0:2]) ;  /* Whole Region */

equ      rho*c*(dt(T)+V..grad(T)) = k*lapl(T)+S ;

bound    T = 0      at   In+X1+Y1 ,
         dz(T) = 0  at   0 ,
         dx(T) = 0  at   X0 ,
         dy(T) = 0  at   Y0 ;
init     T = 0 at R ;

ctr      NT ;     /* Iteration Counter */

scheme ;
    iter NT until NT gt 200;
        T<+1> = T+dlt*((k*lapl(T)+S)/(rho*c)-V..grad(T)) ;
        print T at Y0 ;
        disp T at Y0 every 100 times ;
    end iter ;
end scheme ;
end ;
```

Figure 8. The DEQSOL program to solve a time dependent, three dimensional diffusion problem.

to obtain the linear PDE exhibited in the ELLPACK program of Figure 9. This differentiation is somewhat tedious and a better system would also do that, a MACSYMA program for this is given in [10]. Figure 9 shows an ELLPACK program to implement Newton's method for (1). We do not explain the ELLPACK language here. A simple initial guess is made and the convergence is quite rapid in spite of the fact that the solution has a singularity (the wire has a sharp bend) along one side. The solution is displayed in Figure 10 and Figure 11 shows a contour plot of the difference between the third and fourth iterates.

Our final point in the software and programming area concerns the role *regularity* in data structures, in algorithms and in programs. Clever programmers and hardware designers can do a lot of special things to exploit special situations. This exploitation is usually achieved at the cost of more complex software and hardware. Thus there must be a balance between the execution time costs and the design costs of software and hardware. While it is hard to defend general statements on the matter, we believe that the optimum lies nearer to regularity and its attendant simplicity than it does to irregularity and its attendant complexity. However, we feel *extreme* simplicity is not the best approach either.

This view is illustrated by an example in discretizing a domain. Figure 12 shows a physical domain that has been partitioned in six ways for a problem with difficulties near the right boundary:

(A) A fine triangulation of a common type

(B) A fine, uniform, rectangular overlay grid

(C) Mapping the domain to a rectangle and inducing a logically rectangular partition

(D) Triangulation adapted to the difficulty

(E) Rectangular overlay grid adapted to the difficulty

(F) Logically rectangular partion adapted to the difficulty

We believe that the irregular triangulations do not provide any execution time advantage over the more regular partitions (note that one can do regular triangulation if one wants). On the other hand, we also believe that the uniformly spaced partitions are too simple and have too large an execution time penalty. We believe the adaption will pay off. The logically rectangular one is the simplest to program but the relative execution efficiencies resulting from (E) and (F) are not clear. Thus we believe that the search for the ''best'' method should be concentrated on partitions like (E) and (F) but there are still many undetermined degrees of freedom.

EQUATION.
   (1.+uy(x,y)**2) uxx + (1.+ux(x,y)**2) uyy   &
     - 2.*ux(x,y)*uy(x,y) uxy   &
   + 2.*(ux(x,y)*uyy(x,y) - uy(x,y)*uxy(x,y)) ux   &
   + 2.*(uy(x,y)*uxx(x,y) - ux(x,y)*uxy(x,y)) uy   &
=   2.*(ux(x,y)*uyy(x,y) - uy(x,y)*uxy(x,y))*ux(x,y) &
   + 2.*(uy(x,y)*uxx(x,y) - ux(x,y)*uxy(x,y))*uy(x,y)
BOUNDARY.
   u = bound(x,y) on x = 1.0,   &
           y = 0.5 + p   for p = 0.0 to 3.5
   u = bound(x,y) on x = 1.0 + p,   &
           y = 4.0   for p = 0.0 to 3.0
   u = bound(x,y) on x = 4.0 + .1*p*(p-4.5)**2, &
           y = 4.0 - p   for p = 0.0 to 4.5
   u = bound(x,y) on x = 4.0 - p,   &
           y = -0.5 + p/3.   for p = 0.0 to 3.0

GRID.
   9 x points 1.0 to 5.5 $ 9 y points -0.5 to 4.0

TRIPLE. set (u = gessu)

FORTRAN.
   do 100 it = 1, 5
      call save(r1unkn,i1neqn)
discretization.   **collocation**
solution.   **band ge**
output.   **max(diffu)**

FORTRAN.
   100 continue
OUTPUT. table(u) $ plot(u)

Figure 9. An ELLPACK program for applying Newton's method to solve the Plateau problem.

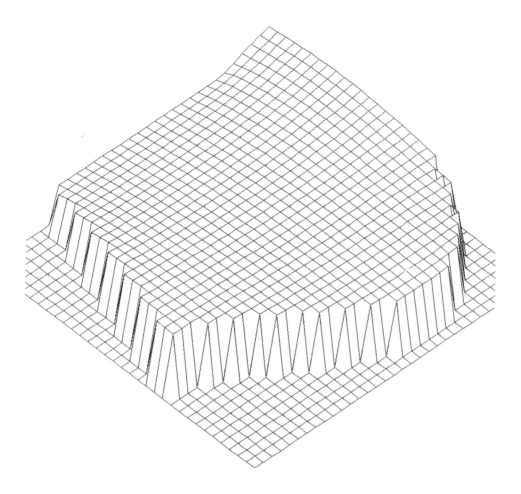

Figure 10. A view of the surface defined by the solution computed by the program in Figure 9.

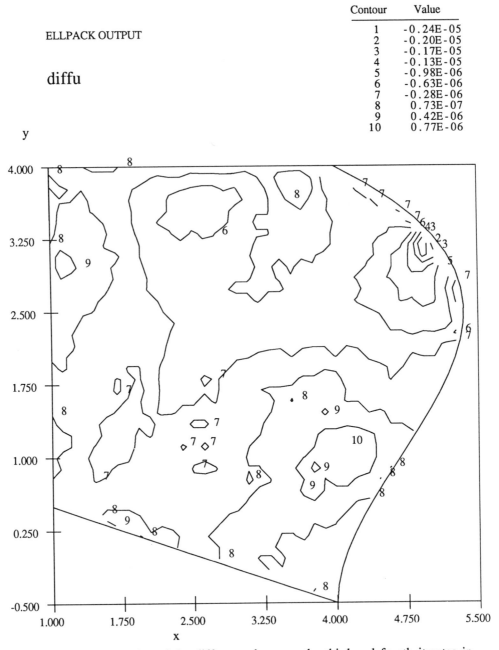

Contour	Value
1	-0.24E-05
2	-0.20E-05
3	-0.17E-05
4	-0.13E-05
5	-0.98E-06
6	-0.63E-06
7	-0.28E-06
8	0.73E-07
9	0.42E-06
10	0.77E-06

ELLPACK OUTPUT

diffu

Figure 11. Contour plot of the difference between the third and fourth iterates in Newton's method. The maximum difference here is 2.4×10^{-5} and that between the fourth and fifth iterates is 5×10^{-7}.

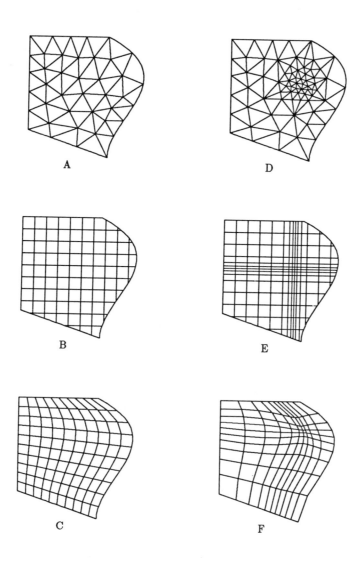

Figure 12. Six ways to partition a domain showing ways to achieve regularity and to adapt to a difficulty. The letters A through F refer to the discussion in the text.

6. Brief Taxonomy Of Parallel Methods For PDEs

This taxonomy assumes that the reader is familiar with the multitude of methods for solving PDEs (see [1], [4] [8] and [9] for recent, systematic treatises). Specific methods are mentioned as examples from larger classes and we do not provide descriptions of these methods here.

We classify methods by the way three basic procedures are used for defining them:

1. **Partitioning**. The unknowns in the problem are divided into groups. The criterion for grouping may be geometric (e.g., substructuring), or a matrix property (e.g., columns of an array) or physical (e.g., neighborhoods of singularities, boundary layers or shock waves). Different partitions may be used at different times (e.g., ADI methods) and partitions may be further partitioned (e.g., nested dissection or multigrid). Partitioning is almost always conducive to the use of parallelism.

2. **Discretization**. The continuous PDE problem is replaced by a finite problem with a set of real numbers as unknowns. The two most common techniques are finite differences and basis functions (e.g., finite elements, polynomials or series expansions). Finite difference methods and basis functions with local support (e.g., piecewise polynomials) are very good for parallel methods in the discretization phase as these computations are highly independent of one another. Discretization are also divisible into high order and low order methods. High order methods are advantageous for parallelism because more of the work can be shifted to the discretization phase of the computation.

3. **Iteration**. Iteration is an all-purpose technique to handle difficulties (e.g., nonlinearities, very large problems or time dependence). Iteration is inherently sequential and thus not very suitable for exploiting parallelism. Yet iteration is essential to solving many, if not most, PDEs, one should concentrate on introducing iteration so that the number of iterations is very small and the work in each iteration is large and easily divisible into parallel components.

There is a fourth basic procedure that is more limited in its applicability and thus usually appears at the "lower levels" of the method:

4. **Solve Directly**. One applies a procedure that exactly solves a problem. By far the most common example in PDEs is some form of elimination to solve a linear system of equations, other examples are FFT methods and symbolic integration.

Figure 13 shows a schematic of our taxonomy of PDE methods. The letters P, D, I and S are used to denote the four procedures, Partitioning, Discretization, Iteration, and Solve, respectively. The classes are defined in terms of the order in which the method is defined. There are some special methods for special problems (e.g., FFT for Laplace's equation) and some methods for linear, steady state PDEs do not use iteration. But, as Figure 13 indicates, most general methods involve all three of the basic procedures. By far the most common methods in current use are those that start with Discretization (corresponding to boxes 3 and 4).

We suggest that there are real advantages to doing the partition or iteration procedures first. It is much easier to use information about the physical domain data structure as this stage of developing the method. It is much easier to see how to apply Newton's method to a PDE than to the thousands of equations generated by a discretization. If Newton's method is applied directly to the PDE it is still easy to see how the physical domain data structure interacts with the linear PDE generated by Newton's methods.

The effective use of parallelism depends in many cases on an astute global organization of the computations. We believe that the most favorable approach to discover such methods is to follow paths 2 or 5, those where the discretization is delayed to the last.

Acknowledgements

This work supported in part by Air Force Office of Scientific Research grant AFOSR-84-0385.

References

[1] G. Birkhoff and R. Lynch, *Numerical Solution of Elliptic Problems*, SIAM, Philadelphia, 1984.

[2] J.J. Dongarra and D.C. Sorensen, "Performance and library issues for mathematical software on high performance computers." In *New Computing Environments: Parallel, Vector and Systolic* (A. Wouk, ed.) SIAM Publications, Philadelphia, 1986, pp. 110–133.

[3] J.A. George, "Nested dissection of a regular finite element mesh." SIAM J. Numer. Anal., 10(1973) pp. 345–363.

[4] I. Gladwell and R. Wait, *A Survey of Numerical Methods for Partial Differential Equations,* Clarendon Press, Oxford, 1979.

[5] R.W. Hockney and C.R. Jesshope, *Parallel Computers,* Chapter 5. Adam Hilger, Bristol, 1981.

[6] K. Hwang and F. Briggs, *Computer Architecture and Parallel Processing,* McGraw-Hill, New York, 1984.

[7] K. Hwang, *Supercomputers: Design and Applications,* IEEE EH0219-6, Silver Spring, 1984.

[8] O.A. McBryan and E.F. Van de Velde, "Parallel algorithms for elliptic equations." In *New Computing Environments: Parallel, Vector and Systolic* (A. Wouk, ed.) SIAM Publications, Philadelphia, 1986, pp. 236–270.

[9] J. Ortega and R. Voight, "Solution of partial differential equations on vector and parallel computers," SIAM Review, 27 (1985), 149–240.

[10] J. Rice and R. Boisvert, *Solving Elliptic Problems Using ELLPACK,* Springer-Verlag, New York, 1985.

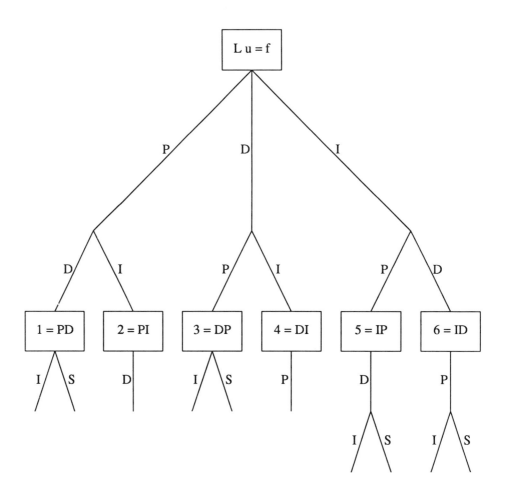

Figure 13. Taxonomy of methods for PDEs. The notation P = Partition, D = Discretization, I = Iterate and S = Solve Directly is used.

[11] A.H. Sameh, "An overview of parallel algorithms for numerical linear algebra," *First Int. Colloquium on Vector and Parallel Computing in Scientific Applications,* Paris, 1983.

[12] A.H. Sameh, "Numerical parallel algorithms - A survey." In *High Speed Computer and Algorithm Organizations*, Academic Press, New York, 1983, pp. 207–228.

[13] Y. Umetami, M. Tsuji, K. Iwasawa and H. Hirayama, "DEQSOL: A numerical simulation language for vector/parallel processors," Technical Report, Hitachi Ltd., Tokyo, 1984.

Parallel Algorithms for Matrix Computations

G. A. Geist, M. T. Heath and E. Ng

Mathematical Sciences Section
Oak Ridge National Laboratory
P.O. Box Y, Bldg. 9207A
Oak Ridge, TN 37831

The fundamental importance of linear algebra problems in science and engineering has placed algorithms for matrix computations in the forefront of research on parallel algorithms. We review the characteristics of parallel architectures that most strongly affect the design and performance of parallel algorithms for various matrix computations. These issues point toward a preliminary taxonomy of parallel matrix algorithms based on problem characteristics, basic solution methods, and specific features of target architectures. More detailed classification schemes require finer analysis of the interplay among algorithms, architectures, and applications.

1. Introduction

Matrix computations are among the cornerstones of scientific computing. Linear algebra problems, including systems of linear equations, linear least squares problems, and algebraic eigenvalue problems, are fundamental to the computational solution of differential equations, optimization problems, and the analysis of various discrete structures. The effective use of parallel computer architectures in scientific computations is therefore critically dependent on exploiting parallelism in matrix computations. Matrix algorithms have been in the vanguard of algorithm development on multiprocessors not only because they are building blocks on which many other scientific computations are based but also because they serve as realistic prototypes that present many of the fundamental challenges of parallel computation in a pure form. Thus, the development of parallel algorithms for matrix computations has received strong emphasis from researchers in parallel computing, both as a tool and as a paradigm for scientific computing in general on parallel architectures.

In this chapter we discuss some of the most important characteristics of parallel algorithms for matrix computations. Our concern here is not so much with the details of specific algorithms, but rather with the general features of parallel architectures that most strongly affect the ease of parallel implementation and determine potential performance of algorithms for various matrix computations. Most of our comments are meant to apply to parallel algorithms and architectures in general, but some of our language is necessarily motivated by particular classes of parallel architectures. Thus, for example, we use the term "communication" to refer not only to the explicit passing of messages in distributed-memory environments, but also to the more implicit interaction of processors through shared-memory.

There is a surprisingly long tradition of research on parallel algorithms for solving computational problems in linear algebra, especially the solution of various types of systems of linear equations. Much of the early research (see [14] for a survey) concentrated simply on attaining the maximum possible concurrency in solving a given type of problem and often employed highly simplified and unrealistic models of parallel computation (e.g., unlimited numbers of processors, no communication costs, no memory contention, no synchronization overhead, etc.). This is understandable, however, in view of the lack of widely available multiprocessor hardware with which to conduct numerical experiments. The advent of cost-effective VLSI components in recent years has made possible the commercial development of multiprocessor architectures having substantial numbers of processors, and this in turn has given new impetus to the development and testing of parallel algorithms in realistic multiprocessor environments.

2. Parallel Architectures

The characteristics of parallel algorithms are intimately intertwined with the characteristics of the problem to be solved and the computer architecture on which the algorithm will be implemented. We use the term "architecture" to include the programming environment and operating system support, as well as machine hardware. Probably the most significant characteristic of parallel architectures is the organization of memory, specifically whether each processor has access only to its own private local memory, or memory is globally shared among all processors (of course, hybrid systems may have both types of memory). In a distributed-memory system, problem data (matrices, vectors, etc.) must be partitioned and distributed across the individual processor memories, while in a shared-memory system all problem data reside in a common global memory where they are accessible by all processors. This means that standard data structures used on traditional serial machines for matrices and vectors usually carry over to shared-memory multiprocessors, whereas distributed-memory systems often require new distributed data structures, which can adversely affect the performance of parallel algorithms. Distributed data structures for sparse matrices, for example, tend to incur a penalty, both in storage overhead and efficiency of access, relative to their serial or shared-memory counterparts due to the loss of context that results from scattering the data across local processor memories.

In a distributed-memory multiprocessor, access to non-local data is provided by passing messages among the processors through an interconnection network. Common topological structures for such networks include rings, meshes, trees, and hypercubes. Analogously, the processors in a shared-memory system are connected to the common global memory by a bus or a switch, such as a crossbar switch or a multistage switching network. Thus, interprocessor communication bandwidth is a key performance parameter for distributed-memory systems, while memory bandwidth is a correspondingly important parameter for shared-memory systems (see, e.g., [16] for a discussion of multiprocessor networks and architectures).

The two main paradigms for memory organization also differ in their methods of coordinating, or synchronizing, the activities of the processors as they collaborate to solve a given problem, and this in turn affects the programming style used on the two types of systems. In distributed-memory systems, the passing of messages tends naturally to synchronize processors, as the availability of necessary data triggers execution of resulting computations. Examples of this approach include dataflow and systolic algorithms, which have been investigated extensively for various linear algebra computations. In this sense, the code running on each processor of a distributed-memory system is an ordinary sequential program with input and output, so that the parallelism is only implicit. Shared-memory systems tend to use more explicit mechanisms for synchronization, such as fork/join, semaphores, and memory locks, which have long been used in multiprogrammed operating systems for serial machines.

Another critical difference between shared-memory and local-memory systems is the manner in which the computational work load is distributed across the processors. Good efficiency requires that the work load be reasonably well balanced among the processors. In a shared-memory system, a good load balance is fairly easily maintained by dynamically assigning work from a common pool of tasks. Such a scheme tends naturally and effectively to accommodate varying numbers of processors and relatively heterogeneous tasks, since processors are assigned new tasks as they become available. Such a common pool of tasks is not easily implemented without access to shared memory, however, so that distributed-memory algorithms usually depend on a static assignment of work to processors that is determined in advance of the computation. This places a significantly greater burden on the algorithm designer to ensure a good load balance. Fortunately, most matrix problems have sufficiently regular computational subtasks that a static preassignment of work is usually quite sufficient to attain good efficiency.

3. Linear Algebra Problems

The most common computational problems in linear algebra are the solution of systems of linear algebraic equations and algebraic eigenvalue problems for various types of matrices (see, e.g., [13]). Significant problem characteristics that affect any computer implementation include whether the matrix is square or rectangular, symmetric or nonsymmetric, dense or sparse, explicitly represented by its entries or implicitly represented by its action on vectors. Such properties determine what algorithm, or family of algorithms, is appropriate for solving the problem in any computing environment, whether serial or parallel, but may have a more dramatic impact in the latter case. For example, the necessity of row or column interchanges for numerical stability can be a much more serious complication in a parallel environment than in a serial one. Another important problem characteristic in any computing environment is the size of the matrix, which determines the computational resources (processor time, storage) that will be required, as well as what classes of algorithms and data structures may be appropriate.

Although linear algebra problems are fairly representative of many other problems in scientific computing, they tend to be rather regular and homogeneous. Vectors and matrices map quite naturally onto many types of multiprocessor architectures, especially one- or two-dimensional mesh-connected arrays of processors, and many standard operations on them (e.g., computing inner products) have very natural parallel implementations.

The solution of a linear algebra problem often has a global data dependence, which implies that some form of global communication is required for solving the problem, and this in turn strongly affects the design of parallel algorithms, especially for architectures that may have

restricted communications. On the other hand, exceptions
to this general rule on data dependence may offer oppor-
tunities to gain additional efficiency in a parallel implementation. For example, one of the most effective techniques for sparse matrices is to decompose the problem
into substructures that are independent and can therefore
be processed in parallel.

4. Parallel Algorithms

The salient characteristics of parallel algorithms for
matrix computations fall under the two main headings of
"concurrency" and "communication." By *concurrency* we
mean the overlapped or simultaneous execution of computations on multiple processors. The degree of concurrency
is determined by the manner in which the overall computation is broken up into subtasks or subunits of computation, specifically their size distribution, scheduling, and
mutual data dependence. In most computations, parallelism can potentially be exploited at any of a number of
levels, depending on the characteristics of the underlying
hardware. The relative size of these subtasks or subunits
of computation that are scheduled for parallel execution is
referred to as the *granularity* of the algorithm. Of course,
it may be possible to exploit parallelism at more than one
level, as in a vector multiprocessor that supports both
inner-loop (fine-grain) parallelism in operating on vectors
and outer-loop (coarse-grain) parallelism in performing
larger tasks. The optimal choice of granularity depends on
the number and type of processors available and on the
overhead associated with communication among tasks,
whether by message passing or through shared memory.

We characterize the exchange of information among
processors in order to satisfy data dependencies among
subtasks as *communication*, whether this is accomplished
explicitly through message passing or implicitly through
shared memory. We have already remarked that many
linear algebra problems inherently require global communication, meaning that results produced by a given processor
must be made available to all other processors. In a

distributed-memory environment, such a communication pattern is typically accomplished by propagating the information in stages through local links in the interconnection network until all processors eventually receive it.

The keys to developing efficient parallel algorithms for a multiprocessor are to maximize concurrency and minimize communication costs. Unfortunately, these two objectives often conflict, and so a compromise must be sought between them. For example, increasing the granularity of an algorithm tends to reduce communication overhead, but it also tends to reduce potential concurrency. The precise trade-off point depends on both the problem and the architecture. The latter can be roughly characterized by the ratio of computation speed to communication speed. Both concurrency and communication requirements are determined by the manner in which the computation and the data are partitioned and distributed over the processors. For matrix problems, the main questions are how to organize the computation and partition the matrix (e.g., by columns, rows, submatrices, or diagonals), how to map the resulting pieces of computation and data onto the processors, and how best to meet the resulting communication requirements.

There are several possibilities for mapping matrices onto a distributed-memory multiprocessor and organizing the corresponding computations. If the network contains or efficiently emulates a two-dimensional grid, then a natural mapping is by submatrices, an approach that has been frequently advocated because of its "minimum perimeter" property (see, e.g., [7], [8]). On the other hand, many existing serial algorithms for matrix computations are column-oriented, as exemplified by such standard packages as LINPACK and EISPACK. This is due both to the convention of interpreting vectors as column vectors and to the column-oriented array storage used by Fortran, which is the most commonly used language for implementing matrix algorithms. Other matrix algorithms are more conveniently implemented by rows, however, and other programming languages have adopted row-oriented array storage. Still another approach is to map by diagonals, which is especially convenient for some implementations of matrix multiplication and for certain matrix structures (e.g., band matrices).

The partitioning of the matrix and mapping of the resulting pieces onto the processors is of critical importance in distributed-memory algorithms because the load balance among processors is determined by this static assignment. Assigning the pieces of the matrix (rows, columns, etc.) to the processors in the same manner that one would deal cards (sometimes called interleaved or scattered storage or wrap mapping) tends to yield better concurrency and load balance in matrix factorization algorithms than assigning contiguous pieces (blocks) to each processor (see, e.g., [10]). On the other hand, keeping contiguous data together in each processor tends to reduce necessary communication in these and many other algorithms. This is another example of a trade-off between concurrency and communication that is dependent on the particular problem, algorithm, and machine characteristics.

Broadcast communication, in which a given processor must make some of its results available to all of the other processors, can be implemented in various ways, depending on the connectivity of the underlying network. Nearest neighbor communication in a ring or grid is a natural scheme in many cases, depending on the mapping employed. Although this approach entails a relatively long path length for propagating a global broadcast, successive broadcasts can be effectively pipelined for good overall efficiency. Wavefront algorithms, in which propagation of broadcasts is interleaved with computation, are a good example of the efficient use of this approach on mesh-connected arrays of processors (see, e.g., [26], [27]). A highly effective approach to broadcasting in a hypercube makes direct use of the recursive structure of the hypercube to embed a minimal spanning tree along which the broadcast fans out, ultimately reaching all processors in a logarithmic number of steps.

5. Matrix Computations

Having established the appropriate background concerning parallel architectures, algorithms, and applications, we are now in a position to discuss some specific problems in computational linear algebra. We briefly overview parallel algorithms for solving systems of linear equations by both direct and iterative methods, computing eigenvalues, and other basic operations on matrices. Although we discuss direct and iterative methods separately, in many modern algorithms elements of both are used in a complementary manner, as in preconditioned iterative methods and multigrid methods. See [28] for a general survey and further references on many of the methods we discuss.

5.1 Direct Methods for Linear Systems

Factorization of dense matrices is an interesting case study for parallel implementation because it has two essential features that tend to inhibit parallel efficiency: serial precedence constraints and global communication. Serial precedence constraints arise because successive columns, rows, or submatrices of the matrix factors must be completed in consecutive order. Global communication is required because each successive column, row, or submatrix of the matrix factors depends on all preceding columns, rows, or submatrices. Nevertheless, many algorithm developers have shown that in practice computations can be sufficiently overlapped and communication sufficiently masked to attain very high processor utilization and parallel efficiency (see, e.g., [5], [10], [11], [17], [18], [25], [26], [27]).

The principal distinctions among parallel algorithms for dense matrix factorization are again concurrency and communication. The degree of concurrency attainable and the communication cost are determined largely by the granularity chosen. A very fine-grain implementation, with subtasks of complexity $O(1)$, as in systolic and wavefront algorithms (e.g., [15], [20], [21], [26], [27]), potentially attains the maximum possible concurrency, but this potential will not be realized unless communication overhead is very small, with the cost of communicating one number comparable to the cost of one arithmetic operation. A medium-grain implementation, with subtasks of complexity $O(n)$, is likely to provide a better trade-off between concurrency and communication cost on most existing general-purpose multiprocessor architectures.

Matrix factorization reduces the problem of solving a general system of equations to the problem of solving triangular systems of equations. The same principles of concurrency and communication apply to the latter problem, but since there is less computation to be done in solving a triangular system compared to computing a factorization ($O(n^2)$ compared to $O(n^3)$), communication overhead tends to be more dominant, and consequently good efficiency is harder to achieve in a parallel algorithm for solving triangular systems. Thus, the cost of the triangular solution phase, while often considered to be negligible in a serial environment, may be significant in a parallel environment.

All of the foregoing discussion of direct factorization methods is predicated on an assumption that the matrix is dense. The picture is quite different if the matrix is sparse. Sparse problems have both advantages and disadvantages relative to the dense case. Their principal advantage is that sparsity potentially leads to greater concurrency and less communication due to the data independence of some parts of the problem from others. Thus, for example, the strict serial precedence constraints may be partially relaxed for a sparse problem, and global communication may no longer be required (see, e.g., [23]). On the other hand, sparse direct methods are inherently fine-grained in that only a small amount of computation occurs between communications, so that the advantages are difficult to exploit unless communication costs on a given architecture are very low (see, e.g., [12]). Sparse matrix algorithms for distributed-memory may also be more sensitive to other hardware and software characteristics. For example, we have found that message buffer space can become saturated due to message production exceeding consumption for sparse problems in which a high degree of independent parallelism permits computations to get too far ahead of communication.

5.2 Iterative Methods for Linear Systems

Iterative methods for solving linear systems also present interesting issues for parallel implementation. Unlike direct methods, in which a fixed set of computations must be performed in order to arrive at the correct solution, the "self-correcting" nature of iterative methods makes it tempting to alter the computations to enhance parallelism (e.g., "chaotic relaxation"). Speeding up individual iterations by increasing concurrency will be of no benefit, however, if the overall convergence rate is slowed down by a corresponding amount, and this is often the case when relatively outdated information is used in advancing an iterative scheme to the next step. See [28] and the many references therein for an extensive discussion of the large body of work on parallel iterative methods for solving linear systems.

Iterative methods are frequently recommended for parallel implementation because of their more limited communication requirements (e.g., corresponding to the stencil of a finite difference operator), in contrast to the global communication required by direct factorization methods. In another sense, however, iterative methods are less amenable to parallel speedup than direct methods: each individual iteration usually requires an order of magnitude less computation than a corresponding factorization algorithm (e.g., an $O(n^2)$ matrix-vector multiplication compared to an $O(n^3)$ matrix factorization in the dense case), and thus there is correspondingly less computation over which to amortize communication costs. A potential solution to this dilemma is to try to achieve concurrency or overlapping in the "time" dimension (i.e., over successive iterations) as well, and some work has been done along this line [34].

Although a basic iterative scheme may involve only local communication, convergence testing necessarily involves gathering global information, and so some of the same global communication issues discussed previously apply here as well. Indeed, if care is not taken, convergence testing can easily dominate the computation in a message-passing environment with relatively slow communication, and therefore techniques for less intrusive convergence checking are imperative [34].

5.3 Eigenvalue Computations

Algorithms for computing eigenvalues of matrices are much less well developed for parallel machines than those for solving linear systems. Of course, algorithms such as the power method that rely solely on matrix-vector multiplication are straightforward to implement, but more sophisticated algorithms, such as the algorithms based on QR and LR iteration found in EISPACK, are more difficult. The latter algorithms have an iterative outer loop around a factorization-like inner loop. The "zero–chasing" often found in the inner loop tends to be a rather sequential and fine-grained computation, and thus it is relatively difficult to implement with good efficiency on a general-purpose multiprocessor. Progress is being made in this area, but much further work is still needed.

One of the most fruitful areas to date in the development of parallel algorithms for eigenvalue problems is the design of special-purpose systolic arrays. The impetus for these potentially very high-performance algorithms is provided by the real-time constraints of signal processing applications. Systolic arrays have been developed for computing eigenvalues and singular values of various types of matrices (see, e.g., [1]). Unfortunately, the practical impact of these algorithms has yet to be felt because of the difficulty of fabricating such devices in silicon.

Another paradigm that has led to new parallel algorithms for eigenvalue computations is that of "divide-and-conquer." This approach has been implemented by Dongarra and Sorensen [6], based on earlier work of

Cuppen [2], to produce a new algorithm that not only is
very natural and effective as a parallel algorithm, but is
actually faster as a serial algorithm than the previously
standard QR algorithm.

5.4 Other Matrix Computations

Other basic manipulations of matrices have also been
investigated on parallel architectures. Matrix transposition, for example, turns out to be an interesting combinatorial problem to implement on distributed-memory multiprocessor networks. Efficient transposition is important
in some implementations of alternating direction (ADI)
methods for partial differential equations ([9], [19]) or for
computing two-dimensional transforms. A number of
efficient transposition techniques have been implemented,
some based on highly pipelined ring or mesh communication, others based on successive pairwise exchanges using
the recursive structure of a hypercube (see, e.g., [24], [32],
[33]).

For many serial machines several Basic Linear Algebra
Subprograms (inner products, norms, etc.) have been
encapsulated into a package called the BLAS [22]. These
have been found to improve both efficiency and portability
across many serial and vector architectures. Most of the
vector-oriented $O(n)$ operations in the BLAS have also
been implemented on many parallel architectures. It has
been found, however, that both efficiency and portability
are further enhanced on many parallel architectures by
implementing higher order ($O(n^2)$) matrix-vector operations, such as matrix-vector multiplication and rank-one
updating [3]. This approach has also been used in developing codes for solving partial differential equations on message passing systems to encapsulate all necessary communication within the matrix-vector subroutine calls so that
serial codes can run on a parallel architecture with no
changes to the user-level source code [24].

6. Concluding Remarks

In this section we address some additional questions regarding parallel algorithms for matrix computations as they relate to such larger issues as identifying commonalities across problem domains, the representation of parallelism, the design of languages and architectures, and taxonomy of parallel algorithms.

In a relatively restricted area such as matrix computations, there are naturally many commonalities across problem domains. For example, similar communication and synchronization patterns occur repeatedly for various problems. Exploiting such commonalities often contributes to greater efficiency both in algorithm development time and in the resulting algorithm. On the other hand, our thinking can sometimes be too narrow in seeking solutions to new problems if we adhere too rigidly to paradigms that have proven useful in the past. For example, the same broadcast communication pattern that arises in matrix factorization can also be used to implement a very efficient row-oriented solution algorithm for triangular systems, but is much less efficient for solving triangular systems with column-oriented storage. Because of this, the development of efficient column-oriented algorithms for triangular solutions lagged far behind that of row-oriented algorithms for largely conceptual reasons. The key observation that broke this conceptual bottleneck was the realization of a deeper kind of duality between row- and column-oriented algorithms. In particular, when the fan-out scatter communication pattern of the row-oriented algorithm is replaced by a fan-in gather communication pattern, an equally efficient column-oriented triangular solution algorithm results [31].

Parallel algorithms for matrix computations have already had an influence on the design of architectures and languages. A classic example is the development of vector computers and array processors, which employ pipelining techniques to provide fast operations on vectors that typically arise in matrix computations. The communication requirements of some of the finer-grained parallel matrix algorithms have made it obvious that hardware designers will have to apply the same effort and ingenuity in designing future communication processors that they have previously applied in developing fast arithmetic processors. In the area of language design, the vector constructs typical in linear algebra have influenced proposals for languages such as Fortran 8x. Looking toward more general types of parallelism, the coarser-grained pipelined execution found in parallel algorithms for matrix factorization has found its way into some parallel programming environments through such constructs as the "do across."

The above ideas lead naturally to the question of how best to represent parallelism in programs. Should it be explicitly specified by the programmer or derived automatically from implicit data dependencies? Certainly there has been a great deal of success in designing optimizing and vectorizing compilers that do an excellent job of producing efficient code for matrix algorithms on serial and vector computers. The inner-loop analysis that these compilers are based on, however, is inadequate to exploit effectively the coarser-grained, outer loop parallelism that is available in many matrix computations. Although much recent progress has been made in this area, it is our feeling that smart compilers would be hard pressed to discover new parallel algorithms, such as Cuppen's algorithm for computing eigenvalues, which was motivated by its excellent potential for parallelism, but has also proven to be a better serial algorithm than the standard QR iteration. It is also our experience that explicitly parallel programs are not particularly difficult to write or debug so long as a reasonable amount of discipline is enforced, especially if one can rely on previously developed and debugged routines or templates for standard communication and synchronization patterns. Of course, additional software tools in this area would be welcome, such as data and control flow analyzers, debuggers, and graphical programming aids.

Finally, we come to the question of a taxonomy of parallel algorithms that initially motivated this book. Parallel algorithms for matrix computations may seem too regular and homogeneous to have much to say about an all-encompassing taxonomy of parallel algorithms. On the other hand, perhaps it is in just such a relatively well-structured area that we should look for the first clues as to what a general taxonomy might look like. First, we observe that for matrix computations the upper levels of such a taxonomy of algorithms are no different than those for the corresponding serial computations. Thus, we would have the usual classifications according to problem class, matrix type, iterative vs. direct, etc., that would be equally applicable to serial algorithms. Second, additional parallel algorithm classes result from largely architectural distinctions: e.g., shared memory vs. distributed memory, static vs. dynamic load balancing, various common synchronization and communication patterns, etc. Thus, the beginnings of a taxonomy of parallel algorithms for matrix computations results from simply taking a "cross product" of existing taxonomies of serial matrix algorithms and parallel architectures.

A further, more detailed taxonomy requires a more penetrating analysis of the interplay between matrix algorithms and parallel architectures. An example of how such an analysis might go is provided by the "*ijk*" classification scheme for matrix factorization algorithms, based on the various ways of rearranging the triple-nested loop defining the factorization. This approach has been used to study the efficiency of factorization algorithms on various vector and parallel architectures [4], [11], [29], [30]. Although it is perhaps a rather modest step toward a general taxonomy of parallel factorization algorithms, the *ijk* classification scheme has proven useful in suggesting the most efficient algorithms for specific types of parallel architectures; such predictive value should be the primary rationale for any taxonomy of algorithms.

Acknowledgements

This work was supported by the Applied Mathematical Sciences subprogram of the Office of Energy Research, U.S. Department of Energy under contract DE-AC05-84OR21400 with Martin Marietta Energy Systems, Inc.

References

[1]   R. P. Brent and F. T. Luk, "The solution of singular-value and symmetric eigenvalue problems on multiprocessor arrays", *SIAM J. Sci. Stat. Comput.*, 6 (1985), pp. 69-84.

[2]   J. J. M. Cuppen, "A divide and conquer method for the symmetric tridiagonal eigenvalue problem", *Numer. Math.*, 36 (1981), pp. 177-195.

[3]   J. J. Dongarra, J. Du Croz, S. Hammarling and R. J. Hanson, "A proposal for an extended set of Fortran basic linear algebra subprograms", *SIGNUM Newsletter*, 20, No. 1 (1985), pp. 2-18.

[4]   J. J. Dongarra, F. G. Gustavson and A. Karp, "Implementing linear algebra algorithms for dense matrices on a vector pipeline machine", *SIAM Rev.*, 26 (1984), pp. 91-112.

[5]   J. J. Dongarra, A. H. Sameh and D. C. Sorensen, "Implementation of some concurrent algorithms for matrix factorization", *Parallel Computing*, 3 (1986), pp. 25-34.

[6]   J. J. Dongarra and D. C. Sorensen, "A parallel algorithm for the symmetric eigenvalue problem", *SIAM J. Sci. Stat. Comput.*, to appear.

[7]   G. C. Fox, "Matrix operations on the homogeneous machine", Tech. Rept. HM-5, California Institute of Technology, Pasadena, CA, 1982.

[8]   G. C. Fox, "Square matrix decompositions - symmetric, local, scattered", Tech. Rept. HM-97, California Institute of Technology, Pasadena, CA, 1984.

[9]   D. B. Gannon and J. Van Rosendale, "On the impact of communication complexity on the design of parallel numerical algorithms", *IEEE Trans. Computers*, C-33 (1984), pp. 1180-1194.

[10] G. A. Geist and M. T. Heath, "Matrix factorization on a hypercube multiprocessor", in *Hypercube Multiprocessors 1986* (ed. by M. T. Heath), pp. 161-180, SIAM, Philadelphia, 1986.

[11] A. George, M. T. Heath and J. Liu, "Parallel Cholesky factorization on a shared-memory multiprocessor", *Linear Algebra Appl.*, 77 (1986), pp. 165-187.

[12] A. George, M. T. Heath, J. Liu and E. Ng, "Sparse Cholesky factorization on a local-memory multiprocessor", Tech. Rept. ORNL/TM-9962, Oak Ridge National Laboratory, Oak Ridge, Tennessee, 1986.

[13] G. H. Golub and C. F. Van Loan, *Matrix Computations*, Johns Hopkins Univ. Press, Baltimore, MD, 1983.

[14] D. E. Heller, "A survey of parallel algorithms in numerical linear algebra", *SIAM Rev.*, 20 (1978), pp. 740-777.

[15] D. E. Heller and I. C. F. Ipsen, "Systolic networks for orthogonal decompositions", *SIAM J. Sci. Stat. Comput.*, 4 (1983), pp. 261-269.

[16] K. Hwang and F. A. Briggs, *Computer Architecture and Parallel Processing*, McGraw-Hill Book Co., New York, 1984.

[17] I. C. F. Ipsen, Y. Saad and M. H. Schultz, "Complexity of dense linear system solution on a multiprocessor ring", *Linear Algebra Appl.*, 77 (1986), pp. 205-239.

[18] S. L. Johnsson, "Communication efficient basic linear algebra computations on hypercube architectures", Tech Rept. YALEU/DCS/RR-361, Department of Computer Science, Yale University, New Haven, CT, 1985.

[19] S. L. Johnsson, Y. Saad and M. H. Schultz, "Alternating direction methods on multiprocessors", Tech Rept. YALEU/DCS/RR-382, Department of Computer Science, Yale University, New Haven, CT, 1985.

[20] H. T. Kung, "Why systolic architectures?", *IEEE Computer*, 15, No. 1 (1982), pp. 37-46.

[21] S. Y. Kung, "On supercomputing with systolic/wavefront array processors", *Proc. IEEE*, 72 (1984), pp. 867-884.

[22] C. Lawson, R. Hanson, D. Kincaid and F. Krogh, "Basic linear algebra subprograms for Fortran usage", *ACM Trans. Math. Software*, 5 (1979), pp. 308-371.

[23] J. W. H. Liu, "Computational models and task scheduling for parallel sparse Cholesky factorization", *Parallel Computing*, to appear, 1986.

[24] O. A. McBryan and E. F. Van de Velde, "Hypercube algorithms and implementations", Tech. Rept. DOE/ER/03077-271, Courant Institute, New York University, New York, 1986.

[25] C. Moler, "Matrix computation on distributed memory multiprocessors", in *Hypercube Multiprocessors 1986* (ed. by M. T. Heath), pp. 181-195, SIAM, Philadelphia, 1986.

[26] D. P. O'Leary and G. W. Stewart, "Data-flow algorithms for parallel matrix computations", *Comm. ACM*, 28 (1985), pp. 840–853.

[27] D. P. O'Leary and G. W. Stewart, "Assignment and scheduling in parallel matrix factorization", *Linear Algebra Appl.*, 77 (1986), pp. 275–299.

[28] J. M. Ortega and R. G. Voigt, "Solution of partial differential equations on vector and parallel computers", *SIAM Rev.*, 27 (1985), pp. 149–240.

[29] J. M. Ortega, "The ijk forms of factorization methods I. Vector computers", Tech Rept., Applied Mathematics Dept., University of Virginia, Charlottesville, VA, in preparation.

[30] J. M. Ortega and C. H. Romine, "The ijk forms of factorization methods II. Parallel systems", Tech Rept., Applied Mathematics Dept., University of Virginia, Charlottesville, VA, in preparation.

[31] C. H. Romine and J. M. Ortega, "Parallel solution of triangular systems of equations", Tech Rept. RM-86-05, Applied Mathematics Dept., University of Virginia, Charlottesville, VA, 1986.

[32] Y. Saad and M. H. Schultz, "Data communication in hypercubes", Tech Rept. YALEU/DCS/RR-428, Department of Computer Science, Yale University, New Haven, CT, 1985.

[33] Y. Saad and M. H. Schultz, "Data communication in parallel architectures", Tech Rept. YALEU/DCS/RR-461, Department of Computer Science, Yale University, New Haven, CT, 1985.

[34] J. H. Saltz, V. K. Naik and D. M. Nicol, "Reduction of the effects of the communication delays in scientific algorithms on message passing MIMD architectures", Tech. Rept. 86-4, ICASE, NASA Langley Research Center, Hampton, VA, 1986.

Parallel Programming with the BLAS

William J. Harrod

Center for Supercomputing Research and Development
University of Illinois
305 Talbot Laboratory
104 South Wright Streeet
Urbana, Illinois 61801

This paper will examine a set of high level modules called the BLAS (Basic Linear Algebra Subroutines) and their use in an algorithm for computing the orthogonal factorization of a matrix. These subroutines consist of the most commonly used vector operations in linear algebra algorithms. In 1984 an extension of the BLAS was proposed these routines consist of matrix_vector operations. Recently another extension of the BLAS has been proposed, in which the routines consist of matrix_matrix operations. In this paper we report on the performance of these three levels of the BLAS and their use in the orthogonal factorization of a matrix on the Alliant FX/8.

1. Introduction

Today's high performance computers use combinations of architectural techniques such as vector, pipeline and parallel processing. With the addition of a multi level memory hierarchical system a software designer faces a difficult task. Portability suffers when a software designer tailors a numerical algorithm to take advantage of the hardware features of an advanced computer. However, it does seem a waste not to take full advantage of a high performance computer. One technique that a software developer can use is to express an algorithm in terms of modules with a high level of granularity. These modules can be designed to include most of the machine dependent details required to achieve the expected high performance of a particular machine. The resulting algorithm may not be the best algorithm for all advanced computers, but the software designer will have an easier task of porting the code to a new machine.

This paper will examine a set of high level modules called the BLAS (Basic Linear Algebra Subroutines) and their use in an algorithm for computing the orthogonal factorization of a matrix. These subroutines consist of the most commonly used vector operations in linear algebra algorithms. In 1984 an extension of the BLAS was proposed (see [DDHH84]), these routines consist of matrix_vector operations. Recently another extension of the BLAS has been proposed (see [GaJM86]), in which the routines consist of matrix_matrix operations. In this paper we report on the performance of these three levels of the BLAS and their use in the orthogonal factorization of a matrix on the Alliant FX/8.

We investigate an algorithm for computing an orthogonal factorization of a matrix that involves a block generalization of the Householder transformation. Research in new formulations of the Householder transformations has been reported by Brenlund and Johnsen [BrJo74] and Dietrich [Diet76]. Recently, a block method has been presented in a paper by Bischof and Van Loan [BiVa85]. The basic idea is to write the product of k Householder transformations as a rank k update of the identity. That is,

$$(I - u_k u_k^T) \cdots (I - u_1 u_1^T)(I - u_1 u_1^T) = I - VU^T,$$

where $U, V \in \mathbf{R}^{m \times k}$. The motivation behind this variation of the classical Householder transformation is to recast the factorization algorithm so that a large percent of the execution time is spent using the BLAS 3 operations instead of BLAS 1 operations.

The Alliant FX/8 is a mini–supercomputer that combines vector and parallel processing. The machine has eight pipelined processors, called computational elements (CE). Each CE has a scalar and vector instruction set. Also, there are eight floating point vector registers. In vector mode, each CE executes double precision floating point operations at the peak rate of 11.8 million floating operations per second (MFLOPS). A cluster of 8 CE's can produce a peak performance of 94.4 MFLOPS. They share a physical memory and a 128 K–byte write back cache that allows up to 8 simultaneous accesses in each cycle of 170 ns. The compiler attempts to detect those sections of the program that have potential for vector and parallel processing and generates code that will use the concurrency and vectorization features of the hardware. For example, when loops are nested, the compiler attempts to generate code so that the inner loop will run in vector mode and the outer loop will run in concurrent mode. The compiler provides the programmer with optimization directives that can be inserted into the program to control optimization at various levels.

2. The BLAS

In 1979 Lawson, Hanson, Kincaid, and Krogh [LHKK79] described a set of commonly used vector routines, called the BLAS (Basic Linear Algebra Subroutines). Most linear algebra algorithms can be expressed in terms of the BLAS. The following are some examples of the BLAS routines:

$$y \leftarrow y \pm \alpha x \qquad \text{vector update}$$

$$\beta = x^T y \qquad \text{dotproduct}$$

$$y \leftarrow \alpha y \qquad \text{vector scale}$$

where $x, y \in \mathbf{R}^n$ and $\alpha, \beta \in \mathbf{R}$. The vector update, dotproduct, and vector scale operations are commonly referred to by their BLAS subroutine names, DAXPY, DDOT, and DSCAL respectively. LINPACK [DBMS79] is an example of high quality mathematical software that is written in terms of the BLAS. A majority of the floating operations are performed in the BLAS routines. Improved performance may be obtained if the standard FORTRAN versions of the BLAS are replaced by a set of highly tuned machine language versions of the BLAS.

The timing results for the original LINPACK routine for solving a linear system of equations on a vector processor, such as the CRAY XMP and CYBER 205, have been disappointing. To improve the performance, one needs to recast LINPACK in terms of matrix_vector operations (see [DoSo86]). In 1984 Dongarra, Du Croz, Hammarling, and Hanson [DDHH84] proposed an extended set of BLAS. We will refer to the original set of BLAS as the BLAS 1 and the additional set of the BLAS as the BLAS 2. The BLAS 2 consist of matrix_vector operations. There are four basic types of operations in the BLAS 2:

$$y \leftarrow y \pm Ax \qquad \qquad \text{matrix} \times \text{vector}$$
$$x^T \leftarrow x^T \pm y^T A \qquad \qquad \text{vector} \times \text{matrix}$$
$$A \leftarrow A \pm yx^T \qquad \qquad \text{rank 1 update}$$
$$x \leftarrow T^{-1}x \qquad \qquad \text{triangular solver}$$

where $A \in \mathbf{R}^{m \times n}$, $x \in \mathbf{R}^n$, $y \in \mathbf{R}^m$, and $T \in \mathbf{R}^{n \times n}$ is a triangular matrix.

Some of the advanced vector computers have the triadic operation $y \leftarrow y + \alpha x$ as a single vector instruction. A few examples of such machines are the Cyber 205, CRAY XMP, and the Alliant FX/8. Thus, when a vector update operation is executed on one of these machines it involves three vector movements and one floating point vector instruction. The BLAS 2 operation $y \leftarrow y + Ax$, $A \in \mathbf{R}^{m \times n}$ can be coded as follows:

$$\text{do } j = 1, n$$
$$y \leftarrow y + \chi_j a_j$$
$$\text{enddo}$$

where χ_j denotes the j–th component in the vector x, and a_j denotes the j–th

column of A. If this routine is implemented as a series of calls to the, vector

update (DAXPY) routine in BLAS 1, then there are 3n vector movements and

n vector operations. If we assume that n is less than or equal to the length of

the vector registers then we could code this routine so that the vector y is

kept in a vector register and the loop is implemented as a series of triadic

operations. In this case there are n+2 vector movements and n vector

operations. Both algorithms involve the same number of vector operations.

However, the second algorithm has approximately 2n fewer vector moves.

Some of the machines with vector capabilities have a mutlilevel memory

system which involves a first level of small fast memory and a second level of

slower larger memory. The Alliant FX/8 has a hardware managed cache, and

the CRAY 2 has a software managed local memory. On these machines

timing results for LINPACK routines written in terms of the BLAS 2 failed to

reach the predicted high rate of performance. The poor results are due to the

poor management of the first level of memory.

The BLAS 2 routine $y \leftarrow y + Ax$ for a square matrix of order n involves

$n^2 + 2n$ floating point numbers and $2n^2$ floating point operations. Thus there

are approximately 2 floating point operations per floating point number. The operation $A \leftarrow A + BC$ for square matrices A, B, C of order n involves $3n^2$ floating point numbers and $2n^3$ floating point operations. Thus there are approximately $\frac{2}{3}n$ floating point operations per floating number. Hence, the algorithm for the operation $A \leftarrow A + BC$ should be designed so that when a floating number is moved into the first level memory it should be used as many times as possible [JaMe86].

An extension of the BLAS 2 has recently been proposed [GaJM86], called the BLAS 3. They consist of matrix_matrix operations, such as:

$$A \leftarrow A \pm BC \qquad\qquad \text{rank k update}$$
$$A \leftarrow A \pm E^T C \qquad\qquad \text{matrix}^T \times \text{matrix}$$

where $A \in \mathbf{R}^{m \times n}, B \in \mathbf{R}^{m \times k}, C \in \mathbf{R}^{k \times n}, E \in \mathbf{R}^{k \times m}$

In the following section we will investigate the performance of some the BLAS operations. Numerical experiments will be reported on an Alliant FX/8.

3. Performance

The results of several timing experiments for the BLAS are presented in this section. The BLAS 1 subroutines are written in Alliant FORTRAN FX/8, while the BLAS 2 and BLAS 3 subroutines are written in assembler language. The experiments were executed on an Alliant FX/8 with 8 computational elements (CE's), a 16 k word cache and a 32 MB physical memory.

Figure 1 shows the performance of the BLAS 1 dotproduct and vector update operations. The vector length ranges from 128 to 16384. The peak performance for the dotproduct is 27.0 MFLOPS, while peak performance for the vector update is 19 MFLOPS. Figures 2a and 2b show the performance of the BLAS 2 matrix $\times$ vector, vector $\times$ matrix, and rank 1 update operations. In Figure 2a the matrices are of dimension n $\times$ n, where n ranges from 100 to 1500. In Figure 2b the matrices are of dimension n $\times$ 32, where n ranges from 100 to 1500. The performance for the matrices used in Figure 2b is higher than those used in Figure 2a. The BLAS 2 operations $y \leftarrow y + Ax$ and $A \leftarrow A + xy^T$, where $A \in \mathbf{R}^{n \times n}$, x, y $\in \mathbf{R}^n$, both involve 2n vector operations on n + 2 vectors each of length n. The algorithms for these operations are similar, in that they both involve vector updates executed in parallel. The primary difference is that each column of A must be read from memory and written back to the memory for the rank 1 update, whereas each column of A must be read, but not written back to memory for the matrix $\times$ vector operation.

Figures 3 and 4 show the performance of the BLAS 3 rank k update and matrix^T $\times$ matrix. They show the superior performance of the BLAS 3 subroutines on the Alliant FX/8 (see [JaMe86]).

MFLOPS

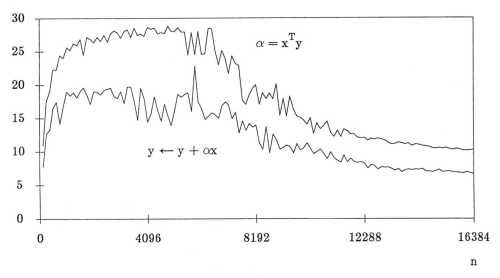

Figure 1: Performance of BLAS 1 operations

$$x, y \in \mathbf{R}^n \quad \alpha \in \mathbf{R}$$

MFLOPS

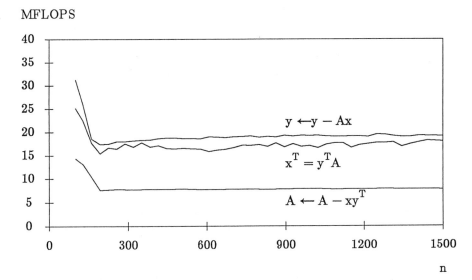

Figure 2a: Performance of BLAS 2 operations

$$A \in \mathbf{R}^{n \times n} \quad x, y \in \mathbf{R}^n$$

MFLOPS

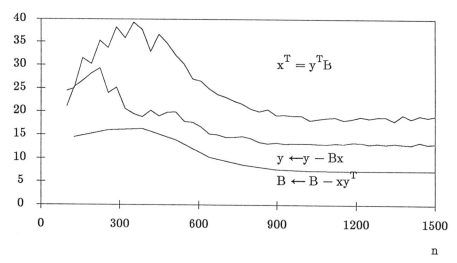

Figure 2b: Performance of BLAS 2 operations

$$B \in \mathbf{R}^{n \times 32} \quad y \in \mathbf{R}^n \quad x \in \mathbf{R}^{32}$$

MFLOPS

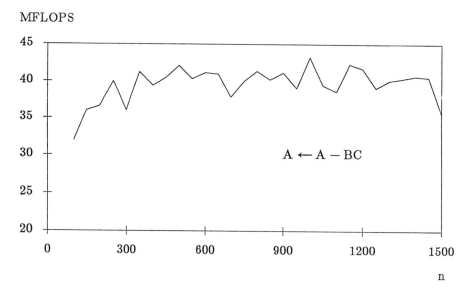

Figure 3: Performance of BLAS 3 operation, rank 32 update

$$A \in \mathbf{R}^{n \times n} \quad B \in \mathbf{R}^{n \times 32} \quad C \in \mathbf{R}^{32 \times n}$$

MFLOPS

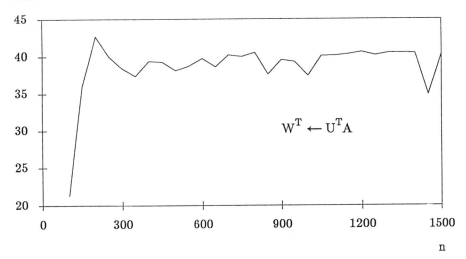

Figure 4: Performance of BLAS 3 operation, $\text{matrix}^T \times \text{matrix}$

$$A \in \mathbf{R}^{n \times n} \quad U, W \in \mathbf{R}^{n \times 32}$$

4. Block Algorithms

A Householder transformation is an orthogonal matrix P of the form

$$P = I - uu^T$$

where $u \in \mathbf{R}^m$ and $u^T u = 2$. If $x \in \mathbf{R}^m$, then Px is a vector in the orthogonal compliment of the set $\{u\}$. In particular, P can be constructed so that Px is a multiple of e_1, the first column of the identity matrix. If

$$u = \frac{1}{\left(\alpha^2 + (e_1^T x)\alpha\right)^{1/2}}(x + \alpha e_1)$$

where $\alpha = \text{sign}(x_1) \| x \|_2$, then $Px = -\alpha e_1$.

Assume that A is partitioned by columns,

$$A = (a_{1,}\, a_{2,} ..., a_n), \qquad a_j \in \mathbf{R}^m.$$

The following is the classical algorithm for computing an orthogonal factorization of A, using Householder transformations. We will assume that $n < m$.

Algorithm: QR
   do j = 1, n
      compute u_j *such that for* $P_j = I - u_j u_j^T$,
      $P_j a_j$ *has zeros in the entries* j+1, ..., m.

      $(a_{j+1}, ..., a_n) \leftarrow (I - u_j u_j^T)(a_{j+1}, ..., a_n)$
      enddo

The above algorithm can be rewritten in terms of the BLAS 1 or BLAS 2 operations. The following is the BLAS 1 version of the classical method.

Algorithm: QR1

   do j = 1, n
      compute u_j *such that for* $P_j = I - u_j u_j^T$,
      $P_j a_j$ *has zeros in the entries* j+1, ..., m.

      do i = j+1, n
         $a_i \leftarrow a_i - (u_j^T a_i) u_j$
         enddo
      enddo

This algorithm is written in terms of three vector operations:

$$u \leftarrow \alpha u \qquad\qquad\qquad \text{vector scale}$$

$$\beta = u^T a \qquad\qquad\qquad \text{dotproduct}$$

$$a \leftarrow a - \gamma u \qquad\qquad\qquad \text{vector update}$$

where α, β, and γ are scalars, and a and u are vectors of length s, $s \leq n$.

There are various ways that this algorithm can be programmed on the Alliant FX/8 to exploit the concurrency in the method. For example, one could use concurrency when computing the dotproducts and vector updates, or one could concurrently apply the transformation P_j to the columns a_{j+1}, a_{j+2}, ..., a_n.

Although this algorithm is fully vectorized, its performance on the Alliant FX/8 was considerably lower than the peak performance of the machine. Algorithm 1 can be rewritten in the following form:

Algorithm: QR2

do j = 1, n
   compute u_j *such that for* $P_j = I - u_j u_j^T$,
   $P_j a_j$ *has zeros in the entries* j+1, ..., m.
   $z^T = u_j^T (a_{j+1}, ..., a_n)$
   $(a_{j+1}, ..., a_n) \leftarrow (a_{j+1}, ..., a_n) - u_j z^T$
   enddo

This version of the classical Householder factorization method is written in terms of the modules:

$$u \leftarrow \alpha u \qquad\qquad\qquad \text{vector scale}$$

$$\beta = u^T a \qquad\qquad\qquad \text{dotproduct}$$

$$v^T = u^TB \qquad\qquad \text{vector} \times \text{matrix}$$

$$B \leftarrow B - uv^T \qquad\qquad \text{rank 1 update}$$

where α and β are scalars, u, v are vectors, and $B \in \mathbf{R}^{\hat{m} \times \hat{n}}$, $\hat{m} \leq m$ and $\hat{n} < n$.

When this algorithm is programmed for the Alliant FX/8, using a set of highly tuned subroutines for the BLAS 2 operations, the performance is lower than the BLAS 1 algorithm for matrices with fewer then 2400 rows.

The Bischof and Van Loan method is based on the observation that the product of k Householder transformations $Q_k = P_k \cdots P_2 P_1$, can be written in the form,

$$Q_k = I - V_k U_k^T, \qquad U_k, V_k \in \mathbf{R}^{m \times k}$$

Assume that $Q_1 = P_1 = I - u_1 u_1^T$ and that $V_1 = U_1 = u_1$. Then

$$Q_j = P_j Q_{j-1} = (I - u_j u_j^T)(I - V_{j-1} U_{j-1}^T)$$

$$= I - (P_j V_{j-1}, \, u_j)(U_{j-1}, \, u_j)^T$$

and thus $V_j = (P_j V_{j-1}, \, u_j)$ and $U_j = (U_{j-1}, \, u_j)$.

The following is the block Householder orthogonal factorization algorithm. In [BiVa85b], it is stated that the floating point operation count is approximately $(1 + 2/p)n^2(m - n/3)$, where $p = n/k$. Also, it is shown that such a block scheme has the same numerical stability properties as the classical Householder factorization method. We will assume that k divides n

and m > n.

Algorithm: QR3

$$p = n/k \quad k_2 = 0$$
$$\text{do } j = 1, p$$
$$\quad k_1 = k_2 + 1, \quad k_2 = k_1 + k$$
$$\quad \text{do } i = k_1, k_2$$

compute u_i *such that for* $P_i = I - u_i u_i^T$,

$P_i a_i$ *has zeros in the entries* i+1, ..., m.

$$\quad\quad V_{i+1-k_1}^j = (P_i V_{i-k_1}^j, u_i)$$
$$\quad\quad U_{i+1-k_1}^j = (U_{i-k_1}^j, u_i)$$
$$\quad\quad \text{if } (i \neq k_2) \text{ then}$$
$$\quad\quad\quad a_{i+1} \leftarrow (I - V_{i+1-k_1}^j U_{i+1-k_1}^{j\,T}) a_{i+1}$$
$$\quad\quad\quad \text{endif}$$
$$\quad\quad \text{enddo}$$
$$\quad \text{if } (j \neq p) \text{ then}$$
$$\quad\quad (a_{k_2+1}, ..., a_n) \leftarrow (I - V_k^j U_k^{j\,T})(a_{k_2+1}, ..., a_n)$$
$$\quad\quad \text{endif}$$
$$\text{enddo}$$

The orthogonal factorization algorithm has now been written in the high

level modules:

$u \leftarrow \alpha u$	vector scale
$\beta = u^T a$	dotproduct
$a \leftarrow a - Vc$	matrix $\times$ vector
$v^T = u^T V$	vector $\times$ matrix
$V \leftarrow V - uc^T$	rank 1 update
$W = U^T B$	matrix^T $\times$ matrix

$$B \leftarrow B - VW \qquad\qquad \text{rank k update}$$

where $B \in \mathbf{R}^{\hat{m} \times (\hat{n} - k)}$, $U, V \in \mathbf{R}^{\hat{m} \times \hat{k}}$, $\hat{m} \leq m$, $\hat{n} < n$, $\hat{k} \leq k$, and k is the block size used in the algorithm.

The numerical results presented in the following section will show the advantage of matrix_matrix operations over matrix_vector operations on a machine such as the FX/8. The implementations of such matrix $\times$ matrix operations are presented in [JaMe86]. The matrix $\times$ vector and vector $\times$ matrix modules are designed to make efficient use of the floating point registers in the CE's.

5. Numerical Experiments

The optimal size of the blocks in the generalized Householder method is difficult to determine. It depends on the dimension of the matrix, the actual declare dimension of the matrix in the computer program, and the subroutines used for the BLAS 3 operations. The following conditions are used to determine the block size k.

```
if (n .lt. 480) then
   k = 16
else if (n .lt. 704) then
   k = 24
else
   k = 32
   endif
```

The floating point operation count used for all three programs is

$$\text{flops}=2n^2\left(m-\frac{n}{3}\right)$$

where m and n are the dimension of the matrix. Thus, the megaflops per second (MFLOPS) is determined by the following equation:

$$\text{MFLOPS}=\frac{\text{flops}}{time\times10^6}$$

where *time* is the execution time. Note that the reported MFLOPS for the program QR3 is less than the actual value.

Figures 5 and 6 show the performance and execution time obtained by the three orthogonal factorization algorithms when applied to a matrix with 1000 rows and n columns. The performance of QR3 ranges from 19.7 MFLOPS for a matrix of order 1000 $\times$ 100 to 36.31 MFLOPS for a matrix of order 1000 $\times$ 1000. Note also that QR3 is 2 to 3.7 times faster then QR2 and 1.7 to 3 times faster then QR1.

It is interesting to observe, for this problem set, that QR1 is faster than QR2. After the Householder transformation has been computed in the QR1, a dotproduct and vector update involving the i–th column of A are concurrently computed. Thus for a matrix of dimension 1000 $\times$ n, the total amount of data involved is 9000 words, which will fit in the cache. Therefore, after the i–th column has been moved into the cache it should still reside in the cache when the vector update is computed. QR2, on the other hand must read into the cache the remaining n–j columns of A to form the vector $\times$ matrix product and then re–read the columns of A into the cache to form the rank 1 update.

Figures 6 and 7 show the performance and execution times for the three orthogonal factorization algorithms when applied to matrices of dimension m $\times$ 200, where m ranges from 200 to 7200. For the larger matrices the QR2 has higher performance than the QR1. Figures 8 and 9 show the performance and execution times for the three orthogonal factorization algorithms applied to square matrices. The performance of QR3 ranges from 13.33 MFLOPS for a matrix of order 100 $\times$ 100 to 38.00 MFLOPS for a matrix of order 1100 $\times$ 1100. Finally, Figure 11 depicts the speedup realized for matrices of order 500 $\times$ 500, 1000 $\times$ 1000, and 1500 $\times$ 1500, vs. the number of CE's.

MFLOPS

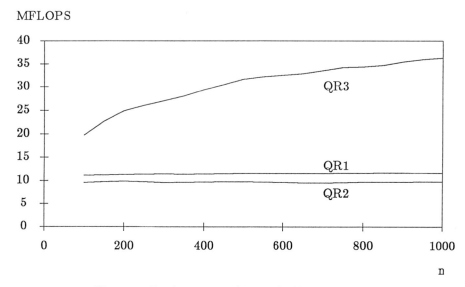

Figure 5: Performance of Householder factorization
of a 1000 × n matrix

seconds

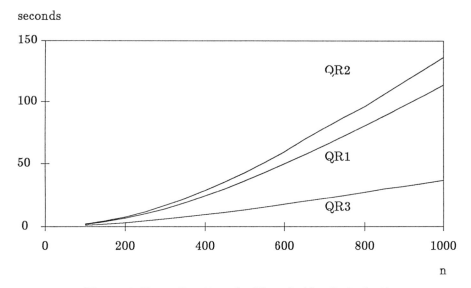

Figure 6: Execution time for Householder factorization
of a 1000 × n matrix

MFLOPS

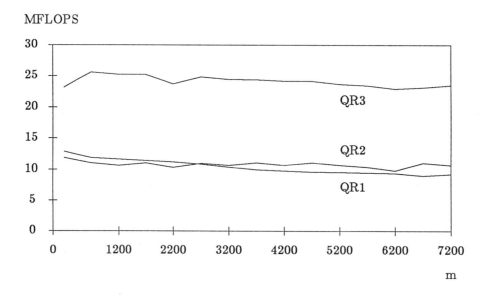

Figure 7: Performance of Householder factorization
of a m × 200 matrix

seconds

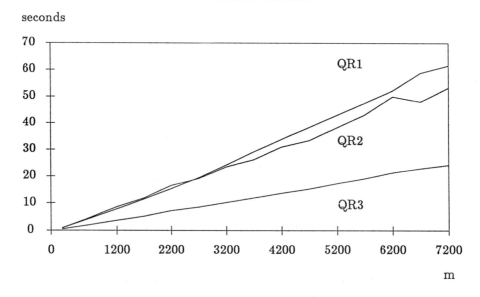

Figure 8: Execution time for Householder factorization
of a m × 200 matrix

MFLOPS

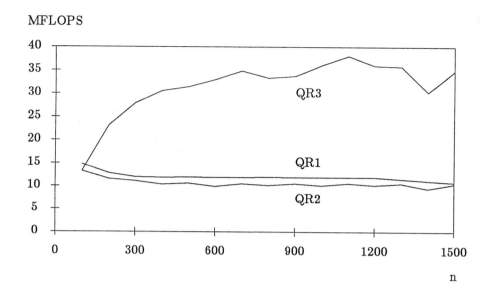

Figure 9: Performance of Householder factorization
of a n × n matrix

seconds

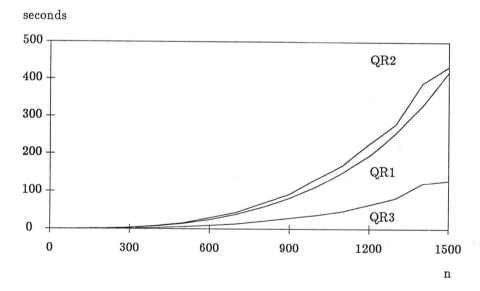

Figure 10: Execution time for Householder factorization
of a n × n matrix

Speedup

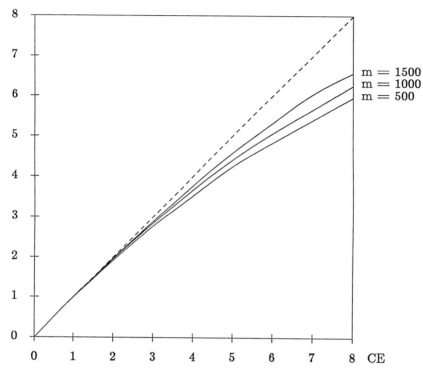

Figure 11: Speedup for the Block Householder factorization
of a m $\times$ m matrix,
that is [time(1CE)/time(k CE's)], $k = 1, ..., 8$

6. Conclusions

We have demonstrated a technique for redesigning an orthogonal
factorization algorithm in terms of the high level modules BLAS 3. The
resulting algorithm achieves high performance on the Alliant FX/8. The
same technique can be used for other factorization algorithms such as; LU-
decomposition, Cholesky factorization, and reduction to upper Hessenberg
form via orthogonal similarity transformations.

Acknowledgements

This work was supported in part by the National Science Foundation under Grant Nos. US NSF DCR84–10110 and US NSF DCR85–09970, the U. S. Department of Energy under Grant No. US DOE DE–FG02–85ER25001, the Air Force Office of Scientific Research Grant No. AFOSR–85–0211, and a donation from the IBM Corporation.

References

[BiVa85]   C. Bischof and C. Van Loan, *The WY Representation for Products of Householder Matrices*, TR 85–681, Department of Computer Science, Cornell University, 1985.

[BrJo74]   O.E. Bronlund and Th. L. Johnsen, *QR–factorization of Partitioned Matrices*, Computer Methods in Applied Mechanics and Engineering 3, 1974, 153–172.

[Diet76]   G.Dietrich, *A New Formulation of the Hypermatrix Householder-QR Decomposition,* Computer Methods in Applied Mechanics and Engineering 9, 1976, 273–280.

[DBMS79]   J. Dongarra, J. Bunch, C. Moler, and G. W. Stewart, *LINPACK User's Guide,* SIAM, 1979.

[DDHH84]   J. Dongarra, J. Du Croz, S. Hammarling, and R. Hanson, *A Proposal for an Extended Set of Fortran Basic Linear Algebra Subroutines,* Argonne National Laboratory Report MCS–TM–41, Revision 1, October 1984.

[DoSo86]   J. Dongarra and D. Sorensen, *Linear Algebra on High-preformance Computers,* Technical Report ANL–82–2, Argonne National Laboratory, 1986.

[GaJM86]   K. Gallivan, W. Jalby, and U. Meier, *The Use of BLAS 3 in Linear Algebra on a Parallel Processor with a Hierarchical Memory,* to appear in SISSC.

[JaMe86]   W. Jalby and U. Meier, *Optimizing Matrix Operations on a Parallel Multiprocessor with a Memory Hierarchy,* Technical Report, Center for Supercomputing Research and Development,

University of Illinois, 1986.

[LHKK79]   C. Lawson, R. Hanson, D. Kincaid, and F. Krogh, *Basic Linear Algebra Subprograms for Fortran Usage*, ACM Trans. Math. Software 5, pp. 308–371, 1979.

The Influence of Memory Hierarchy on Algorithm Organization: Programming FFTs on a Vector Multiprocessor.

Dennis Gannon

Department of Computer Science
Indiana University
Bloomington, Indiana 47401
and
Center for Supercomputer Research & Development
University of Illinois.

William Jalby

INRIA & Center for Supercomputer Research & Development
University of Illinois
Urbana, Illinois 61801.

1. Introduction

Perhaps the most natural taxonomical division in the realm of parallel algorithms is the distinction between the family of algorithms that represent completely new ways to solve a problem and those algorithms that are restructured versions of well-known sequential methods. To make the distinction more formal, we shall say that one algorithm is a parallel restructuring of another if they both have the same underlying data flow graph (even though they may be executed according to a different scheduling strategy). While there are many good examples of "new" parallel algorithms in various disciplines, the vast bulk of program design for parallel machines falls into the latter category of manually "restructured" applications.

The significance of this distinction is the following. If we recognize that the software engineering task involved in "mapping" a computation to a particular parallel machine is, in fact, a process of mechanically reorganizing the control flow of an existing dataflow graph, we have some hope of synthesizing the techniques of transforming the program as an automatic process. In particular, now that we have access to a wide range of parallel systems, the process of designing an algorithm can incorporate an analytic model of the target architecture that can be experimentally calibrated and verified. The problem of "mapping a parallel computation" becomes one of optimization.

An instance where this model of parallel program optimization has proven to be very effective has been the design of optimal synchronous systolic arrays for computations with simple, regular dataflow graphs. Analytical models for the mapping problem have been most effectively applied to computational linear algebra. The reason for this is that the regular structure of linear algebra problems provides natural leverage for finding optimial solutions to the mapping problem.

In this paper we illustrate the basic techniques of such an optimization theory by considering a Fast Fourier Transform package for a parallel vector multiprocessor. We shall show how an algebraic analysis of the computation can be used to guide the programmer to a sequence of transformations that optimizes the use of a memory hierarchy and a vector instruction set.

In section 2 we describe a basic FFT algorithm and the target architecture that we will be considering. In section 3 we consider the general properties of a hierarchial memory system that are important for our study and discuss the general impact this system has on algorithm organization. Section 4 describes the analytical model and shows how it is applied to the FFT problem. The experimental results are summarized in section 5.

2. The Target Architecture and the Basic Algorithm

2.1 The Target.

Our target machine is a multi-vector-processor using a fully shared hierarchical memory system. It consists of a set of P processors (CE = computational element), each being a vector processor. The memory is organized in two levels: a fast small (Cache) memory of size CS words and a slower bigger (Main) Memory. The processors are connected to the cache through a network allowing each of them to read or write any of the cache locations. The set of processors is provided with a synchronization mechanism (e.g. a set of semaphore registers) enabling it to work as an MIMD machine, i.e. the computations of a single program unit may be distributed among the P CE's.

A good example of the architecture described above is the ALLIANT FX/8 (see Figure 2.1). This machine consists of P=8 pipelined CE's, each of which is capable of delivering almost 11 mflops peak speed on 64-bit words, allowing a peak rate of nearly 88 mflops. The CE's are connected by a concurrency control bus (used as a synchronization facility). They share the physical memory as well as a 128K-byte write-back cache that allows up to eight simultaneous accesses in each cycle (of 170 ns). The cache itself is connected to the four-way interleaved main memory through the main memory bus which is able to deliver up to four 64-bit words per cycle. Therefore, the bandwidth between main memory and CE's is half of that between the cache and the CE's. However, from a practical point of view, accessing a vector from main memory may require a time which is larger than twice the time required for accessing the same vector if it is present in cache. This is due to the fact that managing cache misses and possibly writing back words adds a nonnegligible overhead to the memory access time. Another point of practical interest is that each CE has 8 vector registers able to hold up to 32 elements, so for a single CE the maximum vector processing speed will be for vector lengths which are multiples of 32.

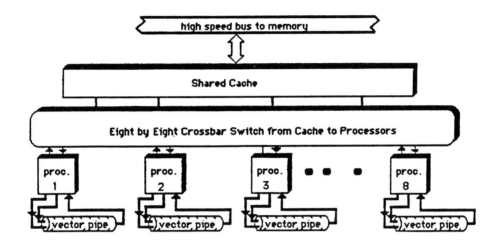

Figure 2.1 Alliant FX/8 Architecture

2.2 The Algorithm

The introduction of the FFT algorithm in 1965 by Cooley and Tuckey [CoTu65], [CoLe70] was a great breakthrough: the number of floating point operations in the DFT was reduced from $O(n^2)$ to $O(n \log n)$, allowing considerable speedup on the sequential computers available at that time. Furthermore, in addition to reducing the floating point complexity the algorithm still exhibited large potential concurrency. The computation is organized in $\log n$ stages, each stage consisting of $n/2$ independent identical elementary operations which may be done in parallel. In fact, a close examination of the data flow will reveal that the maximal degree of parallelism is $O(n)$. Consequently restricting the use of concurrency to one stage involves no loss of potential parallelism.

When implementing this algorithm on a vector machine the main problem was not the discovery of parallelism but how to exploit the limited data addressing capability of these machines to perform the required data reorganization between each stage. The data flow graph for the computation is illustrated in Figure 2.2. Many papers have studied how to rearrange this computation to utilize special features of various machines ([Korn79], [Temp79], [Swar82,84], [Pete83], [Gent66], [Forn81], [Wang80]). The main problems encountered were how to avoid the steady decrease in the vector length at successive stages, how to avoid costly data unscrambling (the famous bitreversal permutation which seriously degrades many implementations) and how to avoid vector access with non-unit strides.

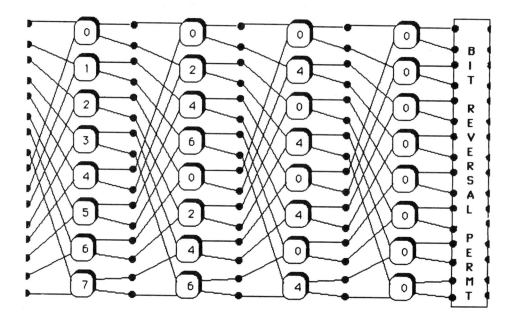

Figure 2.2. 16 Point Complex FFT data flow diagram (based on [Peas68]).

The extensive work for vectorizing the FFT is of little help in implementing it on a multiprocessor, since the optimal parallel scheme is not simply a straight-forward redesign of the best vector algorithm. A number of papers have considered this problem ([Brig85], [Cvet86], [Gann84], [Nort85], [Swar86], [Kwok86]).

We will consider two cases: the computation of a single FFT, and the computation of many FFTs.

2.3. The Single FFT Case.

To start one may begin with the simplest possible vectorization which is given by the following implementation of the Pease algorithm [Peas68].

```
procedure CFT(X,W): returns(X);
parameter     X: array[1..N] of complex;
              W: array[1..N/2,log2(N)] of complex; /* roots of unity */
var           T1,T2: array[1..N/2] of complex;
              N = upper(X);
begin
    for i = 1..log2(N) loop
        T1 = X[1..N/2] + X[N/2+1..N];
        T2 = W[1..N/2,i]*(X[1..N/2] - X[N/2+1..N]);
        X[1..N/2-1 by 2] = T1;
        X[2..N/2 by 2] = T2;
    endloop;
    X = BitRev(X);
end;
```

The function *BitRev*() does the bit reversal permutation and the ith column of the array W contains the roots of unity used in i^{th} stage of the Pease algorithm. This routine was implemented in assembly language on the FX/8 as follows. Each of the 8 CEs was assigned the task of doing the operations on a subrange equal to one eighth of the range of each vector operation. Vector registers in each were used to hold 32 word segments of the subrange. Figure 2.3 shows the performance of the algorithm in a graph measuring effective performance (measured in Mflops) for different values of N. (The horizontal axis is the $\log_2(N)$.)

Looking at these results we can see several interesting and disturbing features. Three main problems are listed below.

1.   Maximum performance is achieved only for relatively large problems.

This is due to a problem associated with vector start-up speeds in multiprocessors. Namely, by dividing a short vector among many processors one is left with many very short vectors. Observe that for $n < 512$ the length of vector operations on each CE is less than 16.

2.   The maximum is less than 25% of the potential of the machine.

This is explained by two characteristics of the machine. First, for a vector register based machine, some of the operands must be loaded into the registers and results must be moved from registers back to the memory. Because these are vector operations that are as costly as vector arithmetic, they are a non-trivial fraction of the computation. Second, the machine only performs at maximum speed when we use the vector operations of the form (v1 <- v2 + s1*v3) where v1,v2 and v3 are vectors and s1 is a scalar. In the algorithm above all operands are vectors.

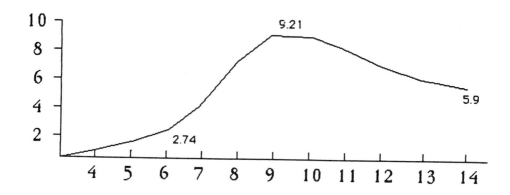

Figure 2.3. Performance for simple CFT algorithm. MFlops vs. $\log_2(N)$.

3. After reaching the maximum the performance decreases as the size of the problem increases.

This problem is related to the cache size. When the problem is small enough to be completely contained in the cache only the initial stage of computation will require a read from memory. All subsequent stages will operate from cache. As soon as the problem exceeds the cache size each stage will have to fetch the data from memory. This, in particular, implies that we will have $\log_2(N)$ reads of the vector X from memory. In section 3 we will describe an analytical model of this behavior and show a way to reformulate the algorithm so that at most $\log_2(log_2(N))$ sweeps of the vector X are needed.

2.4. The Multiple FFT Case.

There are two simple approaches to this problem.

1. Apply all 8 CEs to the the individual FFTs one at a time.

2. Put the parallelism at the outer level and do the FFTs 8 at a time with one FFT per CE.

The first solution will have the same problems and give results identical to the single FFT case described above. In the second case we eliminate one of the problems encountered in the single FFT case. Namely, the vector length is not divided by 8 because each CE is working on a single FFT. However we have aggravated the "cache miss" problem, since now the cache must be able to contain the union of the working sets of all 8 processors. In particular, the amount of X data needed at each stage is multiplied by 8. These points are illustrated in Figures 2.4. In Figure 2.5 we see the effect of the extra cache misses on the multiprocessor speed-up for $N = 1024$.

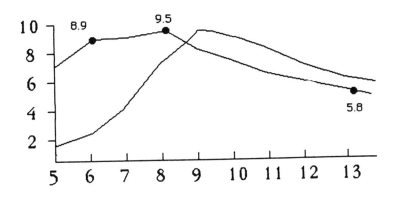

Figure 2.4. Multiple CFT computation.

In the next section we will discuss these issues in much greater detail and show another paradigm for using this machine to implement multiple FFTs.

3. On The Bandwidth of a Hierarchical Memory System.

3.1. A general remark

One easy way to use a multivectorprocessor is just to consider it as a multipipeline unit (as for example the 2 or 4 pipes versions of the CDC 205). This means that we just program the machine as a monolithic vector processor where each vector has been sliced into P pieces, one for each CE. This approach would appear to be well suited to fully vectorizable algorithms such as those of classical linear algebra. Although this approach is very attractive (no redesign of the algorithm is needed), it suffers several major drawbacks:

1. To get near peak performance from the machine, we need to have vectors of length greater than $P*L$, where L is the defined as the length of a vector required to get 90% of peak performance from the vector pipeline. This

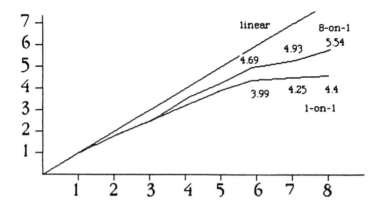

Figure 2.5 Speed-up vs. CEs for $N = 1024$.

requirement may be difficult to meet and, for a given problem size, the speedup obtained by increasing the number of processors P can be very disappointing. As observed in the single FFT case: if v is the speed of a single CE on a problem of size N, P CE's will reach the speed $P*v$ (linear speedup) only if the size of the problem is $P*N$.

2. As a corollary of the previous remark, the increase of the vector length may severely damage the data locality and therefore make poor use of the cache. A good example of this effect may be observed when programming classical linear algebra algorithms in terms of level 1 BLAS [Laws79], [Dong85].

3. A more subtle effect is due to the fact that we are not using the full addressing flexibility of the multiprocessor. One of the important differences between a multiple pipeline unit machine and a multivectorprocessor is that in the first case we generally have only one addressing unit while in the second case we have P independent addressing units, each of which is able to generate its own address stream. This allows the user to access in parallel a rectangular subarray of a large matrix, while with a classical multiple pipeline machine such a subarray cannot be accessed in a single vector operation since the sequence of addresses does not constitute a linear arithmetic progression. Such flexibility in addressing ability is only encountered on systems built from arrays of processors.

3.2. A single vector processor with a memory hierarchy.

As long as the CE accesses vectors with stride one, the memory system offers good performance. If the CE now accesses vectors with non-unit stride, several sources of degradation appear:

1.   At the main memory level:

     The interleaving of the memory will cause disastrous bank conflicts for strides which are multiple of the interleaving factor. Moreover if the main memory is organized by sword (*super word* where in many systems the size of a sword is equal to the size of a cache block), severe performance degradation will occur for every non unit stride access.

2.   At the communication medium level, between the main memory and the cache:

     This medium will have to deal with important traffic, only a fraction of which is useful ($\frac{1}{k}$ where k is the size of the cache blocks) The reason is that if each cache miss causes the sword containing the desired word to be accessed then a non-unit stride sequence may not use much of the remainder of the sword.

3.   At the cache level:

     The cache will be filled with data, only a fraction of which is useful.

3.3. Multivector processor with a memory hierarchy.

In this section, we will show how a multiprocessor can use a memory hierarchy efficiently. This is based on the following remark: although a single stream of data request may achieve a very poor utilization of the memory hierarchy, the combination of several streams of data requests (even if each individually appears to be a disaster for the memory system) can result in a full utilization of the system. More precisely, as argued in the previous section, a single read of a vector by increment of k where k is both the size of a sword and a cache block is the worst case for the single processor case. However, if several processors issue such vector requests of stride k, the global set of requests may appear as contiguous if, for example, there are k processors and processor i starts his request stream at word i. A typical case is the sweep of a two dimensional array; the row sweep (each processor accessing a row in vector mode), can be as efficient as the column sweep (each processor accessing a column in vector mode). This remark may be used to spread the work among the processors so that the global data stream request achieves full use of the memory system.

A second major point is related to data organization. If we consider the parallelization of the FFT, an important problem to be solved is the management of the data permutations between the different stages of the computation. If we consider the algorithms proposed for vector machines, all of them, at some point, need a data permutation which is generally time consuming (namely they often require a scatter/gather primitive to do the bit reversal). On the other hand, a multivector processor has features that may permit us to solve this problem. Let us consider the computation of a group of FFTs. If we use the vectorization across the FFT's and concurrency inside a single FFT, the scatter/gather bottleneck disappears because we are performing a scatter/gather not on scalar entities within a vector, but on vectors handled by different processors. (As we shall see below, this idea can be exploited for a single FFT computation if we can reduce the problem of computing one large FFT to that of computing a set of smaller FFTs.)

4. Computational Organization

4.1. Formulation of the problem.

4.1.1. Notations.

In this section, we just define the notation used throughout the remainder of this paper.

$P = 2^p$ is the number of processors (Alliant FX/8: $p = 3$)

$L = 2^l$ is the vector length for which 90% of the asymptotic performance is reached (Alliant FX/8: $l = 5$). This vector length will be called efficient vector length.

$C = 2^c$ is the size of the cache expressed in complex words (Alliant FX/8: $c = 13$)

$K = 2^k$ is the number of independent FFT to be computed

$N = 2^n$ is the length (number of points) of each FFT to be computed

In the sequel, in order to simplify the notation we will use systematically the base 2 logarithm of the quantities defined above. The "efficient vector length" L is defined to be the vector length for which a CE reaches 90% of its peak performance when performing a triadic operation $V1 \leftarrow V2 op 1 V3 op 2 V4$. For register oriented architectures, L does not depend upon the nature of the arithmetic operations $op 1$ and $op 2$. Rather, it is generally equal to the length of vector register. In the case of memory oriented architectures, the situation is slightly different because L depends more upon the nature of the arithmetic operation involved. However, for this paper, we will express the algorithm only in terms of the triadic instruction $V1 \leftarrow V2 + s1 * V3$ and L will be associated with this instruction.

4.1.2. Assumptions on the cache size.

In this section, we will describe the assumptions to be made about the cache size in terms of the other parameters of the machine. Our idea is to insist that the cache be sized "appropriately" for the rest of the machine. By this, we mean that the cache should be large enough so that the other functional units have a chance of working at full speed. More specifically,

$$c \geq l + p + 1 \qquad \text{(HC1)}$$

$$c \geq 2p + 2 \qquad \text{(HC2)}$$

$$c \geq 2l + 2 \qquad \text{(HC3)}$$

The first inequality expresses the fact that the cache should be large enough to contain a set of vectors (one pair for each processor) for which the peak arithmetic performance can be achieved; this corresponds to the saturation of the one dimensional parallelism ability of the machine.

The second and the third inequalities assumes that the cache is able to contain a pair of 2 dimensional arrays of size k by k (where $k = max(p,l)$) upon which the processors can alternate parallel column sweep (concurrency between columns, vectorization inside a column) and parallel row sweep (concurrency between rows, vectorization inside a row): this corresponds to the saturation of the 2 dimensional parallelism ability of the machine.

It should be noticed that the constraints imposed above are not too drastic. In practice, they can be easily satisfied and are for the Alliant FX/8.

4.1.3. Assumption on the working set.

In order to simplify the analysis, we will suppose that for performing one FFT on 2^n complex points we need a working space of 2^{n+1} of complex words. This, in fact, is an upper bound since we really only need one array of 2^n complex words (for the points) and an array of 2^{n-1} complex words (for the coefficients). However, the extra space can be used for some intermediate results.

4.1.4. Assumption on the problem size.

We will make two assumptions about the problem size.

$$k + n \geq l + p + 1 \qquad \text{(HP1)}$$

$$k + n \geq c \qquad \text{(HP2)}$$

The first inequality expresses that the problem is large enough to saturate the arithmetic ability of the machine. If this inequality is not satisfied we can still apply the algorithms described in the sequel either with a shorter vector length or with fewer processors. (However this choice represents a delicate tradeoff which may be very machine-dependent.)

The second inequality assumes that the problem requires a working space bigger than the cache size; if this not true, no cache management is required and we just apply the algorithms developed below in a straightforward manner.

4.2. Optimization of the arithmetic time.

As mentioned in section 2.2.1 one of the reasons that the first FFT algorithm performed poorly was the fact that the vector triadic operations used were of the wrong form for the machine to exploit at top efficiency. More specifically, let us consider the multiple FFT case and compare the vectorization of a single FFT with the orthogonal notion of using vectorization across many FFT's. In the first case, in all but the last stage, we have to use vector operations of the form: $V1 \leftarrow V2 + W[1..N]*X[i,1..N]$ where the V's denote vectors and $X[i,j]$ denotes the jth component of the ith FFT vector and $W[i]$ is the ith component of the coefficient matrix.. In the second case the multiple FFT's are all using the same W coefficients, so we may vectorize "across problems". In other words, we can use an operation of the form $V1 \leftarrow V2 + W[i]*X[1..N,i]$ where W[i] is a scalar. The first form requires two operands per floating point operation while the second requires only one and half. The difference on some architecture may be very important, for example the CDC 205 is able to perform the second operation twice as fast as the first one. On the Cray X/MP the second form requires two loads and one store which allows optimal use of the three channels to memory (and the chaining capability) of that machine. Because the second form requires only 75% of the bandwidth used by the first it will, in general, yield better performance.

To illustrate how we can improve the locality (and reduce the data bandwidth required by each CE) as well as exploit the property of triadic operations described above, we must show how we can decompose one large FFT problem into a system of smaller FFTs which we can solve in parallel. The "folding" method described below is based on techniques described by Swarztrauber [Swar86].

The basic algorithm for the FFT computes

$$Y_i = \sum_{j=0}^{N-1} W^{ij} X_j$$

If we view X and Y as arrays of size $2^{n-\alpha}$ by $m = 2^{\alpha}$ indexed as

$$X_{r,s} = X_{r \cdot m + s} \quad 0 < r < N/m-1, \ 0 < s < m-1$$
$$Y_{p,q} = Y_{p \cdot N/m + q} \quad 0 < p < m-1, \ 0 < q < N/m-1$$

we can rewrite the computation as

$$Y_{p,q} = \sum_{r=0}^{N/m-1} \sum_{s=0}^{m-1} W^{(pN/m+q)(rm+s)} X_{r,s}.$$

Multiplying out the exponent of W and reorganizing the summations we have (recalling that $W^N = 1$),

$$Y_{p,q} = \sum_{s=0}^{m-1} (W^{N/m})^{ps} W^{qs} \sum_{r=0}^{N/m-1} (W^m)^{qs} X_{r,s}.$$

Notice that the term

$$Z_{q,s} = \sum_{r=0}^{N/m-1} (W^m)^{qs} X_{r,s}$$

is nothing more than the m FFTs of size N/m obtained by operating along the columns of X. Furthermore, the Y array can now be written as

$$Y_{p,q} = \sum_{s=0}^{m-1} (W^{N/m})^{ps} (W^{qs} Z_{q,s})$$

which is the N/m row-wise FFTs of size m applied to the pointwise product of the array $W_{q,s} = W^{qs}$ and $Z_{q,s}$.

Summarizing, let $many\_col\_FFT(m,k,X)$ be a function which takes an m by k array X and returns an m by k array of k FFTs of size m of the columns of X and let $many\_row\_FFT(m,k,X)$ be a function which returns an array of m FFTs of size k of the rows of X. Let Tr() be the function that transposes an array. The FFT of size N of Y is then given by,

$$Y = Tr(\ many\_row\_FFT(\ N/m,\ m,\ (W*many\_col\_FFT(\ N/m,\ m,X))))$$

where the transpose is needed to convert the row-wise ordering of Y to the standard column major order. This composition is illustrated in Figure 4.1.

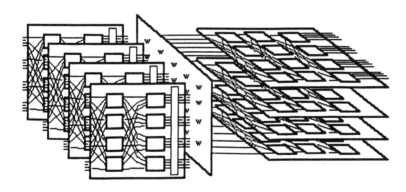

Figure 4.1 Folded computation of FFT as many parallel smaller FFTs.

Commuting the transpose with the row FFTs gives a set of column FFT operating on the transpose, i.e.

$$Y = many\_col\_FFT(\ m,\ N/m, Tr(W*many\_col\_FFT(\ N/m,\ m,X)))$$

Viewed in this way we see that the FFT has a structure that allows us a parameterized way to partition the FFT into subproblems. Norton [Nort85] has observed that if you have p processors the computation can be done as

```
forall i in 1..p do
    for k = (i-1)*m/p +1 .. i*m/p loop
        Z[k, 1..N/m] = W[1..N/m, k]*CFT(X[1..N/m, k], W);
    end;
endforall;
forall i in 1..p do
    for k = (i-1)*N/(m*p) .. i*N/(m*p)-1 loop
        Y[1..m, k] = CFT(Z[1..m, k], W);
    end;
endforall;
```

(where the arrays have been indexed from 1 in each dimension.). The key point here is that the computation is in two phases. During each phase each of the p processors operates on a "private" set of data points which may reside in local memory. No communication is needed between processors except for a barrier between phases.

If we apply vector concurrency within each FFT operation on each processor, we have exploited two levels of concurrency. However, as we have already seen in section 2, this precise formulation is not very well suited to a vector multiprocessor like the Alliant FX/8. To better exploit the vector operations on this machine we formulate the many_col_FFT() function as follows.

```
function many_col_FFT(m, N/m, β, X, W) returns(Y);
    parameter X, Y: array[m, N/m] of complex;
            W: array[1..m/2] of complex;
            m, N/m, β: integer;
        var  s: integer;
    begin
        for i in 1..N/m by β loop
            for j in log₂(m)..1 by -1 loop
                s = 2**(j-1);
                forall k in 1..m/2 loop
                    call RAD2(X[k, i..i+β],
                            X[k+s, i..i+β],
                            W[k])
                endforall;
            endfor;
        endfor;
        /* Y = column_bitrev(X); */
    end;
```

The subroutine RAD2 computes an "in place" CFT of size 2 on an array of size 2 by β, where the vector instructions are used in the β direction so W is used as a scalar in the triadic operations instead of a vector. (One may generalize this program to use a subroutine RADp for any small number p. In fact, the experiments described in the next section use a radix 4 routine RAD4 which was written in assembly language and enabled us to make a few small optimization and better use of the vector registers.) Figure 4.2 illustrates the data flow of this computation. The bit reverse operation was coded as part of the last stage (as a permutation of vectors between processors as described in section 3) so that no additional copy was needed.

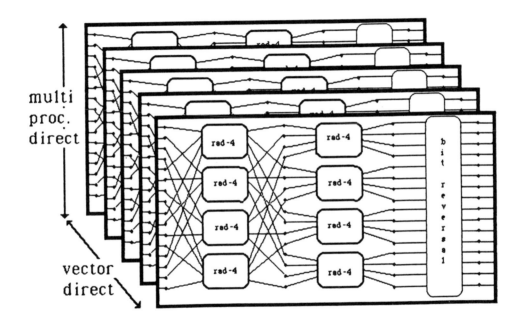

Figure 4.2 Radix 4 based FFT Data Flow

Our final version of the CFT algorithm becomes

function CFT(X,W,N,α, β_1, β_2): returns(Y);
parameter X: array$[1..2^\alpha, 1..2^{n-\alpha}]$ of complex;

```
              W: array[1..2^α, 1..2^{n-α}] of complex;
     var      Y: array[1..2^{n-α}, 1..2^α] of complex;

     begin
              Y = Tr(W* many_col_FFT(2^α, 2^{n-α}, β_1, X, W));
              Y = many_col_FFT(2^{n-α}, 2^α, β_2, Y, W);
     end;
```

The parameters α, β_1 and β_2 can now be chosen to give us the optimal performance.

4.3. Optimization of the data sweep.

The remainder of this section illustrates how an algebraic analysis of the constraints imposed by our hardware assumptions will show how one may best choose these parameters.

4.3.1. The single FFT case: k = 0 and 2(c -l -1) >= n

The first phase imposes the following constraints: The working set for performing the 2^{β_1} FFT's of length 2^α should fit into the cache.

$$c \geq \alpha + \beta_1 + 1 \qquad (3.1)$$

The 2^p processors should be fully employed.

$$\alpha \geq p + 1 \qquad (3.2)$$

Since the inequality (3.2) is between logarithms, it also means that the number of processors divides the length of the FFT's to be performed concurrently, so the load between the processors is perfectly balanced, each processor computing the same number of elementary butterflies.

The vector ability of each processor should be fully employed.

$$\beta_1 \geq l \qquad (3.3)$$

The second phase imposes an identical set of constraints, except with β_2 replacing β_1 and $n - \alpha$ replacing α.

$$c \geq n - \alpha + \beta_2 + 1 \qquad (4.1)$$

$$n - \alpha \geq p + 1 \qquad (4.2)$$

$$\beta_2 \geq l \qquad (4.3)$$

Since, it is always advantageous to fully use the space available in cache, we can suppose that (3.1) and (4.1) are in fact equalities. So, we can express β_1 and β_2 in terms of the other parameter α:

$$\beta_1 = c - \alpha - 1 \tag{4.4}$$

$$\beta_2 = c - n + \alpha - 1 \tag{4.5}$$

Now, we can substitute these values in the inequalities (3.3) and (4.3), so we will have the following system of 4 inequalities with only α:

$$\alpha \geq p + 1$$

$$c - l - 1 \geq \alpha$$

$$n - p - 1 \geq \alpha$$

$$\alpha \geq l + n - c + 1$$

This can be reduced to:

$$min(\; n - p - 1 \; , \; c - l - 1) \geq \alpha \geq max(\; p + 1 \; , \; l + n - c + 1) \tag{4.7}$$

and in fact the only condition for the existence of such α verifying (4.7) is:

$$min(\; n - p - 1 \; , \; c - l - 1) \geq max(\; p + 1 \; , \; l + n - c + 1) \tag{4.8}$$

Let us distinguish two cases:

a) $n + l \geq c + p$: in this case we have

$$min(\; n - p - 1 \; , \; c - l - 1 \;) = c - l - 1$$

$$max(\; p + 1 \; , \; l + n - c + 1) = l + n - c + 1$$

so equation (4.8) becomes:

$$c - l - 1 \geq l + n - c + 1$$

This equation is verified according to our hypothesis on n: $2(c - l - 1) \geq n$. It is easy to check that if we choose α as

$$\alpha = c - l - 1$$

it follows that the constraints are satisfied and we have

$$\beta_1 = l$$

$$\beta_2 = 2c - l - 2$$

b) $c + p \geq n + l$ in this case we have

$$min(\; n - p - 1 \; , \; c - l - 1 \;) = n - p - 1$$

$$max(\; p + 1 \; , \; l + n - c + 1) = p + 1$$

so equation (4.8) becomes:

$$n - p - 1 \geq p + 1$$

or

$$n \geq 2p + 2$$

This equation is verified according to our hypothesis on n and c: $c \geq 2(p + 1)$ (HC2) and $n \geq c$ (HP2). It is easy to check that if we choose α to satisfy

$$\alpha = n - p - 1$$

then the constraints are satisfied and we also have

$$\beta_1 = c - n + p$$

$$\beta_2 = c - p - 2$$

In fact, we have shown that by partitioning the algorithm as described above, we can perform a single FFT of length 2^n, n verifying $2(c - l - 1) \geq n \geq c$, with only two full memory read sweeps the data points. This should be compared to the simple algorithm described in the first section which required $\log_2(n)$ sweeps of the data.

4.3.2. The single FFT case: k=0 and n > 2(c-l-1)

In this case, we recursively use the folding technique in order to reduce the FFT to systems of smaller size that satisfy $n > c$ and $2(c - l - 1) \geq n$.

For example, if we suppose that:

$$4(c - l - 1) \geq n > 2(c - l - 1) \tag{6.1}$$

we fold this FFT in $2^{n/2}$ FFT's; during the two phases of the folding, we will deal with FFT of length verifying the conditions of the previous section. For the case of the equation (6.1), we will sweep the data 4 times.

4.3.3. The multiple FFT case: k > 0 and c >= n

In the multiple FFT case, we have two issues to consider. If $n + k$ is much less than c and k is bigger than l, it may be best to use the $many\_col\_fft()$ algorithm directly. On the other hand, if n is very large it may be best to treat the FFTs one at a time. There are, however, many other possibilities.

First we can restrict ourself to the case where $k + n + 1 \leq c$, because if this condition is not satisfied we can work with a smaller subset of the FFTs that do satisfy this condition and sequentially process these "batches" of problems. Moreover, we just need to study the case where the equality is satisfied i.e. the problem fits in cache. In fact, in this case there is no major difficulty for cache management, we have just to find an efficient organization for the vectorization and concurrency.

Let us distinguish two cases:

a)$k \geq l$:

As stated before, in this case we do not use the "folding" technique, the vectorization takes place across the FFTs.

Let us define $\alpha_1 = k - l$; we can consider that we have now 2^{α_1} groups of 2^l FFT's of length 2^n. If $\alpha_1 \geq p$, the parallelization and vectorization will be used across the FFT's, each CE computing 2^l FFT's in vector mode. On the other hand if $p - 1 \geq \alpha_1$, we divide the 2^p processors in $2^{p-\alpha_1}$ groups of 2^{α_1} processors each. We have logically configured the machine as $2^{p-\alpha_1}$ super processors each of them with an efficient vector length of $2^l 2^{\alpha_1} = 2^k$. With this configuration, since $n - 1 > p - \alpha_1$, we parallelize within the FFT: each FFT of length 2^n is given to one of the $2^{p-\alpha_1}$ super processors.

b)$l \geq k$:

In this case, we again use the folding technique. The computational organization will be similar to that developed in section 4.3.1. The main differences being that now the cache size does not really matter however we have now to compute more than one FFT $(2^k > 1)$. The fact that we have to compute 2^k independent FFT's offers a lot more flexibility for the computation reorganization. For simplifying the analysis, let us suppose that k is a multiple of 2; the folding technique applied to each FFT gives us a 3-D data flow graph, globally we have now 2^k such graphs; we will reorganize them in order to spread the parallelism between this FFT, half for concurrency between the processors, and half for vectorization inside a processor.

Let us define:

$$k' = \frac{k}{2}$$

The first phase impose the following constraints:

The 2^p processors should be fully employed, i.e.,

$$\alpha + k' \geq p + 1 . \tag{7.2}$$

Since the inequality (7.2) is between logarithms, it also means that the number of processors divide the length of the FFT's to be performed concurrently, so the load between the processors is perfectly well balanced, each processor computing the same number of elementary butterflies.

The vector ability of each processor should be fully employed, i.e.,

$$n - \alpha + k' \geq l \tag{7.3}$$

The second phase impose the following constraints:

The 2^p processors should be fully employed.

$$n - \alpha + k' \geq p + 1 \tag{8.2}$$

Since the inequality (8.2) is between logarithms, it also means that the number of processors divide the length of FFT's to be performed concurrently, so

the load between the processors is perfectly well balanced, each processor computing the same number of elementary butterflies.

The vector ability of each processor should be fully employed.

$$\alpha + k' \geq l \tag{8.3}$$

Now, let us rewrite these inequalities explicitly in terms of α:

$$\alpha \geq p + 1 - k'$$

$$n + k' - l \geq \alpha$$

$$n - p - 1 + k' \geq \alpha$$

$$\alpha \geq l - k'$$

This can be reduced to:

$$min(\, n - p - 1 + k'\,,\, n + k' - l\,) \geq \alpha \geq max(\, p + 1 - k'\,,\, l - k'\,) \tag{8.7}$$

and in fact the only condition for the existence of such α verifying (8.7) is:

$$min(\, n - p - 1 + k'\,,\, n + k' - l) \geq max(\, p + 1 - k'\,,\, l - k'\,) \tag{8.8}$$

Let us distinguish two cases:

a) $l \geq p + 1$:
in this case we have

$$min(\, n - p - 1 + k'\,,\, n + k' - l) = n + k' - l$$

$$max(\, p + 1 - k'\,,\, l - k'\,) = l - k'$$

so equation (8.8) becomes:

$$n + k' - l \geq l - k'$$

$$n + k \geq 2l$$

This just means that the amount of work must be large enough. In other
words, the total size of the problem should be at least as big as the effective vector length. Anything smaller will never achieve good performance.

b) $p + 1 \geq l$ In this case we have

$$min(\, n - p - 1 + k'\,,\, n + k' - l) = n - p - 1 + k'$$

$$max(\, p + 1 - k'\,,\, l - k'\,) = p + 1 - k'$$

so equation (8.8) becomes:

$$n - p - 1 + k' \geq p + 1 - k'$$

$$n + k \geq 2(\, p + 1)$$

Again, this means that the amount of work must be large enough, but in this case the size of the problem should be as big as the number of processors.

5. Experimental Results.

There are sets of experimental results that go along with the algorithm formulations in the previous section. In the first case we consider the performance of the many_col_FFT() function implemented as described above. Figure 5.1 shows the performance of the algorithm superimposed over the two simple methods described in section two. The X-axis is the $\log_2()$ of the size of each FFT (which is also the $\log_2()$ of the number of FFTs.) The value of β was set to 32 in this experiment. The Y-axis is again the effective Megaflop rate.

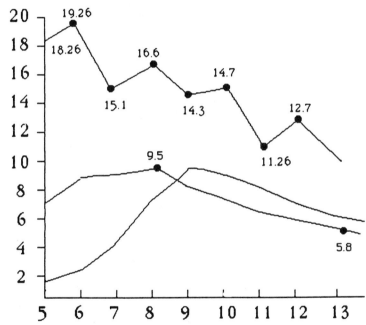

Figure 5.1 Multiple FFTs using many_col_FFT(). $\beta = 32$.

There are several important points to notice here. First, if we compare the many_col_FFT() function to the more simple manyCFT() function we see the improved cache use and better vector instruction use gives us a speed up ranging from 2 to 4. Notice that many_col_FFT() performs better when $\log_2(N)$ is even. This is due to the fact that a complete radix 4 formulation is not possible when n is odd and an extra radix 2 stage is needed. The problem is that the radix 2 stage is not as efficient in vector instruction use as the radix 4 stages.

The most important feature of this graph is that, like the earlier experiments, the performance advantage drops off when n becomes large. This is because we have exceeded the constraints described in 4.3.3 and we are better with a "folded" technique.

Figure 5.2 describes the performance of the folded algorithm for a single FFT. As can be seen we have performance increase of nearly a factor of 3 in the single vector oriented fft algorithm (in the range $N = 512$ to $N = 4096$.) For larger N performance is dropping off because we have entered the range in which a recursive application of the folding technique is needed (section 4.3.2).

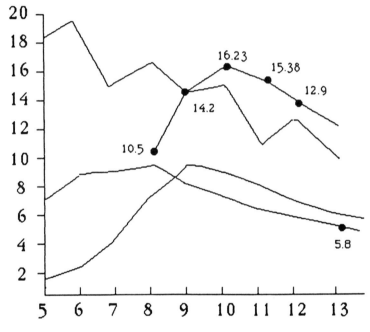

Figure 5.2. Folded CFT Algorithm.

Figure 5.3 gives the measured multiprocessor speed-up for p processors vs. 1 processor for this algorithm. This was done in the case $N = 1024$, but the results do not change much with other values of $N < 8K$. These numbers should be compared with the speed-up figures given earlier. The improvement can be attributed to the fact that we no longer have as critical a memory cache conflict problem. By improving locality, the the cache misses are less frequent. Because processors share the data that is in cache we see that more of them can be busy at one time.

Speed-up as a function of Processors							
p = 1	2	3	4	5	6	7	8
1.0	1.9	2.8	3.7	4.1	4.5	5.9	6.6

Figure 5.3 Multiprocessor Speed-up for Folded CFT for $N = 1024$

6. Conclusions.

Though we have only considered one example algorithm, that of the FFT, the analysis performed here was in terms of a general vector-multiprocessor. By changing the parameters we should be able to use the model to characterize performance for the Cray and ETA multiprocessors. A similar analysis has been carried out for a number of other problems including many basic linear algebra routines [Jalb86].

There are two key ideas presented here. First, real performance improvements are not simply a matter of deriving a means to use as many processors as possible. While this is important, many other issues are of nearly equal concern. Specifically, machines with a complex memory hierarchy require that a careful eye be kept on the way in which data is moved from one level to another. For parallel systems with no shared memory such as the many "hypercube" designs or systolic arrays, the data management strategy is an essential part of the programming process. In fact we believe that there is a strong link between this type of systolic array data management and the process of blocking a computation for a parallel system so that we optimize memory system use.

The idea that we have tried to emphasize is that our mathematical model has been designed not to give us performance estimates, but to measure the number of data movement operations incurred by a given formulation of an algorithm. In this way the data is used to pick between alternate ways to restructure the computation. It is our feeling that these techniques could be intergrated into an automatic restructuring package at some future time.

7. Acknowledgements

The authors would like to thank Ahmed Sameh and Paul Swartztrauber for thier outstanding help on this project. We would also like to thank the AFOSR, DOE and NSF for thier generous support of this research.

8. References

[Brig85]   W. L. Briggs, L. Hart, A. O'Gallagher and R. A. Sweet, "Parallel FFT Methods," Tec. Report RP85-0407, Dept. of Mathematics, Univ. of Colorado, Denver.

[Cvet86]   Z. Cvetanovic, "Performance Analysis of the FFT Algorithm on a Shared-Memory Parallel Architecture," Tec. Rep. RC11749 (52739) IBM T. J. Watson Research Center, 1986.

[CoTu65]   J.W. Cooley, and J.W.Tuckey, "An algorithm for machine calculation of complex Fourier series," Math. Comput., no. 19, pp. 297-301, 1965.

[CoLe70]   J. W. Cooley, P.A.W. Lewis and P.D. Welch, "The Fast Fourier Transform Algorithm: Programming Consideration in the Calculation of Sine, Cosine and Laplace Transforms," J. Sound Vib. Vol. 12(1970) pp.315-337.

[Dong85]   J. Dongarra, J. Du Croz, S. Hammerling and R. Hanson, "A Proposal for an Extended Set of Fortran Basic Linear Algebra Subprograms," ACM SIGNUM Newsletter, vol. 20, No. 1, Jan. 1985, pp. 2-18.

[Forn81]   B. Fornberg, "A Vector Implementation of the Fast Fourier Transform Algorithm," Math Comput. no. 36, pp. 189-191, 1981.

[Gann84]   D. Gannon and J. Van Rosendale, "On the Impact of communication Complexity on the Design of Parallel Numerical Algorithms," IEEE TOC, C-33(1984), pp. 1180-1194.

[Gent66]   W. M. Gentleman and G. Sande, "Fast Fourier Transforms for fun and profit," 1966 Fall Joint Comp. Conf., AFIPS Proc. 29 (1966), pp. 563-578.

[Korn79]   D. Korn and J. Lambiotte, "Computing the Fast Fourier Transform on a Vector Computer," Math. Comput., Vol. 33, pp252-264, 1968.

[Kwok86]   A. Y. Kwok, "The Multiprocessor Modified Pease FFT Algorithm,"
Tech. Report. Center for Supercomputer Research and Development, Urbana, Illinois, 1986.

[Laws79]   C. Lawson, R. Hanson, D. Kincaid and F. Krough, "Basic Linear
Algebra Subprograms for FORTRAN usage," ACM TOMS 5(1979),
pp. 308-323.

[Nort85]   V.A. Norton and A. Silberger, "Parallelization and Performance
Prediction of the Cooley-Tukey Algorithm for Shared-memory
Architecutres," Tec. Report. IBM T.J. Watson Research Center,
Yorktown Heights, N.Y. July, 1985. Presented at second Siam
Conf. on Parallel Processing, Norfolk, Va., Oct. 1985.

[Peas68]   M.C. Pease, "An Adaption of the Fast Fourier Transform for
Parallel Processing," J. ACM Vo. 15, pp. 252-264, 1968.

[Pete83]   W. P. Peterson, "Vector Fortran for Numerical Problems on
CRAY-1," Communication of the ACM, Vol. 26, No. 11, pp.1008-
10021, Nov. 1983.

[Swar82]   P. N. Swartztrauber, "Vectorizing the FFT's," in Parallel Computation, G. Rodrigue, Ed. Academic Press, 1982, pp. 51-85.

[Swar84]   P.N. Swarztrauber, "FFT algorithms for vector computers," Parallel Computing, no. 1, pp. 45-63, 1984

[Swar86]   P.N. Swarztrauber, "Multiprocessor FFT's," Tech. Report. No. 608,
Center for Supercomputer Research and Development, Urbana, Illinois. Also presented at LOEN, Norway, 2-4 June 1986.

[Temp79]   C. Temperton, "Fast Fourier Transforms and Poisson Solvers on
the Cray 1," Infotech State of the Art Report: Supercomputers: 2,
C. Jessope, ed., Infotech Int'l Ltd. 1979, pp. 359-379.

[Wang80]   H. H. Wang, "On Vectorizing the Fast Fourier Transform," BIT
Vol. 20, pp.233-243, 1980.

[Wino78]   S. Winograd, "On computing the discrete Fourier transform,"
Math. Comput., no. 32, pp. 175-199, 1978.

Section III. Software Tools

One simple and practical reason for attempting to understand the general characteristics of parallel algorithms is to see where an idea that worked well on one problem can be applied to another. As we gain a clearer picture of where parallel algorithms differ in their structure from each other and from "serial" algorithms, we start to see where common structures emerge in the organization of concurrent software. These common control and data structures become the idioms of expression for parallel computation. Furthermore, as we gain more experience with different parallel architectures, we start to see which of these idioms of expression are suitable for describing algorithms for given classes of machines and which are not efficiently implemented (or meaningless) on others. This process of abstracting the model of computation and the idioms of expression not only gives us a way to characterize different modes of parallel computation, it also gives us the basis for designing software tools to better use the machines under study.

Because the technology of parallel computation is so new, most commercial systems have been delivered with very little software support other than a few primative extensions to standard programming languages like C and FORTRAN. With the sole exception of the Alliant FX/8, no commercial parallel system provides a reasonably automatic means of converting a sequential program to a parallel one. In the case of research parallel systems, the situation for the applications programmer is usually much worse. This situation has had the effect of forcing the applications programmers and algorithm designers into putting into practice the principle of idiomatic abstraction described above. They have been forced into becoming the designers of the first software tools to use these machines. But if parallel architectures can be characterized by the classes of algorithms for which they are well suited, it is only natural that algorithm designers are the first to recognize which tools are needed to express algorithms in that class.

A further motivation to build tools is that many of the researchers in the parallel algorithm community have access to more than one parallel architecture to experiment with. Consequently, they have been confronted with the problem of moving software from one machine to another. This has led to a movement to design tools that represent a model of parallelism that will allow software to be written for a wide class of machines so that some portability of code is maintained.

In this section five different projects are presented. In the first, Fran Berman describes the Prep-P software system. The objective of this project is to provide a tool to help automate the "mapping" problem for non-shared memory parallel processors. In particular, it is targeted at machines based on a fixed or configurable communication network between processing elements. One common model of parallel algorithm abstraction is to view a computation as a network of communicating processes. The topology of the communication graph may not be a natural subgraph of the topology of the processor interconnection graph or, more commonly, the process communication graph may be much larger. The mapping problem is the process of making a static assignment of processes to processors.

The second chapter, by Babb and DiNucci, describes a parallel programming software engineering methodology based on large-grain data flow principles. This system (as well as those described in all the following chapters) is based on a layered language where the bottom language is standard FORTRAN and the top layer is set of special macros. The LGDF methodology is designed to properly encapsulate the segments of code that belong to one data flow block and to help the programmer isolate the data dependences that exist between blocks. The system has been successfully used on the Cray-1 and on the Hep and Cyberplus parallel systems.

The chapter by Lusk and Overbeek describes an adaptation of a well known operating system concept, Hoare's Monitor, to the task of structuring parallel algorithms. This has been used by a number of people and has proven to be very easy to port to a shared memory environment. In its current form a problem first written on the Hep using this package can now run on the parallel systems from Alliant, Sequent, Encore, the Cray-2, and the IBM 3033-AP.

A system with very similar objectives, but different organization, is the Schedule package by Dongarra and Sorensen. This set of tools provides a way for users to structure FORTRAN programs as a system of static or dynamically allocated, "lightweight" processes. The user specifies communication between processes by defining a precedence graph of data dependences. The system then provides an automatic scheduling mechanism to execute the program subject to these constraints.

The last chapter describes the FORCE system designed and implemented by Jordan and his students. Again we have a macro package for the parallel execution of shared memory FORTRAN programs, but a completely different model of parallel execution has been designed. Based on the concept of a large number of processes concurrently executing the code, it is the job of programmer to decide which sections of code are sequential rather than where to find concurrency.

It should be noted that there are several other systems and tool packages in use that are not represented here, for example, the Domino system from Maryland and the refined language concept of Klappholtz and Dietz. These systems are all first attempts to implement a parallel execution model based on a coherent view of parallel algorithm organization. As such, they also serve as testbeds for ideas used in later generations of tool systems based on new languages and compilers as well as attempts to automate the process of extracting and mapping the concurrency from high level programs.

Experience with an Automatic Solution to the Mapping Problem

Francine Berman
University of California, San Diego

Prep-P project members: *Mark Anderson, Jonathon Buchwald, Jeffrey Conroy, Barbara Donovan, Michael Goodrich, Patricia Haden, Shanti Hofshi, Darin Johnson, Dino Karabeg, Charles Koelbel, Clay Mayers, Dwight Newton, Denise Neufer, Brian Parent, William Robison III, Daniel Rose, Karen Showell, Bernd Stramm, Joel Strickland, Shouhan Wang, Anne Wilson.*

Introduction

In theory, it is easy to argue that a multiprocessor is faster than a single processor computer: A task can be divided into several subtasks which may be executed simultaneously. The parallel program terminates in the time it takes the longest subtask to complete execution. In the limit, the speedup for the parallel execution of a problem is the time complexity of the fastest sequential algorithm divided by the number of processors in the multiprocessor. The implication is that the overhead of decomposing the algorithm and mapping it into the target machine has a negligible effect on the implementation process.

In reality, the process of implementing a parallel algorithm on a multiprocessor is usually machine-specific and rife with low-level detail. The parallel algorithm must be expressed in a language supported by the multiprocessor. This language may require explicit references to the processor memory organization, processor interconnection structure, synchronization and I/O protocols, etc. The advantage of computing in a multiprocessor environment is often offset by the complexity of the implementation process. Moreover, programs for one multiprocessor are rarely portable to other multiprocessors, so that for each algorithm and each parallel architecture, the process must be repeated.

Such difficulties in implementing parallel algorithms on multiprocessors render multiprocessor computing inaccessible to all but a dedicated user community. This contrasts sharply to the uniprocessor environment where users can program algorithms relatively efficiently in any one of a number of supported high-level languages, and generally depend upon existing system software to allocate memory, swap resources with other jobs, communicate with I/O devices, etc. Multiprocessor computing will be accessible to a wider community when it can offer not only faster and more powerful machines but also programming tools with which to use them.

The implementation problem described above is often called the **mapping problem.** This term can be used for almost any aspect of the implementation of parallel algorithms on multiprocessors with some legitimacy. For this report, we will limit ourselves to mapping problems which arise when there is a mismatch between the number of processes (size) of a parallel algorithm and the target multiprocessor, or between the communication requirements (structure) of a parallel algorithm and the processor interconnection structure of the multiprocessor, or both. More informally, we are interested in automatic solutions for mapping parallel algorithms which are "too big" for their intended parallel machines.

This problem is fundamental to parallel computing: applications often require large-sized instances of an algorithm and multiprocessors are hardwired architectures which cannot be changed according to program size or structure. For special-purpose or dedicated multiprocessors, an instance of the mapping problem can be solved by letting the algorithm domain determine the design and size of the architecture. For general- or broad-purpose multiprocessors, this is not possible. What is needed are analogues to the system software of the uniprocessor environment: software which maps and multiplexes parallel algorithms into multiprocessors regardless of size, available memory, or communication structure.

Recent work on the mapping problem can be divided into two categories: research on mappings in which the algorithms are allowed to vary but in which the architecture type remains fixed, and research on mappings in which both the algorithms and architectures are allowed to vary.

There is a sizable body of interesting work on mappings in which the algorithms are allowed to vary. In particular, this approach includes a large number of papers on mapping algorithms into array processors. One of the first and most well-known papers of this type was by Bokhari ([9]) who considered mappings of algorithm communication graphs into the finite element machine architecture. Also representative of this category is the work of Moldovan ([24]) who described an approach to designing VLSI array processors specialized to the communication graphs of high-level language programs. Other researchers who have taken this approach include Ramakrishnan, Fussell and Silberschatz ([26]), Kung and Stevenson ([21]), Kuhn ([20]), DeGroot ([12],[13]), Heller ([15]), and others.

There is a less sizable body of work when both the algorithms and architectures are allowed to vary. One approach was taken by Fishburn and Finkel who mapped a small class of commonly used algorithm communication graphs into smaller-sized architectures of the same graph type ([14]). Preparata ([25]) also discussed the mapping of several algorithm paradigms into "hypercube equivalent" networks such as the shuffle-exchange, cube-connected cycles and butterfly. Another approach is taken by Jamieson ([16]) who uses algorithm and architecture characteristics to model the mapping process. More theoretical research on the embedding of algorithm communication graphs into silicon includes the work of Valiant ([34]), Leiserson ([22]), and others.

Our interest has focused on determining general automatic mappings of algorithms into architectures. One fruitful approach has been to divide the mapping problem into several component parts: algorithm partitioning (contraction), layout (placement and routing), and multiplexing. This allows the investigation of performance-efficient solutions for each of these parts independent of the constraints of the others. The investigation of this approach has produced an automatic mapping software system called **Prep-P** ([5]), and we report on its general design and development in

Section 1. In Section 2, we detail the algorithms used in the software and our experience with them; and in Section 3, we discuss the current status of the project.

Section 1: The Design of the Prep-P Mapping Strategy

We began by studying the mapping problem at a theoretical level ([7]). The first task was to determine an appropriate abstraction for the parallel algorithm, multiprocessor and mapping process. Following the literature, we represent an instance of a parallel algorithm as a *communication graph* G_i whose nodes represent processes and whose edges represent communication links between processes ([10],[33]). The parallel algorithm is then a *family* of communication graphs $\{G_i\}$, one for each problem instance. The graph family representation is a natural model which does not constrain the algorithm representation to any specific set of parallel constructs. (The problem of developing such constructs and more generally parallel programming languages is a complex and fundamental one which we do not attempt to solve.) This representation presupposes that the algorithm has already been decomposed into a set of processes which run concurrently. A large and useful class of parallel algorithms can easily be described within this framework.

To represent the target multiprocessor, we use an undirected *computation graph* H in which the nodes are processors and the edges are data paths. Each edge meets its incident nodes at unique *ports*. For many architectures, the number of ports is bounded for each processor and influences the geometry of the processor interconnection structure ([33]).

Using these abstractions, a family of communication graphs $\{G_i\}$ for the algorithm and a computation graph H for the multiprocessor, the mapping process can then be abstracted as an embedding problem from $\{G_i\}$ into H. We began by investigating different strategies for performing the embedding. One fruitful approach was to divide the embedding process into the tasks of partitioning or contracting algorithm processes into groups, placing these groups at processors in the target multiprocessor, routing paths between process groups in the target machine, and multiplexing the processes within each group to execute the original parallel algorithm. This approach describes a series of transformations *(contraction, placement* and *routing)* which can be applied to the algorithm communication graph and which result in a mapping of the algorithm into the multiprocessor. The last task, *multiplexing,* ensures that the transformations preserve the termination and output behavior of the original algorithm following execution of the multiplexed code.

To evaluate the efficiency of mappings based on this approach we considered a diverse group of benchmark algorithms and architecture interconnection structures ([8]). We measured complexity in terms of the amount of sequential simulation during multiplexing and the amount of edge expansion (communication delay) resulting from a given set of contraction, placement and routing transformations. The mappings for these benchmark examples were generally efficient and in many cases optimal. Encouraged by the theoretical performance of this approach, we began to design and implement a mapping preprocessor, called Prep-P, based on these ideas for non-shared memory multiprocessors. We discuss this software in detail in the next section.

To illustrate the mapping strategy described above, consider the following example. We want to compute the maximum of n=8 integers. One approach is to use a parallel version of the tournament method algorithm ([1]):

> *Initially, place the data at the leaves of a complete binary tree. In the first step, each leaf will send its data value to its parent at height 1. In subsequent steps, each node at height k will compare the values from each of its children and send the maximum of the two values to its parent at height k+1. After O(logn) steps the algorithm will terminate with the maximum of all of the input data at the root.*

The representation for this algorithm is a family of communication graphs, each of which is a complete binary tree. The communication graph for the parallel max-finding algorithm with n=8 nodes is shown in Figure 1. Each node is identified with a process describing its function. For this algorithm, a possible set of process codes is given in Figure 2. (The programming language used for processes is XX ([31])).

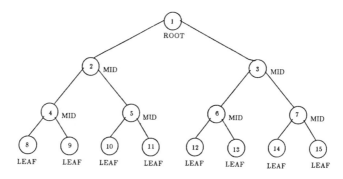

Figure 1
Communication graph for the parallel max-finding algorithm.

Continuing with the example, suppose that we have on hand a small array multiprocessor with 9 processors (Figure 3). Each processor has its own local memory and can communicate with its nearest neighbors only through message-passing. In addition, assume that each processor can also communicate with the "outside world". There are two immediate difficulties with implementing the parallel tournament method algorithm on this array processor. First, the algorithm requires 15 (non-distinct) processes and the multiprocessor has only 9 processors. Second, the communication structure of the algorithm is a tree and the multiprocessor interconnection configuration is a grid. To implement the algorithm on the multiprocessor, some form of mapping must be done.

We may apply the transformations described above. First contract the communication graph of the algorithm so that the number of partitions does not exceed 9. One

```
code ROOT;                          code MID;
ports LSON, RSON;                   ports PARENT, LSON, RSON;
begin                               begin
int x, y, z;                        int x, y;
x<-LSON;                            x<-LSON;
y<-RSON;                            y<-RSON;
if (x > y) then z:=x else z:=y;     if (x > y) then PARENT<-x else PARENT<-y;
end.                                end.


                                    code LEAF(m);
                                    ports PARENT;
                                    begin
                                    integer m;
                                    sint x;
                                    x:=m;
                                    PARENT<-x;
                                    end.
```

Figure 2
Process codes for the parallel max-finding algorithm.

way of doing this is shown in Figure 4. The left subtree of the root is folded onto the right subtree and the root is coalesced with its two children. The resulting contracted graph is a 7 node tree. (Although in this example, both the contracted and uncontracted graphs are trees, the contracted graph need not be of the same graph type as the uncontracted graph). Next, the contracted tree is placed on the multiprocessor as shown in Figure 5. In this case, the routing transformation is the identity function.

We have now defined a mapping (contraction, placement and routing) of the tree into the array. The algorithm is then executed by multiplexing the original algorithm on the array processor, sequentially executing the process codes of nodes mapped to the same processor and executing in parallel the multiplexed code of distinct processors. Following execution, the multiplexed algorithm should have the same output and termination behavior on the 9 node array processor as the original parallel tournament method algorithm would have had on a 15 node binary tree multiprocessor.

The mapping given in the preceding example is similar to the mapping produced by the Prep-P software. In the next section, we describe this software in detail and discuss our experience developing mapping protocols in a cooperative software project.

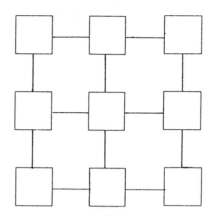

Figure 3
Array multiprocessor.

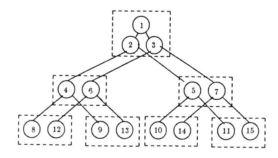

Figure 4
"Folding" contraction of the 15 node complete binary tree.
(Boxed nodes are mapped to the same processor).

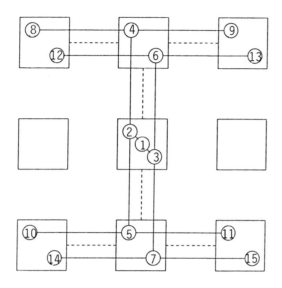

Figure 5
*Mapping of the contracted 15 node
complete binary tree into the array multiprocessor.
(Some array nearest-neighbor connections not shown).*

Section 2: The Implementation of Prep-P

The overall organization of Prep-P is based upon the mapping approach described
in Section 1. The algorithm is input as an undirected graph, each of whose nodes is
identified with a process (written in **XX**); the output of the system is Intel 8051 assembly code which, when run, executes the algorithm communication graph on a fixed-size
parallel architecture simulator. The Prep-P system contracts, places, routes and multiplexes the communication graph automatically. We describe the design and implementation of each of the modules of the Prep-P system in the following pages.

The User Interface: A Graph Description Language

The goal of the user interface is to provide a representation for the algorithm communication graph which can be processed into files used by subsequent modules of the system. Since the communication graph is given as an undirected graph with associated processes, we developed a simple graph description language to represent the adjacency list and process assignment. Since each pair of processes communicates via a unique path in the graph, the incident input/output ports of each node are also labeled.

As an example, consider the communication graph given in Figure 1 with process codes given in Figure 2. A representation for this graph is given in Figure 6.

```
irregular
nodemin=1
nodecount=15

procedure ROOT
  nodetype: {i=1}
  port LSON: {2*i}
  port RSON: {2*i+1}
procedure MID
  nodetype: {i>1 && i<(15+1)/2}
  port PARENT: {i/2}
  port LSON: {2*i}
  port RSON: {2*i+1}
procedure LEAF
  nodetype: {i≥(15+1)/2}
  port PARENT: {i/2}
```

Figure 6
Representation of a 15 node complete binary tree
in the graph description language.

The first line is a description of the graph type. Although the graph is a complete binary tree, we call this graph *irregular* since we are explicitly providing the adjacency list and we wish to use the general contraction and placement algorithms rather than library mappings. The second line gives the minimum integer label a node may have, and the third line gives the number of nodes in the graph. Each *procedure* segment gives the name of the process code which will be associated with the nodes whose labels are described by the expression *nodetype*. Distinct adjacencies of a node are distinguished by assigning each nexus of a node a distinct *port* label. For example, nodes whose process code is MID have 3 ports: PARENT, LSON and RSON. If a MID node is labeled i, its PARENT port is connected to a unique port of node i/2 (integer division), its LSON port is connected to a unique port of node 2i and its RSON port is connected to a unique port of node 2i+1. Note that although each adjacency on the adjacency list may be given explicitly, regular structure in the communication graph may be indicated using C-like arithmetic expressions, substantially abbreviating the representation.

Some communication graphs are used frequently in the literature. These include complete binary trees, linear and two-dimensional arrays, hypercubes and others. For these graphs, we have added a set of library descriptions which may be used with a specified labeling. (The library routines also may be used to define graphs constructed from several regular graphs). In practice, for the communication graph and codes given in Figures 1 and 2, we would most likely use the representation given in Figure 7 rather than that given in Figure 6.

```
tree
nodemin=1
nodecount=15
use tree_lib
```

Figure 7
Library representation of the complete binary tree.

Note that to use the library placement and contraction of the tree, we must also use library procedure names and the library labeling of the tree. (There is a facility which allows the ports to be renamed). As before, *nodemin* is the minimum integer label, and *nodecount* is the number of nodes or size of the graph.

Given a communication graph represented using the graph description language, Prep-P creates internal files which represent the process and adjacency information in a more convenient form. These include *AdjLst,* which lists the adjacent nodes in the graph, and *ProcLst,* which associates with each node in the graph the name of its process code. An excerpt from the AdjLst and ProcLst for a complete binary tree with 15 nodes is given in Figure 8.

If the number of nodes in the algorithm communication graph does not exceed n, the number of processors in the multiprocessor, an additional file, *SmallAdjLst,* is generated, the contraction module is bypassed, and control is passed to the placement module. If the number of nodes in the communication graph is greater than n, control is passed by Prep-P to the contraction module.

AdjLst

9
15
4
:1:(RSON,PARENT)3:(LSON,PARENT)2
:2:(RSON,PARENT)5:(LSON,PARENT)4:(PARENT,LSON)1
:3:(RSON,PARENT)7:(LSON,PARENT)6:(PARENT,RSON)1
:4:(RSON,PARENT)9:(LSON,PARENT)8:(PARENT,LSON)2
:5:(RSON,PARENT)11:(LSON,PARENT)10:(PARENT,RSON)2
.
.
.
:14:(PARENT,RSON)7
:15:(PARENT,LSON)7

ProcLst

:1:ROOT
:2:MID
:3:MID
:4:MID
.
.
.
:8:LEAF
:9:LEAF
.
.
.
:15:LEAF

Figure 8
Excerpts from AdjLst and ProcLst
for a complete binary tree with 15 nodes.

Partitioning the Communication Graph: The Contraction Module

The goal of the contraction module is to take the representation of an input communication graph with more than n nodes, and to produce an intermediate contracted graph with no more than n nodes. (In the Prep-P system, n is generally 64). We have experimented with several different designs for the contraction module during the development of Prep-P. Each design served to define our focus and provide more intuition about the contraction problem. This perspective was useful since the determination of good solutions for the contraction problem is a complex and challenging problem. In particular, both "optimal" and "good" solutions vary with the model chosen for the algorithm communication graph and multiprocessor interconnection structure. For Prep-P, we considered several different models for contraction and varied the algorithms used for each one. We describe this process in the next few paragraphs.

Our first approach to the contraction problem was to describe the algorithm communication graph using *edge grammars* ([4]), a type of graph grammar which promotes automatic contraction for regularly structured graphs. The graph was specified using the graph description language and identified as a member of a library of frequently used edge grammar-defined graphs. ("Irregular" graphs were not permitted). The edge grammar representation for the graph was used to define a set of library contractions which were indexed by graph type. Since the module essentially consisted of library routines, performance was fast, and the contracted graphs produced by the module were of the same class of regularly structured graphs used as input. On the down side, the contraction module was not general: Although the graph description language permitted the definition of any algorithm which could be represented as an undirected graph, the contraction module could only map those which could be represented by a class of regularly structured edge grammars. In addition, the input communication graphs were required to be labeled in precisely the same way as the library graphs. Since the problem of identifying distinct sets of labels for the same graph is NP-complete, the description process could permit little freedom. The narrow graph domain and restrictive description process encouraged us to investigate more general strategies for contraction.

The next approach was to find good algorithms for the following version of the general contraction problem: *Given an undirected graph with m nodes, find a partition into at most n≤m groups such that a given cost function is minimized.* For our purposes, we chose the cost function to be the number of edges between distinct partitions in the induced (contracted) graph. To simplify the design process, we made several assumptions:

1) that the processes identified with each node perform roughly the same number of reads and writes,

2) that parallelism is maximized when the m processes are distributed roughly equally over the n partitions, and that

3) intra-processor communication is more performance-efficient than inter-processor communication (i.e. wires should be mapped "inside" processors rather than "between" them when possible).

Note that assumptions 2 and 3 are conflicting: in the extreme, intra-processor communication is maximized by mapping all processes to one of the partitions while under assumption 1, parallelism is maximized by mapping m/n processes to each partition. This conflict influences the identification and weighting of cost functions.

With this general approach, the graph domain is the set of undirected graphs and the contraction algorithms we chose to test contract any element in this set. (For algorithms in which assumption 1 is not true, the contraction may not be as performance-efficient as for algorithms which are more homogeneous). In contrast to the edge grammar approach, the contracted graph need not be of the same graph type as the original communication graph (i.e. if the input graph is a hypercube, the contracted graph need not be a smaller hypercube).

Under assumptions 1-3, we tested several algorithm techniques to perform contraction: simulated annealing, local neighborhood search, branch-and-bound, greedy, etc. The most promising of these algorithms were simulated annealing ([19]) and local neighborhood search ([2]). We "fine-tuned" each of these algorithms extensively and compared their performance on a small suite of test graphs. The following is a description of the algorithms and our conclusions.

Local Neighborhood Search

The local neighborhood search algorithm for contraction can be described as follows: Start with a random contraction of m processes into n partitions ($m \geq n$). While the exit criterion has not been achieved, generate a new configuration. If the new configuration cost is greater than the old configuration cost, then restore the old configuration, otherwise accept the new configuration. We experimented with different methods for generating an initial configuration, generating new configurations, defining the exit condition and calculating the cost of a given contraction. Pilot studies on the first three algorithm parameters gave clear results of the effect of these parameters on the cost of the final solutions and the response time of the algorithm ([6]). The last parameter, cost, requires a function which is minimized when assumptions 1-3 are satisfied. These assumptions are conflicting and the determination of a cost function which accurately represents their intent is an open problem. At this writing, we are using the cost function

$$\textit{(number of distinct edges in the contracted graph)} + sqrt(\Sigma_i |Rp_i|^2)$$

(where Rp_i is the number of processes mapped to partition i) for local neighborhood search. We discuss possibilities for better cost functions at the end of this subsection.

Simulated Annealing

Simulated annealing is a probabilistic modification of traditional neighborhood search techniques. Both approaches find a solution to a combinatorial optimization problem by starting with a random solution and making a series of modifications to the solution. In neighborhood search algorithms, modifications which improve the solution by some given cost criteria are accepted and others are rejected. The acceptance criterion in simulated annealing is more complex. All modifications which lead to a better solution are accepted. All modifications which result in a poorer (higher cost) solution are accepted with probability exp(-D/T) where D is the difference between the costs of the solutions before and after the update, and T is an algorithm parameter known as temperature. Over time, the parameter T is slowly reduced, causing a reduction in the probability that a modification which results in a poorer solution will be accepted. The term simulated annealing is derived from an analogy between combinatorial optimization problems and the behavior of physical systems (such as solids) during cooling.

The simulated annealing algorithm corrects a major flaw of neighborhood search algorithms -- the tendency to get stuck in local minima in the solution space. Since neighborhood search techniques can accept only modifications that improve the solution, they are unable to escape from a local minimum. Simulated annealing has a nonzero probability of making a modification that worsens the solution and thus may permit an escape from a local minimum.

We experimented with varying or "tuning" the simulated annealing algorithm with a set of five variables: starting temperature, temperature schedule, acceptance condition, exit condition, and cost ([6]). We instantiated each of these variables based on guidelines from the literature ([3],[23],[35]) and considered algorithm performance in a fully crossed factorial design. In comparison with the local neighborhood search algorithm, the performance of the simulated annealing algorithm was unpredictable: the resulting contractions tended to be either much better or much worse than the contractions resulting from local neighborhood search. In addition, simulated annealing produced good contractions primarily on trials having prohibitive runtimes. Based on these pilot studies, we chose local neighborhood search as the contraction algorithm for the current version of Prep-P.

For convenience, we call the processes of the original algorithm **Vips** or virtual processes, and the nodes of the contracted communication graph **Rps** or real processes. After execution of the local neighborhood search algorithm, files are created which encode the assignment of Vips to Rps and the adjacencies of the contracted graph. These files are then passed to the placement module.

Under the assumptions 1-3, the contractions obtained using the local neighborhood search heuristic described above are generally good. However, many algorithms are not homogeneous, i.e. are not decomposable into a communication graph all of whose processes perform roughly the same function. We are currently considering a more sophisticated set of assumptions for the contraction problem. These assumptions include 2 and 3 given above and use more explicit information about the I/O behavior of the algorithm communication graph. The strategy is to perform a small amount of preprocessing on the communication graph in order to obtain information about algorithm behavior which will promote better contractions. For example, it is possible to preprocess the communication graph to determine whether two processes will communicate using a bounded number of reads and writes, or to determine whether communication is between processors, or between a processor and an I/O pad. New contraction heuristics using this information are currently under investigation and we hope to report on them in the near future.

Placing the Small-sized Graph: The Placement Module

The goal of the placement module is to take a small-sized (no more than n Rps) graph and to produce an embedding into n PEs (processing elements) where n is the number of processors in the target multiprocessor. If the algorithm communication graph was small to begin with, the contraction module is bypassed and files processed through the graph description language are used as input to the placement module. For algorithm communication graphs with greater than n nodes, the contraction module produces a contracted intermediate graph. This contracted graph is then given as input to the placement module.

Our version of the placement problem is: *Given a graph G with at most n Rps, find an embedding of G into a grid with n PEs so that a given cost function is minimized.* For Prep-P, we chose the cost function to be the total wirelength (using the manhattan or "taxicab" metric) between Rps in the embedded graph. The maximum wirelength, average wirelength, or some other cost function could also be used.

As was the case with contraction, we tested several algorithms for inclusion in the placement module. The most promising of these algorithms were an adaptation of simulated annealing and the Kernighan and Lin algorithm.

Kernighan and Lin

The Kernighan and Lin algorithm ([18]) is a divide-and-conquer algorithm for circuit placement. The adapted version we use in Prep-P can be described as follows: Starting with a random placement of Rps on PEs, partition the grid into halves. (Assume that each PE is assigned some Rp. PE's which have not been assigned a Rp in the initial placement are assigned a null or dummy Rp). Compute the number of edges crossing each partition boundary resulting from the exchange of each pair of Rps (including dummy Rps). Pairs whose exchange results in a reduction of the number of edges between partitions are swapped. Next, each partition is halved and the process is repeated. The resulting four subgraphs are then assigned randomly to the four quadrants of the grid. The cost of this assignment, measured as the sum of the manhattan distances between each pair of nodes in distinct subgraphs, is computed. (Since individual nodes in each of the four subgraphs have not been placed, we measure the distance between nodes in distinct quadrants as the distance between the centers of those quadrants). All 24 possible assignments are computed and the assignment of subgraphs to quadrants which minimizes the cost is accepted. The procedure is then iterated for each (subgraph, quadrant) pair until the resulting subgraphs only contain single items.

Unlike simulated annealing, the only parameter which may be varied in Kernighan and Lin is cost. Since we believe that cost measured as total wirelength predisposes the algorithm in our environment towards better routings, we did not test other cost functions.

Simulated Annealing

For placement, we adapted the simulated annealing algorithm described above for contraction. Using a parameter suite appropriate to placement, we performed the same test set as for contraction to evaluate the effect of various parameters on the resulting placements. As before, the results were unpredictable in comparison to the Kernighan and Lin embeddings, and good results required multiple runs (and thus prohibitive runtimes) to achieve. Consequently, we chose the Kernighan and Lin algorithm to include in the placement module.

Library Contractions and Placements

Just as the graph description process can be abbreviated for some frequently used, regularly structured graphs, we can also use library routines within Prep-P to contract and place this same class of graphs.

The advantages of using library routines are twofold: First, the runtime of using the contraction and placement libraries for library-defined graphs is much faster than invoking the local neighborhood search and Kernighan and Lin algorithms. Second, the contractions and placements resulting from the library routines are more aesthetically appealing, i.e. the contractions are generally small-sized versions of the original communication graph, and the placements utilize the regularity and symmetries of the graphs' topology. Note that to use this feature of the Prep-P software, the graph type cannot be given as "irregular". As an example, the library placement for the 63 node complete binary tree is given in Figure 9a; the Kernighan and Lin placement for the same tree is given in Figure 9b. (Both graphs have also been routed).

The output of both the library placement and the Kernighan and Lin placement routines includes files which describe the assignment of Rps to PEs. After placing the contracted or small-sized graph on the grid, the next task is to route "wires" between adjacent PEs.

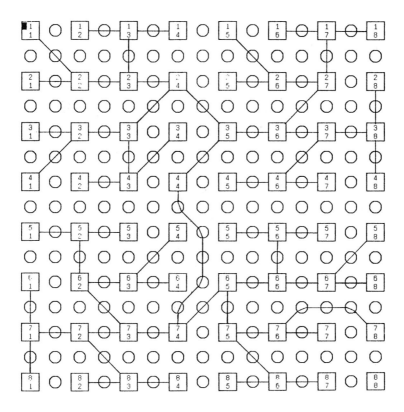

Figure 9a)
Library placement of the 63 node complete binary tree.

Connecting the Rp grid: The Routing Module

The goal of the routing module is to connect the Rp grid on a target parallel architecture. Unlike contraction and placement, routing is constrained by the physical interconnection structure of the target machine. In the Prep-P project, the target parallel architecture is a CHiP machine.

The CHiP machine ([30]), designed and developed by Lawrence Snyder, is a configurable, message-passing parallel architecture which may be used to simulate the processor interconnection structure of a variety of message-passing parallel machines. The CHiP architecture consists of a lattice of processing elements, or PEs, interleaved

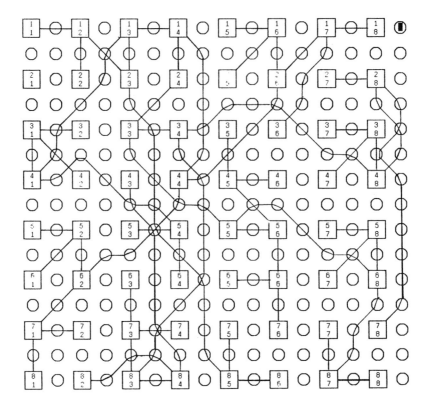

Figure 9b)
Kernighan and Lin placement of the 63 node complete binary tree.

with rows and columns of switches (Figure 10a). The lattice can be configured to represent a given interconnection structure by designating paths between the PEs using the switches (Figures 10b, 10c). Note that the switches are programmable and "wires" may intersect at switches without confusing their data. We chose the CHiP as our target architecture because of its flexibility: the CHiP can be configured to simulate many frequently used message-passing architectures (including complete binary trees, arrays, hypercubes, butterflies, shuffle-exchange and omega networks, rings, cube-connected cycles, etc.). In addition, the CHiP is available in the form of both a hardware emulator (Pringle, [17]) and a distribution software simulator (Poker, [31]). We currently run Prep-P as a front-end to Poker and plan to run on Pringle when possible.

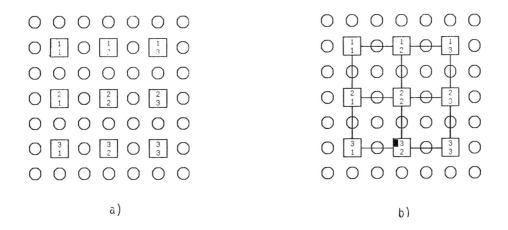

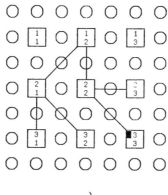

Figure 10
*a) CHiP architecture. (Boxes are PEs; circles are switches).
b) CHiP architecture configured as an array.
c) CHiP architecture configured as a complete binary tree.*

After contraction and placement, the Rps of the contracted graph have been identified with PEs in the CHiP lattice. The next step is to route "wires" between adjacent PEs. To accomplish this, Prep-P uses a simple breadth-first search algorithm ([1]) on switches in the CHiP lattice. For example, suppose we want to route an adjacency between PE1 and PE2 in Figure 11a. Note that some paths between switches have already been used. The routing algorithm examines the free switches immediately neighboring PE1 searching for PE2. These switches are placed at the end of a queue. (Consider a switch free if its nearest port is not full). Next, the first element of the queue is examined, looking for PE2. If PE2 is an immediate neighbor of a free (unused) port of the first element of the queue, then PE2 has been found and the search terminates. Otherwise, all free immediately neighboring switches of the first element of the queue are placed on the end of the queue and the first element is deleted. When a path to PE2 has been found, the wiring is accomplished by walking backwards up the designated path, marking each element encountered as having a wire out of the port to its neighbor on the path. The breadth-first search and wiring is performed for each pair of adjacent processors. The routing produced by this algorithm for Figure 11a is given in Figure 11b.

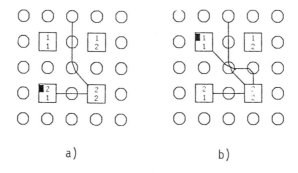

Figure 11

a) Unrouted path from PE1 to PE2. b) Routed path from PE1 to PE2.
(PE1 is 1 1; PE2 is 2 2).

The routing algorithm is simple and seems to be robust. In an attempt to improve the quality of the routings and the runtime of the routing code, we added the following shortest path heuristic: Before routing, sort the pairs of adjacent PEs by the manhattan distance between them placing the pairs with the shortest manhattan distance first on the list. Then perform the routing algorithm on the list. Preliminary studies have shown that while the sorting modification may not improve response time, it leads to more "regular" routings. However, the test suite was small, and it remains an open question as to whether the sorting heuristic does improve the running time on average and shortens the wirelength of the resulting routed graph.

Executing the Embedded Graph: The Multiplexing Module

The goal of the multiplexer is to produce code which simulates the execution of the algorithm communication graph on a fixed-size CHiP machine. The code may then be run on the Poker system and should exhibit the same output values and termination behavior as if the original communication graph were run on an architecture which matches its communication structure. Multiplexing is generally accomplished in Prep-P by concatenating the Vip codes assigned to each PE and pairing the concatenated Vip codes with a series of support routines to manage intra- and inter-PE communication. The support routines ensure accurate addressing between the source and destination Vips of messages and also manage the context switching for each PE so that all Vips may execute. The following example illustrates how this "mini-operating system" functions.

Consider the communication graph shown in Figure 12a. During execution of the non-multiplexed algorithm, communication consists of simple reads and writes along unique paths between distinct Vips. However in the multiplexed system (Figure 12b), communication between Vips 1 and 2 will take the form of *intra-writes* or *intra-reads* (communication between Vips mapped to the same PE) and communication between Vips 1 and 3 and Vips 1 and 4 will take the form of *inter-writes* and *inter-reads* (communication between Vips mapped to distinct PEs).

To simulate correct execution of the message-passing portion of the original communication graph, the support routines must provide message-passing protocols which tag the source and destination Vips of each message, and buffer messages which have not yet been requested. To ensure that all Vips assigned to a PE are executed, the support routines must also contain a mechanism for context-switching between them. In the example illustrated by Figure 12b, Vip codes 1 and 2 are concatenated, paired with support routines and mapped to PEi (Figure 12c). After preprocessing, the code associated with PEi may be run on the Poker system. The code associated with Vip 1 begins executing first. After the first context switch, the code associated with Vip 2 will begin executing. During Poker execution, PEi will run Vip 1's (2's) code until it encounters a read or write instruction. Execution of Vip 1 (2) is interrupted, its state is saved, and all I/O which may be processed by PEi at this time is performed. The context switcher then resumes execution of Vip 2 (1) at the saved state. This approach is straightforward and seems to be robust. The following paragraph describes the support routines in more detail.

The multiplexer is a series of 5 routines called *premux, s2a51, As51, postmux* and *HxToO*. These routines perform the multiplexing as follows: Before multiplexing, files have been generated which describe the assignment of Vips to PEs, the process codes associated with each Vip, the routing of the contracted graph on the CHiP lattice, etc. *Premux* creates for each PE a file of its assigned compiled Vip process codes. The I/O calls of this intermediate code are intercepted, and identified as intra-PE or inter-PE reads and writes. *s2a51* takes the modified intermediate code created by premux and generates an 8051 assembly code file for each PE (the Intel 8051 processor

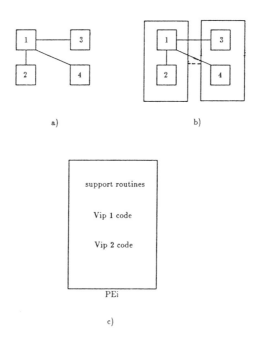

Figure 12
a) Unmultiplexed communication graph.
b) Multiplexed communication graph.
c) Multiplexed PE code.

is the PE architecture of the Pringle, simulated by Poker). These assembly code files are then run through the Intel 8051 assembler *As51* which produces hexadecimal PE files. The next routine, called *postmux,* links the hexadecimal files for each PE with the multiplexing support routines. These support routines include the message-passing protocols as well as the context-switcher which enables an individual PE to execute several different Vip codes in succession. Finally, each PE has a code file to execute. The program *HxToO* then converts the hexadecimal files to executable object code. Poker may then be called to execute this object code, simulating the execution of the original algorithm.

Upon completion of the multiplexing module, the Prep-P script terminates. An organizational chart of the routines which make up the Prep-P system is given in Figure 13.

Uncontracted Graphs

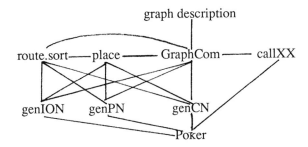

Contracted Graphs

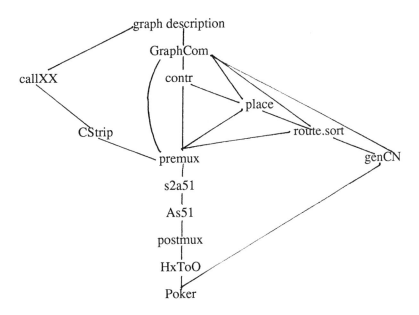

Figure 13
Prep-P organizational chart.

Special Poker Features: Phases and External I/O

The Poker system has some special facilities which provides its users with a flexible and natural parallel programming environment. We have been modifying the Prep-P system to process algorithms which incorporate these features. In particular, we are currently extending the Prep-P system to handle algorithms which use the Poker facilities of phases and external I/O.

Phases

The Poker system allows an algorithm to be decomposed into a linear sequence of subroutines or *phases*. Each phase has its own PE interconnection structure and associated code. For example, if an algorithm begins by broadcasting data from a single data stream to its PEs, and then pipelines the results of PE computations, the algorithm may be decomposed into a broadcast phase and a pipelining phase. The user may choose a complete binary tree as the interconnection structure for the broadcast phase and a linear or two-dimensional array for the interconnection structure of the pipelining phase. The code and PE interconnection structure for each of these phases are treated separately by Poker although values of user-declared variables are shared between phases. The sharing of variables between phases is achieved for each Vip through the use of a "common data area" which serves as the only internal means of communication and data transfer between distinct phases.

Prep-P treats the contraction, placement, routing and multiplexing of each phase as a separate activity. The phases extension of the Prep-P script must orchestrate the progress of each phase through the system modules and must ensure that the common data areas are maintained between phases. The first task, the processing of each phase through the Prep-P system modules, is conducted analogously to the processing of a single phase program. (The current software does perform some optimization, however, for graphs which need not be contracted). The second task, the maintainance of the common data areas between phases, is accomplished as follows: Prior to the execution of any phase, the contraction, placement and routing routines are conducted for the communication graphs specified for each phase. The first phase is processed in the conventional way by the multiplexing module. Poker is then called to execute phase 1. When phase i terminates, the Poker output is saved in a file which includes the instantiated data areas for each Vip in the original communication graph. The Prep-P phases script extracts the common data area for each Vip and saves it. The common data areas saved from phase i are inserted in the appropriate PE codes along with the support routines and compiled process codes for the Vips associated with phase i+1. Prep-P then transfers control to the Poker system for execution of phase i+1.

External I/O

The designs of Poker and Pringle are modeled after the "silicon" design of the CHiP architecture. In the spirit of this model, I/O pads have been added to the CHiP lattice to direct data on- and off-chip. These I/O pads are represented in Figure 14 by solid black boxes.

External I/O in Poker is manifested in the form of streams. Each I/O pad is responsible for one data stream which is either read from or written to exclusively by a single PE. An I/O pad resides at the periphery of the CHiP and may be routed to a PE anywhere in the lattice but not, of course, to another I/O pad.

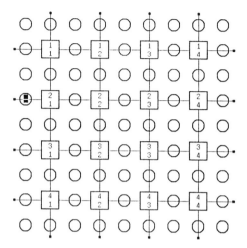

Figure 14
CHiP architecture with external I/O pads.

The multiplexing of a contracted graph with external I/O creates several interesting and challenging problems. The first problem is to determine whether I/O pads may be contracted. Consider the communication graph in Figure 15a. If we do not allow I/O pads to be contracted, and we contract Vips 1 and 2 in PEi as shown in Figure 15b, then the contracted graph may have some nodes with degree higher than 8, the maximum degree in the CHiP lattice. This approach generates some contracted graphs which cannot be placed on the CHiP lattice; however no special handling of external I/O is required since the data streams associated with I/O pads adjacent to PEs in the contracted graph are themselves not contracted. If we do allow I/O pads to be contracted, one possible contraction of Vips 1 and 2 and their I/O pads is given in Figure 15c. This approach requires special handling for the multiplexed I/O streams. In particular, Vip 1 may request data many times before Vip 2 requests data, and data items from the stream associated with the I/O pad adjacent to PEi must be tagged with the individual stream name and/or buffered appropriately.

The choice of a multiplexing strategy for the I/O pads and the extension of the contraction algorithm to include external communication are interdependent, and will follow from further analysis of the contraction model and empirical studies of Poker communication delays. Once we have chosen a contraction and multiplexing strategy, we will also extend the placement and routing algorithms to include external I/O. This will involve load-balancing between external, intra-processor and inter-processor communication, and identifying heuristics which promote good placements and routings. Finally, we will extend the support routines to include protocols which manage off-chip communication. We expect that this will involve some modification to the Poker system as well.

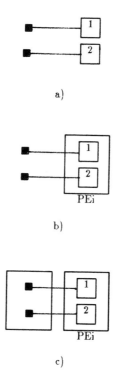

Figure 15
a) Uncontracted communication graph with I/O pads.
b) Contracted communication graph with uncontracted I/O pads.
c) Contracted communication graph with contracted I/O pads.

Section 3: Prep-P -- Work in Progress

At present, we are continuing to design and develop the Prep-P system. Our goals are threefold: 1) To complete, test and distribute the Prep-P software to interested users in the scientific community, 2) To port appropriate Prep-P protocols to other message-passing architectures, and 3) To continue investigating models and methodologies for the mapping problem using the insights gained in designing and developing Prep-P.

Currently, we are progressing towards the completion of the first goal: the general distribution of the Prep-P system. Most of the software has been designed and written, and we are optimizing and extending the current system to include the new contraction heuristics, external I/O and phases. The new contraction algorithm is still in the design stage. The phases extension has been designed ([11]), mostly coded, and is currently being completed and tested. The largest of the development tasks which remain is the Prep-P external I/O extension. This involves changes to almost every existing Prep-P routine (including the user interface and graph description language, contraction, placement, routing and the support routines). At this writing, the I/O extension has been designed ([27],[28]) and is in the process of being implemented.

Our second goal is the porting of Prep-P protocols to other architectures. We are planning to begin this process by porting Prep-P to the hypercube ([29]). This seems to be a reasonable place to start since there already is an implementation of the Poker system on the hypercube ([32]). For this architecture, the contraction, placement and routing heuristics will require a reorientation towards the hypercube rather than CHiP interconnection structure. Additionally, the Prep-P modification of some aspects of Poker (particularly those routines which deal with external I/O) may conflict with the Poker implementation on the hypercube. The conversion therefore will not be completely straightforward. We expect that the experience gained from porting Prep-P to the hypercube will help distinguish the CHiP-dependent aspects of the system from the CHiP-independent aspects, and provide insight on the problems of porting Prep-P algorithms and protocols to other message-passing architectures.

The third goal, the investigation of models and methodologies for the mapping problem, is the more general focus of our continuing research. Although the Prep-P software solves a specific instance of the mapping problem for a specific architecture, the algorithms and protocols were developed with the general mapping problem in mind. At present, we are continuing the development of models and methodologies for the mapping problem which include explicit information on communication and computation loads, bandwidth, memory constraints, etc. Using these models, we hope to provide a framework for the comparison of mapping strategies with one another and with the "optimal", the determination of deadlock, and the analysis of performance/overhead trade-offs.

The experience of designing and developing Prep-P has provided a focus for a theoretical modeling of the mapping problem, and the theoretical research in turn has provided a framework for the software development. It is the interplay between the theoretical work and the software development that has been a rich source of insights and questions, and the most challenging and intriguing aspect of our work on this project.

Acknowledgements

Much of the work on Prep-P has been funded by the Office of Naval Research under Contract No. N00014-86-K-0218. We would like to thank Charles Holland and Philip Hwang at ONR for their interest and support of this project. We are also grateful to the following people for their helpful discussions, encouragement, and support: Mark Anderson, Robert Grafton, Greg Hidley, David Mizell, Steve Rose, and Lawrence Snyder.

REFERENCES

[1] Aho, A., Hopcroft, J. and J. Ullman, *The Design and Analysis of Computer Algorithms,* Addison Wesley, 1974.

[2] Aho, A., Hopcroft, J. and J. Ullman, *Data Structures and Algorithms,* Addison Wesley, 1983.

[3] Aragon, C., Johnson, D., MeGeoch, L., and C. Schevon, *"Optimization by Simulated Annealing: An Experimental Evaluation,"* manuscript in preparation, 1984 draft.

[4] Berman, F., "Edge Grammars and Parallel Computation," *Proceedings of the 21st Annual Allerton Conference on Computing, Control and Communication,* 1983.

[5] Berman, F., Goodrich, M., Koelbel, C., Robison III, W. and K. Showell, "Prep-P: A Mapping Preprocessor for CHiP Architectures," *Proceedings of the 1985 International Conference on Parallel Processing.*

[6] Berman, F. and P. Haden, "A Comparative Study of Mapping Algorithms for an Automated Parallel Programming Environment," Technical Report CS-088, Department of Computer Science, University of California, San Diego.

[7] Berman, F. and L. Snyder, "On Mapping Parallel Algorithms into Parallel Architectures," *Proceedings of the 1984 International Conference on Parallel Processing.*

[8] Berman, F. and L. Snyder, "On Mapping Parallel Algorithms into Parallel Architectures," submitted to the Journal of Parallel and Distributed Computing.

[9] Bokhari, S., "On the Mapping Problem," *IEEE Transactions on Computers,* March, 1981.

[10] Chazelle, B. and L. Monier, "A Model of Computation for VLSI with Related Complexity Results," *Proceedings of the 13th Symposium on the Theory of Computing,* 1981.

[11] Conroy, J., "Phases," Prep-P internal documentation.

[12] DeGroot, D., "Expanding and Contracting SW-Banyan Networks," *Proceedings of the 1983 International Conference on Parallel Processing.*

[13] DeGroot, D., "Partitioning Job Structures for SW-Banyan Networks," *Proceedings of the 1983 International Conference on Parallel Processing.*

[14] Fishburn, J. and R. Finkel, "Quotient Networks," *IEEE Transactions on Computers,* April, 1982.

[15] Heller, D., "Partitioning Big Matrices for Small Systolic Arrays," Technical Report CS-83-02, Department of Computer Science, Pennsylvania State University, 1983.

[16] Jamieson, L., "Dimensions for Describing Parallel Algorithms," position paper, *NSF-CMU Workshop on Performance-Efficient Parallel Computing,* September, 1986.

[17] Kapuan, A., Wang, K.Y., Gannon, D., Cuny, J. and L. Snyder, "The Pringle: An Experimental System for Parallel Algorithm and Software Testing," *Proceedings of the 1984 International Conference on Parallel Processing.*

[18] Kernighan, B. and S. Lin, "An Efficient Heuristic Procedure for Partitioning Graphs," *Bell System Technical Journal,* 49 (2), February, 1970.

[19] Kirkpatrick, S., Gelatt Jr., C. and M. Vecchi, "Optimization by Simulated Annealing," *Science,* May, 1983.

[20] Kuhn, R., "Efficient Mapping of Algorithms to Single Stage Interconnections," *Proceedings of the 7th Annual Symposium on Computer Architecture,* 1979.

[21] Kung, H.T. and D. Stevenson, "A Software Technique for Reducing the Routing Time on a Parallel Computer with a Fixed Interconnection Network," *High Speed Computer and Algorithm Organization,* D. Kuck et al., editors, Academic Press, 1977.

[22] Leiserson, C., "Area Efficient Graph Layouts (for VLSI)," *Proceedings of the 21st Annual Symposium on the Foundations of Computer Science,* 1980.

[23] Mitra, D., Romeo, F. and A. Sangiovanni-Vincentelli, *"Convergence and Finite-Time Behavior of Simulated Annealing,"* Technical Report UCB/ERL M85/23, Electronics Research Laboratory, University of California, Berkeley.

[24] Moldovan, D., "On the Design of Algorithms for VLSI Systolic Arrays," *Proceedings of the IEEE,* Vol. 71 (1), January, 1983.

[25] Preparata, F., "VLSI Algorithms and Architectures," *Proceedings of the Conference on Mathematical Foundations of Computer Science,* 1984.

[26] Ramakrishnan,I., Fussell, D. and A. Silberschatz, "Systolic Matrix Multiplication on a Linear Array," *Proceedings of the 20th Annual Allerton Conference on Computing, Control and Communication,* 1982.

[27] Rose, D., "Implementing External I/O in the Prep-P System (Stage I: Uncontracted Graphs)," Prep-P internal documentation.

[28] Rose, D., "Implementing External I/O in the Prep-P System (Stage II: Contracted Graphs)," Prep-P internal documentation.

[29] Seitz, C., "The Cosmic Cube," *Communications of the ACM,* January, 1985.

[30] Snyder, L., "Introduction to the Configurable, Highly Parallel Computer," *Computer,* January, 1982.

[31] Snyder, L., "Introduction to the Poker Parallel Programming Environment," *Proceedings of the 1983 International Conference on Parallel Processing.*

[32] Snyder, L. and D. Socha, "Poker on the Cosmic Cube: The First Retargetable Parallel Programming Language and Environment," *Proceedings of the 1986 International Conference on Parallel Processing.*

[33] Thompson, C., "A Complexity Theory for VLSI," Ph.D. Dissertation, Department of Computer Science, Carnegie-Mellon University, 1980.

[34] Valiant, L., "Universality Considerations in VLSI Circuits," *IEEE Transactions on Computers,* February, 1981.

[35] White, S., "Concepts of Scale in Simulated Annealing," *Proceedings of the IEEE International Conference on Computer Design: VLSI in Computers,* 1984.

Design and Implementation of Parallel Programs with Large-Grain Data Flow

Robert G. Babb II and David C. DiNucci

Dept. of Computer Science and Engineering
Oregon Graduate Center
19600 NW Von Neumann Drive
Beaverton, OR 97006

Large-Grain Data Flow (LGDF) is a model of computation that combines sequential programming with dataflow-like program activation. LGDF applications are constructed using networks of program modules connected by datapaths. Parallel execution is controlled indirectly via the production and consumption of data. With the aid of the LGDF Toolset, LGDF programs are automatically transformed for efficient implementation on a particular (parallel) machine.

This paper provides an overview of the current state of development of the Large-Grain Data Flow computation model, and summarizes experience gained during the course of using LGDF to implement parallel versions of several small and medium-sized scientific application codes.

1. Introduction

Implementing parallel versions of scientific programs has in general proved much more difficult than most computer scientists expected[1]. Parallel programming errors can be nearly impossible to diagnose because they are often non-repeatable. Debugging can be difficult since tracing execution sequences usually affects the behavior of the parallel program being debugged by making the error disappear! These difficulties seem especially frustrating in the case of scientific programs, since even large scientific programs typically use simple data structures (e.g. matrices) and implement algorithms that are readily parallelized.

A possible explanation for this situation may be that the complexity of parallel processing is fundamentally different from that of sequential processing. In sequential processing, the global state of a computation at any point in time is completely defined by the values of all variables in memory, plus the current value of the program counter. All subsequent computation states can be predicted by analysis of the interaction between program instructions and data values:

$$State_1 + PC_1 \rightarrow State_2 + PC_2$$
$$State_2 + PC_2 \rightarrow State_3 + PC_3$$
$$\vdots$$

Note that the sequence of states is uniquely determined whether or not the program is correct.

Now consider the case of a shared memory parallel computer with N processors. Assume for simplicity that the parallel execution is serialized such that only one processor is allowed to execute an instruction at a time. At each execution step, we have N possible next states, depending on which processor's instruction is chosen next for execution:

$$State_{1,2,\ldots,N} + PC_{1,2,\ldots,N} \rightarrow \begin{bmatrix} State_{1',2,\ldots,N} + PC_{1',2,\ldots,N} \\ State_{1,2',\ldots,N} + PC_{1,2',\ldots,N} \\ \vdots \\ State_{1,2,\ldots,N'} + PC_{1,2,\ldots,N'} \end{bmatrix}$$

Since this pattern is repeated for each instruction executed, we have a combinatoric explosion of possible execution histories. In general, some of the possible "next states" at each step will represent parallel programming bugs (unintended state changes due to synchronization errors).

One way to improve the odds of correct execution for parallel programs is to restrict the number of possible parallel execution patterns. This can lead to improved program reliability, generally at the expense of a decrease in potential parallelism. The difficulty in achieving reliable, efficient, maintainable, and portable parallel programs argues strongly for the use of such higher level models of parallelism. In particular, it would seem to be a good idea to restrict access by programmers to basic machine mechanisms for controlling parallelism. Programmers would use the parallel constructs of the higher level model to design applications, and rely on automatic means, such as a preprocessor, to generate a concrete implementation tailored for a particular parallel machine[1].

Our experience has shown that programming within the restrictions of a suitable higher level parallel model can make parallel program design more reliable and in many cases even easier than design using a sequential computation model. (For an example of such a program, see [2]).

---

[1]These arguments are similar to those originally used to encourage the use of FORTRAN in preference to assembly languages in the late 50's.

Large-Grain Data Flow is implemented as a *layered language*. A variety of layered languages have been used to implement higher level models of parallel programming on top of FORTRAN and other languages [3] [4] [5] [6]. To use these programming environments, programmers code parallel operations using macros. The macros are expanded to produce low level synchronizing operations suitable for a particular target parallel machine. The macro preprocessing step tends to lead to fewer bookkeeping errors in the coding of process synchronization, as well as providing for portability of code between different parallel processors. Ideally one need only modify the macro definitions to get layered language programs running on a new target machine.

2. Overview of LGDF

The overall goals of the LGDF computation model are to provide a restricted model of parallelism that is:

1) implementable efficiently on a variety of parallel machines

2) easily ported from sequential to parallel machines, and between parallel machines

3) simple to understand

4) visual, allowing use of graphic design and monitoring tools

5) a useful software engineering tool for implementing large scale parallel programs

For more details on the LGDF software tools, and a complete example of the use of LGDF for parallel processing see [7].

2.1. LGDF Networks

A Large-Grain Data Flow *network* consists of a set of *nodes* (represented graphically by circles or "bubbles") connected by *datapaths*. Datapaths are represented graphically by various kinds of arrows. A datapath that enters a bubble is termed an *input datapath* while one that leaves a bubble is termed an *output datapath*. Datapaths can also be *dot-shared* so that one datapath can be an input to or an output from more than one node. Large-Grain Data Flow networks must be acyclic.

A bubble can represent either an LGDF program or an LGDF network. The meaning of a network containing a subnetwork node is the same as if the lower level network were substituted graphically for the higher level node. Any set of hierarchically related LGDF networks is therefore semantically equivalent to a single network containing only LGDF programs and datapaths.

A datapath represents an interface mechanism for communicating data values from one LGDF program to another. Datapaths can assume two states: *empty* and *full*. The data associated with a datapath can be thought of as residing in a single fixed-length data structure, access to which is controlled by the empty/full property. The empty/full state is an attribute that is manipulated explicitly by the programs that the datapath is connected to. This state signifies whether there is valid data to be read on an input datapath and/or whether an output datapath is available for writing new data. In general, it is illegal for an LGDF process to write values on an input datapath, or to read values that it did not write from an output datapath.

2.2. LGDF Process Activation

An LGDF program's eligibility for activation depends only upon the state of its associated input and output datapaths, according to following rule:

> **Execution Rule**—A program may start an execution cycle if and only if all of its input datapaths are in the *full* state and all of its output datapaths are in the *empty* state.

Once a process wakes up, it may read the data associated with any of its input datapaths and modify the data associated with any of its output datapaths. When a program has finished accessing an input datapath, it can change the state of the path to empty by using the LGDF *clear_* command. (In LGDF programs, the command refers to a specific datapath by its unique tag, e.g. *clear_(d06)*). The usual reason an LGDF program clears an input datapath is to allow an upstream process to wake up and write the next set of data values onto the datapath. Likewise, when a program has finished writing values onto an output datapath, it can change the state of the path to full by using the LGDF *set_* command.

After a process has changed the state of a datapath, it cannot access the data on that datapath for the remainder of the current execution cycle. An LGDF process, then, is limited in the amount of work it can do in a single execution cycle. When it is finished with that work, it puts itself to sleep using the *suspend_* command, and it will remain suspended until its datapaths again satisfy the LGDF Execution Rule. At the time a process suspends, it checked for conformance with the

> **Data Flow Progress Rule**—Upon suspension of an execution cycle, a program must have cleared at least one input or set at least one output datapath. Otherwise, it is terminated.

A process that is in the *terminated* state can never re-awaken, regardless of the empty/full state of its datapaths.

2.3. Update in Place

The LGDF model extends the idea of a simple input datapath to include capabilities for updating values on the datapath that will persist across program activations. The model also includes facilities for various kinds of shared datapaths. A shared datapath is one that is both an input and an output from the same LGDF program, or one that is input to or output from more than one LGDF program. The latter type is referred to as *dot-shared*, in reference to the dot that is placed wherever the datapath arrow splits or merges.

If a datapath functions as both input and output for the same process, the datapath state cannot be the same for both since it would be impossible for the process to wake up. Therefore, the "two" datapaths have independent empty/full states even though there exists only one copy of the associated data values. The output datapath tag is typically given a different suffix from the input datapath (e.g. d01 and d01a).

The ability to update a datapath is explicitly shown on LGDF network diagrams by a dashed line (update in place allowed) or solid line (no update permission, read only) passing into the bubble itself as shown in Figure 1. A datapath that functions as both an input and output is diagramed by continuing the input datapath through the bubble (with a solid or dashed line depending on the updatability of the datapath) directly to the output datapath. Some combinations of input, update, and output are obvious, in which case the update or lack thereof is not drawn in the diagram, as shown in Figure 2.

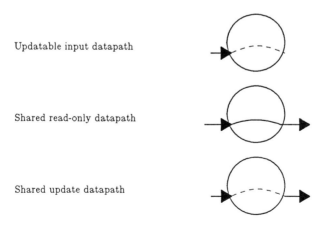

Updatable input datapath

Shared read-only datapath

Shared update datapath

Fig. 1. Update permissions on datapaths.

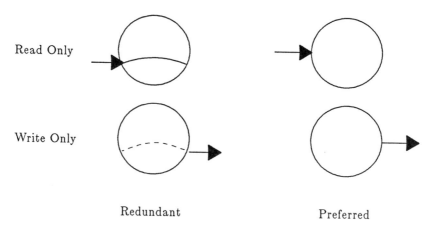

Read Only

Write Only

Redundant   Preferred

Fig. 2. LGDF update permission graphic conventions.

2.4. Clearable, Nonclearable, and Side-effect Datapaths

LGDF allows specification of several different permissions for clearing datapaths, as shown in Figure 3. These are shown by the type of arrowhead entering the process bubble. An arrowhead shown as an empty triangle indicates that the associated datapath cannot be explicitly cleared by that process, while a filled triangle arrowhead means that it can be cleared. If a non-clearable datapath is continued through a bubble as a shared datapath, any clear of the output datapath by a downstream process will be propagated backward through the bubble to the input datapath. A third type of datapath, termed a side-effect, is shown as an open V shape and is used to show non-dataflow inputs and outputs (e.g. a hardware device which must be queried for data such as a clock). This type of datapath does not have an empty/full condition associated with it, and therefore can not affect the scheduling of LGDF processes.

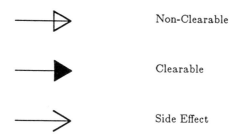

Non-Clearable

Clearable

Side Effect

Fig. 3. Graphics for various types of datapaths.

2.5. Competition for Exclusive Access to Shared Datapaths

When a dot-shared datapath splits to become an input to several different processes, any one of them can start executing after the datapath becomes full. Without a further control mechanism, it would be possible for one of the reading processes to access the associated data values and clear the path before a second process starts to read the data. In such a case, a writing process (on the other end of the datapath) could start running while the second read process was running, causing a read/write conflict. To guard against this possibility, a reader must be able to obtain exclusive read access permission to a datapath before reading values from it or clearing it. This is accomplished in LGDF by executing the special command *aread_*, which allows processes to compete for exclusive access to dot-shared datapaths. Exclusive access to the path is relinquished when the process clears the path, using the *aclear_* command, or it can be relinquished explicitly (without clearing the path) by executing an *unread_* command.

The same kinds of conflicts can occur when several processes write to the same output datapath. The analogous commands *awrite_*, *aset_* and *unwrite_* are available for resolving such conflicts.

2.6. Rules and Semantics

This section will define the semantics of the LGDF computation model formally. It can be skipped by readers who do not need an in-depth understanding of the definition of LGDF mechanisms.

Each datapath has two binary semaphores associated with it which will be labeled R and W. (We are using Dijkstra's formulation for semaphores [8]). Initially, all W semaphores are set to their "after V" state (i.e. non-zero) and all R semaphores are set to their "after P" state (i.e. zero). The LGDF command primitives can be defined by the following operations on W and R:

LGDF Command	Semaphore Definition
*awrite_*	P(W)
*aset_*	V(R)
*unwrite_*	V(W)
*aread_*	P(R)
*aclear_*	V(W)
*unread_*	V(R)

The *clear_* operation is defined as an *aread_* followed by an *aclear_*, and *set_* is defined as an *awrite_* followed by an *aset_*.

In terms of semaphore operations, a more precise definition of the LGDF Execution Rule is:

A process is eligible for execution if and only if the sequence P then V can be performed on all W semaphores associated with its output datapaths and all R semaphores associated with its input paths. The semaphores can be checked in any order.

The possible states and transitions for an LGDF process are shown in Figure 4. An LGDF process that is executing can transfer to one of the four next states shown. The transfer to the "latched" state occurs after a process sets a sequentially shared datapath whose corresponding input arrow is open (non-clearable). This effect is termed "autolatching". The process blocks waiting for the shared output to clear, at which time it propagates the clear signal backwards and transfers to the suspended state. An LGDF process is allowed at most one of these shared autolatch datapaths.

3. Design of Parallel Programs with LGDF

This section discusses the effect of various features of the basic LGDF model on parallel program design strategies.

3.1. Execution Rule

The LGDF execution rule would be classified as a *strict firing rule* in traditional (fine-grain) dataflow terminology. An LGDF program wakes up at the beginning of each execution cycle with a completely protected data environment. If a program is running, it is guaranteed safe read access to its input datapaths, and safe write access to its output datapaths. The only

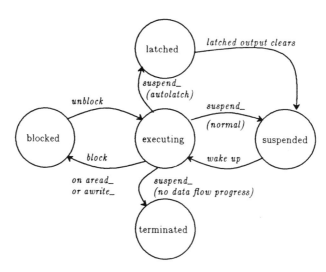

Fig. 4. State transition diagram for an LGDF process.

possible exceptions to this can occur if a program:

1)   accesses values on an input datapath after executing a *clear_* on the datapath, or

2)   writes values on an output datapath after executing a *set_* on the datapath, or

3)   fails to observe the proper protocol for accessing dot-shared datapaths, i.e.:

   (*aread_*, access, *aclear_*) or (*aread_*, access, *unread_*) for input datapaths

   (*awrite_*, access, *aset_*) or (*awrite_*, access, *unwrite_*) for output datapaths

The LGDF execution rule allows construction of safe producer/consumer pipelines. LGDF program pipelines in their simplest form resemble Unix[2] *pipes*[9]. (See Figure 5a). Of course, the possibility of specifying more than one input and output datapath allows construction of more complicated (non-linear) pipeline structures than standard Unix systems allow, such as shown in Fig. 5b. Program pipelines, in which different phases of data can be processed simultaneously, are a common way of specifying potential parallelism in LGDF network designs.

Instead of calling subroutines, LGDF programmers must arrange to send input data to them. Of course for the "call" to be effective, previous subroutine output data values must have been consumed by other LGDF programs.

LGDF programs can make use of local variables during an execution cycle, but are not allowed to save internal state between execution cycles. Thus the simplest type of LGDF program resembles the functional style of programming. The same sequences of input data values will always yield the same sequences of output data values. Of course, with update-in-place datapaths, history sensitive computations can also be easily implemented.

3.2. Data Flow Progress Rule

The requirement that LGDF programs must make data flow progress on every execution cycle forces networks to be defined in such a way that "useless" program activations are avoided. Since every program activation must do something to affect the overall "data flow state" of the system, the common situation of mysteriously "hung" parallel program executions can be prevented. The result is analogous to prevention of ineffective infinite looping in sequential programs.

---

[2]Unix is a trademark of Bell Laboratories

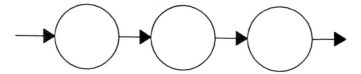

(a) Simple LGDF program pipeline.

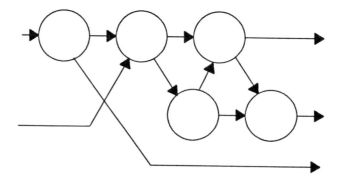

(b) Non-linear LGDF program pipeline network.

Fig. 5. LGDF program pipelines.

The data flow progress requirement has the effect of forcing "busy wait" cycles up to the level where they are visible to the LGDF scheduling mechanism. As a simple example, consider a program whose job it is to check for the occurrence of an asynchronous event, such as a keystroke. Suppose also that the keystroke event must be modelled as a side-effect (non data flow) input datapath because the only way to tell if a keystroke has occurred is to poll the keyboard. If the keyboard monitor program uses an internal loop to poll the keyboard, it may never make data flow progress by producing the character. (See Figure 6a). In a parallel implementation, the processor executing such a program could be tied up indefinitely polling for the event.

In order to make data flow progress whether or not a keystroke has occurred, we can split the program into two parts connected by a datapath that encodes either the latest keystroke or a "null" value indicating the absence of a new keystroke, as shown in Figure 6b. The second design allows the LGDF scheduler to re-assign the processor executing the poll keyboard program to another LGDF program between polling cycles. This kind of design is particularly crucial if the LGDF scheduler is emulating parallel execution on a uniprocessor, since processing in other parts of the network can be interleaved with the keystroke monitoring process.

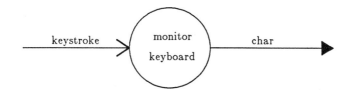

(a) Keyboard monitor as a single LGDF program.

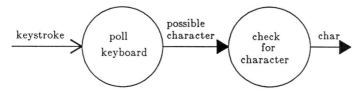

(b) Keyboard monitor as two LGDF programs.

Fig. 6. Two designs for an asynchronous event monitor.

3.3. Control and Data Feedback

Since LGDF networks are restricted to being acyclic, control feedback can not be constructed in the obvious way, by routing an output datapath back around as an input to the same network. The effect of control feedback can be constructed, however, by blocking a producer on a full output. A consumer process can indirectly activate (and thus give control to) the producer process by choosing when to clear the intervening "interlock" datapath. An example of control feedback is shown in Figure 7. Suppose it is desired to prevent the LGDF program p10 from running again until after programs p11 and p12 have finished executing (to prevent unsafe overlapped updating of a shared array on datapath d01, for example). After p10 has started an update cycle by setting d02, it can block itself from executing by setting the "interlock" datapath d05. After it is safe for p10 to run again, p12 can set the "done" datapath d04. Program p13 can wake up and clear both of its inputs (d04 and d05) allowing p10 to run again, effectively feeding back control. The scope of control feedback can be extended indefinitely by constructing chains of such interlocks.

Data feedback can be implemented in two ways within the LGDF model. One way is shown in Figure 7. The first time program p10 executes, data values on datapath d01 will be determined by the program (not shown) that initialized it "upstream". Each successive time p10 wakes up, it will have available the latest values on d01 (as updated by p11 and p12). The results of processing by "downstream" programs p11 and p12 are effectively fed back to p10. Note that the solid line through p10 for the shared

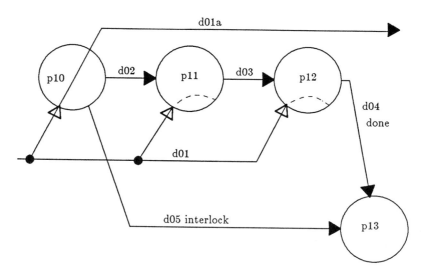

Fig. 7. An example of control feedback.

datapath d01-d01a indicates that p10 can choose when to make the updated values available to the outside world via datapath d01a, but it is not allowed to change the values itself.

Another way to achieve the effect of data feedback that avoids the use of dot-shared data paths is shown in Figure 8. The open arrowheads on datapaths d10a and d10b indicate that programs p21 and p22 are not allowed to explicitly clear those inputs. Program p23, on the other hand, can choose either to clear (datapath d10c) or set (datapath d10d). If it clears the datapath, the datapath automatically clears back to the first previous solid arrowhead on the chain. The programs p21 and p22 do not wake up except

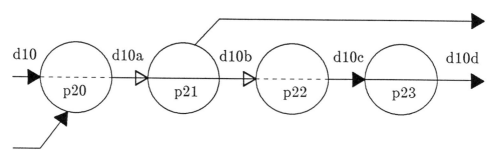

Fig. 8. Data feedback via a sequentially shared datapath.

to propagate the clear backwards. In this example, a clear by p23 would have the effect of feeding the updated values on d10 back to program p20, which could then feed them forward again to programs p21 and p22. Note that p21 has read-only access to datapath d10a-d10b, while p22 is allowed to update datapath d10b-d10c.

While both forms of data feedback are useful in designing LGDF networks, feedback involving a dot-shared datapath such as that shown in Fig. 7, generally has greater potential for parallelism at the risk of possibly hazardous simultaneous updates.

4. Implementation Considerations

Several steps are involved in using LGDF for programming parallel processors. First, a hierarchical set of LGDF network data flow graphs is designed based on logical data dependencies inherent in the application and as required by the parallelization strategy adopted. The structure of the resulting network is encoded in a *wirelist* file. Data declarations for each datapath in the network are packaged separately. LGDF programs, written in a combination of a base language such as FORTRAN and LGDF macro calls, are preprocessed to produce compilable code for a particular (parallel) machine. The generated code includes "scheduler" code and mechanisms to efficiently simulate LGDF on the particular target machine.

An important goal in designing LGDF schedulers is to minimize scheduling overhead. To achieve this goal, it is necessary to minimize the time when one processor must lock out other processors from accessing shared scheduler tables and to minimize the amount of resources (CPU time, bus cycles, and/or memory access cycles) used by processes waiting to run. On some parallel architectures, efficient mechanisms must also be designed for the transfer of data among LGDF processes to simulate shared datapaths. For a further discussion of these issues, see [10]. Other design considerations include providing the ability to:

- schedule arbitrarily large LGDF networks
- schedule processes efficiently with any number of processors
- simulate LGDF parallelism on a single processor.

The last capability allows debugging of applications in a more controlled (less parallel) setting, and provides greater flexibility in assigning processes to processors.

Two basic strategies have been used to implement the basic LGDF process activation mechanism:

1) self-scheduling, and
2) queued-scheduling

Self-scheduling is attractive on parallel architectures in which:

- a large number of semaphores are available
- a process that blocks on a semaphore is automatically (and cheaply) suspended by hardware or software mechanisms
- later re-activation after the semaphore has changed value is efficient

For such architectures, we can generate code for *suspend_* that checks for data flow progress, then branches back to a barrier at the beginning of the program that implements the LGDF Execution Rule. The overall effect is of multiple distributed (almost free) busy-waits, one for each LGDF program.

Queued-scheduling is appropriate for architectures where a busy-wait on a semaphore is an expensive (and possibly non-terminating) operation. In this case, a more centralized scheduler can be designed that uses semaphores only to interlock parallel scheduling actions (such as simultaneous access to datapath empty/full state tables). The scheduler code can maintain queues of LGDF programs that are eligible to execute but are not currently executing. The parallel processors compete to advance the data flow state of the computation by executing LGDF programs. Bottlenecks are possible in gaining access to the scheduler tables, but this effect can be counteracted somewhat by distributing the work queues and/or by subsetting the protection for the LGDF state tables. For a description of such a scheduler for the CRAY X-MP, see [11].

Acknowledgements

This work was supported in part by Los Alamos National Laboratory Computer Research and Applications Group (contract 9-Z-34-P-3915-1), Lawrence Livermore National Laboratory Computation and Information Services Directorate, the National Science Foundation Office of Advanced Scientific Computing (grant no. ASC-8518527), and Cray Research, Inc.

References

[1]   J. R. McGraw and T. S. Axelrod, "Exploiting multiprocessors: Issues and options", Lawrence Livermore National Laboratory Preprint UCRL-91734, Oct. 31, 1984.

[2]   R. G. Babb II, "Data-driven implementation of data flow diagrams," in *Proc. 6th Int. Conf. on Software Engineering,* Tokyo, Japan, Sept. 1982, pp. 309–318.

[3] H. F. Jordan, "Structuring parallel algorithms in an MIMD, shared memory environment," in *Proc. 18th Hawaii Int. Conf. on System Sciences,* Jan. 1985, vol. 2: *Software,* pp. 30—38.

[4] E. L. Lusk and R. A. Overbeek, "An Approach to Programming Multiprocessing Algorithms on the Denelcor HEP," Argonne National Laboratory, Mathematics and Computer Science Division, Technical Report No. ANL—83—96, Dec. 1983.

[5] E. L. Lusk and R. A. Overbeek, "Implementation of Monitors with Macros: A Programming Aid for the HEP and other Parallel Processors," Argonne National Laboratory, Mathematics and Computer Science Division, Technical Report No. ANL—83—97, Dec. 1983.

[6] W. F. Appelbe and C. E. McDowell, "Anomaly detection in parallel Fortran programs", in *Proc. of the Workshop on Parallel Processing using the Heterogeneous Element Processor,* Norman, OK, March 1985, pp. 389—399.

[7] R. G. Babb II, "Programming the HEP with Large-Grain Data Flow techniques," in *MIMD Computation: HEP Supercomputer and Its Applications,* (ed. by J. S. Kowalik). Cambridge, MA: The MIT Press, 1985.

[8] E. W. Dijkstra, "Cooperating sequential processes," Technological University Eindhoven, The Netherlands, 1965. (Reprinted in *Programming Languages,* ed. by F. Genuys. New York: Academic Press, 1968)

[9] B. W. Kernighan and R. Pike, *The UNIX Programming Environment.* Englewood Cliffs, NJ: Prentice-Hall, 1984.

[10] R. G. Babb II, L. Storc, and W. C. Ragsdale, "A Large-Grain Data Flow scheduler for parallel processing on CYBERPLUS", in *Proc. 1986 International Conference on Parallel Processing,* pp. 845-848.

[11] R. G. Babb II, "Parallel processing on the CRAY X-MP with Large-Grain Data Flow techniques", in *Supercomputers: Class VI Systems, Hardware, and Software* (ed. by S. Fernbach). Amsterdam: North-Holland, 1986, pp. 239-251.

A Minimalist Approach to Portable, Parallel Programming

Ewing L. Lusk and Ross A. Overbeek

Mathematics and Computer Science Division
Argonne National Laboratory
Argonne, IL 60439

A wide variety of approaches to expressing parallel algorithms are currently being taken. We argue here that one of the needs not being met by this diversity is the need for a portable, clear semantics for the mechanisms being implemented by hardware and software vendors. We describe monitors, a well-studied, machine-independent technique for expressing and controlling parallelism, and show how monitors can be implemented in a portable yet efficient way. Our experiences with this approach have led us to believe that the needs of programmers of parallel systems can be met by having the operating environment supply only a small number of simple primitives.

1. Introduction

In this paper we discuss the current search for mechanisms by which parallelism can be expressed and controlled, with particular attention to the issue of portability. We argue that portable concepts as well as portable programming constructs are needed, and we offer *monitors* as an approach to this problem. We demonstrate how monitors can be implemented both efficiently and portably for a variety of parallel computers, and report on some experiences with this approach. We describe what types of primitives this type of approach requires from the hardware, operating system, and/or compiler, and conclude that it is more important for vendors to provide a small set of well-implemented, low-level primitives that to supply extensive multiprocessing libraries or complex, machine-specific operations.

2. The Problem

The relative success of vectorizing compilers has led many to expect similar success in the area of automatic parallelization. The hope is that the compiler, either alone or aided by a few directives inserted into the code by the programmer, will be able to discover data dependencies and synchronization requirements in the algorithm, and generate machine instructions appropriate for utilizing multiple processing elements of the hardware and providing the appropriate protection for operations on shared data items.

It is our belief (not argued here at length) that this hope will not be realized except in a narrow range of applications. It will succeed where automatic vectorization has succeeded–in algorithms with a regular, uniform structure–and fail precisely in those areas where vectorization is obviously inappropriate but parallelization should help–in problems with an irregular and dynamically changing structure.

Thus we assume that compilers won't do it all; that some means of expressing and controlling parallelism will have to be provided to the programmer of a parallel algorithm. Our focus in this paper is the issue of how to use such mechanisms, while constructing portable implementations of algorithms.

2.1. The Current State of Affairs

A programmer with a parallel algorithm and a target machine in mind wants to understand the mapping of his algorithm onto at least the computational model offered by his target machine. What tools are currently being provided to help him do this?

2.1.1. New Languages

Certainly one approach is that offered by completely new languages that either include some notion of parallelism within the language or are at a high enough level of abstraction to allow effective exploitation of parallelism. Examples are ADA, experimental languages like SISAL, and semantic modifications of existing high-level languages, like PARLOG. These may be the wave of the future, but for the present (and near future) a large community will rely on vendor-supplied mechanisms that work in conjunction with existing conventional languages like C and FORTRAN.

2.1.2. Language Extensions

In some cases, popular languages are being extended to provide features designed to exploit parallelism. Within the Fortran community a number of constructs have been proposed are are being considered by the standards committees. Given the lack of experience with multiprocessors, it seems likely (and proper) that such committees will proceed with caution. Hence, it may well be several years before such constructs are actually included in the language.

2.1.3. Parallel Subroutine Libraries

In order to make no change to the semantics of an existing language, and thus retain current compilers, some vendors have adopted the subroutine library approach, in which synchronization operations are carried out by calling a system routine. This is probably the most common approach. It has the disadvantage of requiring the overhead of a subroutine call to get access to what may be only one or two machine instructions.

An intermediate approach is to retain current language structure but have certain subroutine calls recognized by the compiler and generate in-line code.

2.1.4. Compiler Directives

An increasingly popular technique is to provide compiler directives that allow the compiler to generate machine instructions to control parallelism from ordinary code. The directives typically have the syntax of comments, so that the program remains a syntactically correct sequential program.

In some cases, for example Alliant and Sequent, a number of approaches can be used simultaneously. The compiler will automatically parallelize when it can. Optional compiler directives supplied by the programmer can help the compiler find more parallelism than it could find by itself, and explicit calls to system routines can also be made.

2.2. The Portability Issue

There is already a wide variety of multiprocessors available. However, it does seem possible to identify broad classes of such machines and to write software that is portable between machines in the same class. For example, one might group a majority of the existing MIMD multiprocessors into one of the following three broad classes:

a.   The class of machines that support a single, shared memory.

b.   The class of machines that do not support shared memory. In this case, communication takes place via message-passing.

c.   The class of machines that support "clusters". A cluster is a set of processors that do share memory. Processors in distinct clusters must communicate via message-passing.

One can achieve portability between machines in all three classes by utilizing properly implemented message-passing primitives. However, for many applications access to shared memory offers substantial advantages in terms of programming convenience and performance.

The first major decision facing someone implementing an application on MIMD multiprocessors is to choose one of the three memory models. The next step is to determine a portable means of expressing algorithms for execution on machines that support that model.

2.2.1. Portable Programs

Many of the early conferences devoted to parallel computing were dominated by talks of the form "How I got program X to run on machine Y". These were interesting talks, representing hard work, and valuable in establishing that parallelism could indeed be found in useful algorithms and that the new machines could indeed be programmed to exploit it. But these talks were disquieting in that they seemed to be reporting isolated, disconnected experiments instead of progress on a broad front toward concepts and tools central to parallel computing.

It is quite understandable that these presentations were so machine-specific. Each parallel computer (in some cases each release of the operating system) provided a

different set of mechanisms for expressing and controlling parallelism. It was important to try these various mechanisms out and report on how well they matched a given
application, regardless of how non-portable they were.

And yet the need for portability is now greater than ever. As the lifetime of
hardware shortens and that of software lengthens, it becomes more and more likely that
one will want to move parallel programs from one environment to another. In fact, it is
almost inevitable that the production environment for any large system will be different
from its development environment. Thus it is crucial that one be able to move programs from one environment to another with only minimal changes to the source code.
We have long expected high-level languages to provide this capability for sequential
programs; we should not lower our expectations for parallel programs.

2.2.2. Portable Concepts

Even more important than portable code, however, is the portability of the *concepts* in which a programmer thinks about his algorithm. The various mechanisms
currently being provided by vendors are even more of a problem here than they are for
portable code. As an extreme example, imagine the hopelessness of trying think out a
complex algorithm in terms of the semantics of an Alliant "carry-around variable" or a
Denelcor "asynchronous variable." Both of these concepts are the bread and butter of
the "How I got program X . . ." discussion, but are too machine-specific, low-level, and
complicated to *think* with. And parallel algorithms require more thinking about than
sequential ones.

What we need are clean, well-understood semantics for our parallel programming
primitives. The literature of operating systems theory fortunately provides a number of
approaches, many of which have been extensively studied. Examples are *wait* and *signal* primitives, *P* and *V* operations, monitors, rendezvous operations, etc. For a survey,
see [1].

Some of these, like *P* and *V*, while theoretically adequate, are lacking in expres-
sive power. For algorithms in today's applications, one needs a somewhat higher-level
set of constructs. On the other hand, too high a level of content in a primitive may act
to prevent efficient implementation. (It remains to be seen, for example, whether the
rendezvous mechanism of Ada falls into this category).

The approach that seems best to fit our needs as we have defined them is the classical (early 1970's) one of *monitors[2,3]*.

3. Monitors

Monitors were originally invented in the context of operating systems. Here we
give a relatively informal presentation.

A *monitor* is an abstract data type consisting of:

1. data to be shared among concurrently executing processes
2. initialization code for the data

3. a set of operations on the data

The behavior guaranteed by the monitor is that

1. the initialization code is performed before contention for the shared data begins

2. only one of the operations may be performed at any one time

3. the code for the operations may include *delay* and *continue* operations.

A simple example illustrating a specific monitor is given by the SEND and RECEIVE operations by which two processes might communicate via a single shared buffer.

```
SEND (message):
    enter
    if FULL is true then
        delay(SENDQ)
    endif
    move message to BUFFER
    set FULL to true
    continue(RECEIVEQ)
    exit

RECEIVE (message):
    enter
    if FULL is false then
        delay(RECEIVEQ)
    endif
    move contents of BUFFER to message
    set FULL to false
    continue(SENDQ)
    exit

Initialization:
    set FULL to false
```

3.1. Implementing Monitors

The logic above expresses the meaning of the SEND and RECEIVE operations in a machine-independent way. In order to implement them in a specific environment we need only a few fundamental operations, which we expect to find *some* way to implement in any specific environment (hardware–operating system–compiler):

1. In order to specify the operations, we need to implement the mutual exclusion instructions `enter` and `exit`, and the `delay` and `continue` operations.

2. We need a way of creating multiple processes and identifying the program each process will execute.

3. We need a mechanism for identifying which program variables are shared among processes.

Most environments on current multiprocessors provide a lock mechanism of some kind, which is all that is required for the implementation of the `enter`, `exit`, `delay`, and `continue` operations. Many environments also provide an explicit `create` mechanism, although in others the creation of processes is handled in some implicit, declarative way. Similarly, there are various, more or less convenient, ways in which shared variables can be declared.

3.2. A Portable Implementation of Monitors

While most environments, particularly those on machines with a concept of shared memory, make it possible to implement monitors, the implementations typically look very different in each environment. This is not due solely to different names for the primitives, but also is due to the difference in the *method* of implementation (compiler extension, library routines, and in some cases even compiler directives).

In order to achieve portability, we have found it useful to hide machine dependencies in macros. There are macros for all of the primitive monitor-building operations, for the creation of processes, and for the declaration and initialization of shared variables.

3.3. Higher-level Operations

At this point we have described a means of implementing a set of low-level operations portably. These operations (like *P* and *V*) are too low level to be a useful tool. The choice of these operations to be exactly those required to construct monitors means that we can now implement any specific monitor portably, by expressing it in terms of the low-level macros. That is, we have constructed a small set of macros that must be recoded for each target machine. In addition, we have defined a number of higher-level monitors. Since these are defined in terms of the low-level macros, they are completely portable.

1. The named lock: The simplest monitor is a lock. Its purpose is to simply guarantee exclusive access to some resource as long as a process owns the lock.

2. The barrier: It is often useful to mark a certain place in the code which a certain number of processes must reach before any of them can proceed. Usually, but not always, this number is the total number of processes. The barrier monitor holds the processes with a delay operation until the required number have arrived.

3. The self-scheduling loop: One of the most common forms of parallelism in numerical programs is the parallel execution of independent loop iterations. The shared data item owned by the monitor in this case is the next available subscript. Each process executes a loop in which it enters the monitor to "ask for" a loop index; then it executes the loop body for the value of the index that it receives. There is an implied barrier at the end of the loop: no process is allowed to finish until all of them do.

4. The askfor monitor: The self-scheduling loop is really a dispatcher in which the
   tasks being dispatched are all represented by loop index values. This gives rise to
   a natural generalization, which we call the "askfor" monitor. The monitor manges
   a pool of work units described by user-supplied data structures. All processes typically execute a loop in which they request a unit of work, do it, and then request
   another. The units of work may be of different sizes and types, and the pool of
   work units may dynamically grow and shrink during the course of the run.

We have found this general model–a pool of work units being dynamically
dispatched to a pool of processes–to be an extremely useful paradigm for a wide
range of parallel algorithms. Our particular implementation handles several subtle
synchronization patterns that may occur within this model.

It is important to realize that, since all of these synchronization patterns are formulated as monitors and therefore implemented in terms of the monitor-building primitives, they are portable. In our implementation they are packaged as macros for
efficiency (no subroutine calls) and ease of use. There is one version of these macros
for each language (Fortran and C) but not for each parallel machine. On the other hand,
if one machine supports a particularly efficient implementation of one of these patterns,
a machine-specific version of the macro for that pattern can be created and used without
impact on the source code (by invoking the machine-specific primitives in the definition
of the monitor, rather than building it strictly upon the lower-level macro package).

3.3.1. Developing other synchronization patterns.

We have found the above set of high-level synchronization patterns to be surprisingly stable over the last few years. Originally conceived in order to deal with programming a small set of problems on the Denelcor HEP, they have not changed as they
have been applied by ourselves and others to a wide range of parallel algorithms. There
can be no claim that they are complete, however, and programmers will find new, useful synchronization patterns, particularly application-specific ones, and will require
more general versions of the above patterns. The usefulness of the monitor mechanism
is that it provides a paradigm for expressing these new patterns in an environment-
independent way, and implementing them in a portable way (since they will be implemented in terms of the monitor-building primitives.)

3.4. Efficiency Issues

3.4.1. Efficiency of Monitor Operations

It will always be difficult for any general implementation of anything whatsoever
to match a machine-specific, vendor-supplied one. And yet the macro implementation
of monitors does not *necessarily* add any overhead of its own. If the primitives upon
which the fundamental operations are built are efficiently accessible, then the portable
code can be as efficient as the non-portable code.

The Denelcor HEP was an example of a machine with an extremely efficient locking mechanism (the asynchronous variable, indicated by a "$" prefix in the early versions of the Fortran compiler.). The barrier was recognized as an fundamental synchronization pattern, and HEP manual gave the following code to express it (this is being offered as a horrible example; don't study it):

```
      IF (WAITF($INLOCK)) CONTINUE
      N = $NP + 1
      IF (N .NE. IP) GO TO 5
      PURGE $INLOCK
      $OUTLOCK = .TRUE.
    5 $NP = N
      IF (WAITF($OUTLOCK)) CONTINUE
      N = $NP - 1
      IF (N .NE. 0) GO TO 10
      PURGE $OUTLOCK
      $INLOCK = .TRUE.
   10 $NP = N
```

We illustrate our approach with the three levels at which the barrier is expressible. The version seen and used by the programmer is just the invocation of the `barrier` macro instruction:

```
      barrier(B1,IP)
```

where `B1` is the barrier's name and `NPROCS` is the number of processes to wait here before proceeding.

This macro is expanded into a version in some host language (say Fortran) in terms of the basic primitives. The following code is portable and can be read as a precise definition of the barrier.

```
      enter(B1)
      IF (B1C1 .LT. (IP - 1)) THEN
          delay(B1,1)
      ENDIF
      continue(B1,1)
      exit(B1)
```

Finally, by expanding the basic primitives using their HEP-specific definitions, we get code ready for the HEP Fortran compiler:

```
      B1 = $B1
      IF (B1C1 .LT. (IP - 1)) THEN
          B1C1 = B1C1 + 1
          $B1 = 0
          B1D1 = $B1D1
      ENDIF
      IF (B1C1 .EQ. 0) THEN
          $B1 = 0
      ELSE
          B1C1 = B1C1 - 1
          $B1D1 = 0
      ENDIF
      GO TO 800
      $B1 = 0
  800 CONTINUE
```

The point of this is not that this code is more readable that the version from the HEP manual (although perhaps that could be argued) but rather that it is just as efficient in terms of number of Fortran instruction executed on the various branches through the code. Thus we have what we would have wished for: portability at the programmer's level and efficiency at the machine-specific level.

3.4.2. Creating Processes

The overhead of creating processes can vary greatly. On the HEP it was quite inexpensive; on machines in which process creation amounts to a unix *fork* operation it can take many thousands. We adopt the conservative approach of creating processes once at the beginning of a program, each with the capability (code) to process the all of the different kinds of work units that will be created during the run. When there is no work for them to do, they are delayed in a monitor rather than destroyed and then recreated.

3.4.3. Efficiency of Locks

The most common implementation of *enter*, *exit*, *delay*, and *continue* operations is in terms of whatever mechanism the environment supplies for obtaining and releasing locks. The monitor paradigm makes it clear that two somewhat contradictory flavors of lock are desirable.

It is usually the case that a process will hold a lock associated with an *enter* operation for a short time (Programmers know enough to keep critical sections of code–the monitor operations–short). A process can be expected either to exit quickly or to execute a *delay* operation, which gives up the lock associated with the *enter* operation. On the other hand, a delayed process may be held up for a relatively long time. If the only type of lock available in a given environment is a spin lock, then these long delays waste processor cycles, but if the only type of lock available is one that causes a context switch (dispatching of another process by the operating system) then the *enter* and *exit* operations may be needlessly expensive. It is inefficient to force a context switch when all that the monitor operation does is update a counter, for example.

4. Experiences with this Approach

We have implemented versions of the low-level macros on the Denelcor HEP, the Encore Multimax, the Sequent Balance, the Alliant FX/8, the Cray 2, and the IBM 3033-AP. (We have found it useful to have a version for uniprocessors as well, in which many of the macros generate no code. If a program is structures so that it will execute properly with any number of processes, including one, then one level of debugging can be done on a unprocessor). There are C and Fortran version for most environments.

We use the m4 macro processor, although the macros as we have implemented them make no unusual demands on a macro processor and it would not be difficult to recode them for a different macro processor. The m4 macro processor has the advantage of being widely available. Of course the macro processor used need not run on the target multiprocessor; the source code can be expanded (macro-processed) on one machine and then compiled on another.

As mentioned above, the high level patterns have remained relatively stable. Experience on machines with more processes, however, have demonstrated the bottlenecking effect of the single dispatching pool in the original *askfor* monitor. This has led to the development of a distributed version with multiple dispatching pools to eliminate the global critical section while maintaining for the programmer the convenience of a single dispatcher[4].

5. Minimal Requirements from the Operating Environment

Many vendors of multiprocessors today are offering a variety of complicated mechanisms for expressing and controlling parallelism. It is the thesis of this paper that a usable, efficient, and portable mechanism (monitors) can be developed using only a minimal set of basic primitives. In this section we describe exactly what is needed for an implementation monitors.

5.1. Process Creation

It is important that there be an explicit mechanism for creating processes, one at a time. It is not important that it be particularly fast; we expect to create processes only once at the beginning of a program.

5.2. Shared Memory

There must be way of declaring that certain variables are to be shared among processes. There are two approaches to this, static and dynamic.

The purely static approach has its origins in the Fortran COMMON statement, and many systems still use this mechanism for specifying shared memory. This has left no standard notation for memory local to a process but global to a collection of procedures; the name TASK COMMON is beginning to be used in the Fortran community for this class of memory.

The dynamic approach is more natural in C, where dynamic allocation of memory is a fundamental feature. Here, some systems provide a special form of *malloc* which returns a pointer to a shared data structure. Other systems stick to the declarative approach even in C, decreeing that all external memory is to be shared.

5.2.1. Locks

As mentioned above, no single implementation of the *lock* primitive is entirely adequate. Both short-term locks and long-terms locks are required. The short-term lock needs to be capable of being acquired in a few instructions; in particular, the overhead of a subroutine call to a library routine is undesirable, although not as fatal as a process switch. It is acceptable for short-term locks to be implemented via a busy wait precisely because it is not expected that they will be held for more than a few instructions.

On the other hand, a busy wait is inefficient as a mechanism for delaying a process, which may be waiting a relatively long time. Thus a lock that causes a process switch if it cannot be obtained is also a desirable primitive to have available.

A hybrid approach may be best, in which a process first attempts to obtain a lock using the efficient, busy-wait mechanism, and if it fails after some number of attempts, executes the higher-overhead process-switch operation.

6. Summary

We have presented here our experiences with an approach to programming multiprocessors that leads to portable yet efficient programs. This approach is based on a small number of system-specific primitives. We believe that it is premature to standardize on high-level language extensions for parallelism; at the same time the variety of mechanisms currently being proposed by vendors are not only non-portable, but lack the clear semantics required of a useful programming paradigm. We believe that a better approach is for manufacturers to concentrate on providing efficient user access to the low-level synchronization operations of the system, in order that various user communities will be able to build the tools they need. We have demonstrated a set of useful tools emphasizing generality, portability, and efficiency.

Acknowledgement

This work was partially supported by the Applied Mathematical Sciences subprogram of the Office of Energy Research of the U.S. Department of Energy under Contract W-31-109-Eng-38.

References

1. M. Ben-Ari, *Principles of Concurrent Programming,* Prentice-Hall, Inc., Englewood Cliffs, New Jersey (1982).

2. Per Brinch Hansen, *The Architecture of Concurrent Programs,* Prentice-Hall, Inc., Englewood Cliffs, New Jersey (1977).

3. C. A. R. Hoare, "Monitors: an operating system structuring concept," *Communications of the ACM*, pp. 549-557 (October 1974).

4. E. L. Lusk and R. A. Overbeek, "The tradeoffs among portability, complexity, and efficiency in multiprocessing environments," pp. 245-260 in *Proceedings of the Workshop on Parallel Processing Using the Heterogeneous Element Processor*, Norman, Oklahoma (March, 1985)

SCHEDULE: Tools for Developing and Analyzing Parallel Fortran Programs *

J. J. Dongarra and D. C. Sorensen

Mathematics and Computer Science Division
Argonne National Laboratory
Argonne, Illinois 60439-4844

Center for Supercomputer Research and Development
University of Illinois
Urbana, Illinois 61801-2932

1. Introduction

Many new parallel computers are now emerging as commercial products[9]. Exploitation of the parallel capabilities requires either extensions to an existing language such as Fortran or development of an entirely new language. A number of activities[16, 17] are under way to develop new languages that promise to provide the ability to exploit parallelism without the considerable effort that may be required in using an inherently serial language that has been extended for parallelism. We applaud such activities and expect they will offer a true solution to the software dilemma in the future. However, in the short term we feel there is a need to confront some of the software issues, with particular emphasis placed on transportability and use of existing software.

Our interests lie mainly with mathematical software typically associated with scientific computations. Therefore, we concentrate here on using the Fortran language. Each vendor of a parallel machine designed primarily for numerical calculations has provided its own parallel extensions to Fortran. These extensions have taken many forms already and are usually dictated by the underlying hardware and by the capabilities that the vendor wishes to supply the user. This has led to widely different extensions ranging from the ability to synchronize on every assignment of a variable with a full empty property[13] to attempts at automatically detecting loop-based parallelism with a preprocessing compiler aided by user directives[6]. The act of getting a parallel process executing on a physical processor ranges from a simple "create" statement[13] which imposes the overhead of a subroutine call, to "tskstart" [1] which imposes an overhead on the order of 10^6 machine cycles, to no formal

* Work supported in part by the Applied Mathematical Sciences subprogram of the Office of Energy Research, U.S. Department of Energy, under Contracts W-31-109-Eng-38, DE-AC05-840R21400, and DE-FG02-85ER25001.

mechanism whatsoever[6]. These different approaches reflect characteristics of underlying hardware and operating systems and to a large extent are dictated by the vendors view of which aspects of parallelism are marketable. It is too early to impose a standard on these vendors, yet it is disconcerting that there is no agreement among any of them on which extensions should be included. There is not even an agreed-upon naming convention for extensions that have identical functionality. Program developers interested in producing implementations of parallel algorithms that will run on a number of different parallel machines are therefore faced with an overwhelming task. The process of developing portable parallel packages is complicated by additional factors that lie beyond each computer manufacturer supplying its own, very different mechanism for parallel processing. A given implementation may require several different communicating parallel processes, perhaps with different levels of granularity. An efficient implementation may require the ability to dynamically start processes, perhaps many more than the number of physical processors in the system. This feature is either lacking or prohibitively expensive on most commercially available parallel computers. Instead, many of the manufacturers have limited themselves to providing one-level loop-based parallelism.

This paper describes an environment for the transportable implementation of parallel algorithms in a Fortran setting. By this we mean that a user's code is virtually identical for each machine. The main tool in this environment is a package called SCHEDULE which has been designed to aid a programmer familiar with a Fortran programming environment to implement a parallel algorithm in a manner that will lend itself to transporting the resulting program across a wide variety of parallel machines. The package is designed to allow existing Fortran subroutines to be called through SCHEDULE, without modification, thereby permitting users access to a wide body of existing library software in a parallel setting. Machine intrinsics are invoked within the SCHEDULE package, and considerable effort may be required on our part to move SCHEDULE from one machine to another. On the other hand, the user of SCHEDULE is relieved of the burden of modifying each code he disires to transport from one machine to another.

Our work has primarily been influenced by the work of Babb [2], Browne [3], and Lusk and Overbeek [14]. We present here our approach, which aids in the programming of explicitly parallel algorithms in Fortran and which allows one to make use of existing Fortran libraries in the parallel setting. The approach taken here should be regarded as minimalist: it has a very limited scope. There are two reasons for this. First, the goal of portability of user code will be less difficult to achieve. Second, the real hope for a solution to the software problems associated with parallel programming lies with new programming languages or perhaps with the "right" extension to Fortran. Our approach is expected to have a limited lifetime. Its purpose is to allow us to exploit existing hardware immediately.

2. Terminology

Within the science of parallel computation there seems to be no standard definition of terms. A certain terminology will be adopted here for the sake of dialogue. It will not be "standard" and is intended only to apply within the scope of this document.

Process - A unit of computation, an independently executable Fortran
      subroutine together with calling sequence parameters, common
      data, and externals.

Task   - A main program, processes, and a virtual processor.

Virtual Processor - A process designed to assume the identity of every
      process within a given task (through an appropriate subroutine call).

Processor - A physical device capable of executing a main program or a
      virtual processor.

Shared Data - Variables that are read and/or written by
      more than one process (including copies of processes).

Data Dependency - A situation wherein one process (A) reads any shared
      data that another process (B) writes. This data dependency
      is satisfied when B has written the shared data.

Schedulable Process - A process whose data dependencies have all
      been satisfied.

3. Parallel Programming Ideas

When designing a parallel algorithm one is required to describe the data dependencies, parallel structures, and shared variables involved in the solution. Typically, such algorithms are first designed at a conceptual level and later implemented in Fortran and its extensions. Each manufacturer provides a different set of extensions and targets these extensions at different implementation levels. For example, some manufacturers allow only test-and-set along with spawn-a-process, while others allow concurrent execution of different loop iterations.

Our attempt here is to allow the user to define the data dependencies, parallel structures, and shared variables in his application and then to implement these ideas in a Fortran program written in terms of subroutine calls to SCHEDULE. Each set of subroutine calls specifies a unit of computation or process which consists of a subroutine name along with the calling parameters and the data dependencies necessary to coordinate the parallel execution.

The basic philosophy here is that Fortran programs are naturally broken into subroutines that identify self-contained units of computation which operate on shared data structures. This allows one to call on existing library subroutines in a parallel setting without modification, and without having to write an envelope around the library subroutine call in order to conform to some unusual data-passing conventions imposed by a given parallel programming environment.

A parallel(izable) program is written in terms of calls to subroutines which, in principle, may be performed either independently or according to data dependency requirements that the user is responsible for defining. The result is a serial program that can run in parallel given a way to schedule the units of computation on a system of parallel processors while obeying the data dependencies.

4. Parallel Programming Using SCHEDULE

The package SCHEDULE requires a user to specify the subroutine calls along with the execution dependencies in order to carry out a parallel computation. Each of these calls represents a process, and the user must take the responsibility of ensuring that the data dependencies represented by the graph are valid. This concept is perhaps difficult to grasp without some experience with writing parallel programs. We shall try to explain it in this section by example; in the following section we shall describe the underlying concepts and the SCHEDULE mechanism.

To use SCHEDULE, one must be able to express (i.e., program) an algorithm in terms of processes and execution dependencies among the processes. A convenient way to view this is through a computational graph. For example, the following graph denotes five subroutines A, B, C, D, and E (here with two "copies" of subroutine D operating on different data). We intend the execution to start simultaneously on sub-

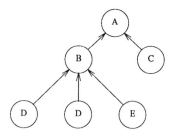

Figure 4.1.
Data Dependency Graph

routines C,D,D, and E since they appear as leaves in the data dependency graph (D will be initiated twice with different data). Once D,D,and E have completed then B may execute. When B and C have completed execution then A may start and the entire computation is finished when A has completed. To use SCHEDULE, one is required to specify the subroutine calling sequence of each of the six units of computation, along with a representation of this dependency graph.

For each node in the graph, SCHEDULE requires two subroutine calls. One contains information about the user's routine to be called, such as the name of the routine, calling sequence parameters, and a simple tag to identify the process. The second subroutine call defines the dependency in the graph to nodes above and below the one being specified, and specifies the tag to identify the process. In this example, after an initial call to set up the environment for SCHEDULE, six pairs of calls would be made to define the relationships and data in the computational graph.

These concepts are perhaps more easily grasped through an actual Fortran example. A very simple example is a parallel algorithm for computing the inner product of two vectors. The intention here is to illustrate the mechanics of using SCHEDULE. This algorithm and the use of SCHEDULE on a problem of such small granularity are not necessarily recommended.

Problem: Given real vectors a and b, each of length n, compute $\sigma = a^T b$.

Parallel Algorithm:

Let $a^T = (a_1^T, a_2^T, \ldots, a_k^T)$ and $b^T = (b_1^T, b_2^T, \ldots, b_k^T)$

be a partitioning of the vectors a and b into smaller vectors a_i and b_i.

Compute (in parallel)

$$\sigma_j = a_j^T b_j \ , \ j = 1,k \ .$$

When all done

$$\sigma = \sigma_1 + \sigma_2 + \cdots + \sigma_k .$$

Each of the parallel processes will execute code of the following form:

```
      subroutine inprod(m,a,b,sig)
      integer m
      real a(*),b(*),sig
      sig = 0.0
      do 100 j = 1,m
          sig = sig + a(j)*b(j)
  100 continue
      return
      end
```

The following routine is used to accumulate the result:

```
      subroutine addup(k,sigma,temp)
      integer k
      real sigma,temp(*)
      sigma = 0.0
      do 100 j = 1,k
          sigma = sigma + temp(j)
  100 continue
      return
      end
```

The first step in constructing a code is to understand the parallel algorithm in terms of schedulable processes and a data dependency graph. Then the algorithm is expressed in a standard (serial) Fortran code. This code consists of a main program which initializes the shared data and a "parallel" subroutine **parprd** to compute the inner product by invoking the parallel processes **inprd** and **addup**. The program and associated data dependency graph are shown below.

Serial Code:

```
      program main
      integer n,k
```

```
        real  a(1000),b(1000),temp(50),sigma
        read  (5,*)  n,k
        do  100  j  =  1,n
           a(j)  =  j
           b(j)  =  1
  100  continue
c
        call  parprd(n,k,a,b,temp,sigma)
c
        write(6,*)  '  sigma  =  ',sigma
        stop
        end


        subroutine  parprd(n,k,a,b,temp,sigma)
c
c       declare  shared  variables
c
        integer  n,k
        real  a(*),b(*),temp(*),sigma
c
c       declare  local  variables
c
        integer  m,indx,j
c
        m  =  n/k
        indx  =  1
        do  200  j  =  1,k
c
           call  inprod(m,a(indx),b(indx),temp(j))
c
           indx  =  indx  +  m
           if  (j  .eq.  k-1)  m  =  n  -  indx  +  1
  200  continue
c
        call  addup(k,sigma,temp)
c
        return
        end
```

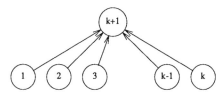

Figure 4.2.
Depedency Graph for Parallel Inner Product

In this data dependency graph we have identified k processes

$$inprod(\,m\,,a\,(indx\,),b\,(indx\,),temp\,(j\,))\,,\quad j=1,2,...,k\quad,\quad indx=1+(j-1)*m$$

which are not data dependent. Each of them reads a segment of the shared data a , b and writes on its own entry of the array *temp*, but none of them needs to read data that some other process will write. This fact is evident in the graphical representation shown in Figure 4.2 where they are *leaves*. One process,

$$addup\ (k\ ,sigma\ ,temp\ ),$$

labeled $k+1$ is data dependent on each of the processes $1,2,...,k$. This is because
addup needs to read each entry of the array *temp* in order to compute the sum and
place it into σ.

From this data dependency graph we may proceed to write the parallel program.
Once we have understood the computation well enough to have carried out these two
steps, the invocation of SCHEDULE to provide for the parallel execution of schedulable processes is straightforward. Calls to **parprd**, **inprod**, and **addup** are replaced
by calls to SCHEDULE to identify the routines to be executed as well as the information relating to the dependency graph. The modified code follows.

Parallel Main:

```
          program main
          integer n,k
c
          EXTERNAL PARPRD
c
          real a(1000),b(1000),temp(50),sigma
          read (5,*) n,k,NPROCS
          do 100 j = 1,n
             a(j) = j
             b(j) = 1
    100 continue
c
          CALL SCHED(nprocs,PARPRD,n,k,a,b,temp,sigma)
c
          write(6,*) ' sigma = ',sigma
          stop
          end


          subroutine parprd(n,k,a,b,temp,sigma)
c
c         declare shared variables
c
          integer n,k
          real a(*),b(*),temp(*),sigma
c
c         declare local variables
c
          integer m1,m2,indx,j,jobtag,icango,ncheks,mychkn(2)
c
          EXTERNAL INPROD,ADDUP
          save m1,m2
c
          m1   = n/k
          indx = 1
          do 200 j = 1,k-1
c
c             express data dependencies
c
```

```
                JOBTAG = j
                ICANGO = 0
                NCHEKS = 1
                MYCHKN( 1 )  =  k+1
c
                CALL   DEP ( j obt ag , i c ango , nc hek s , my c hkn )
                CALL   PUTQ( j obt ag , INPROD, m1 , a ( i ndx ) , b ( i ndx ) , t emp ( j ) )
c
                indx  =  i ndx  +  m1
    200  c ont i nue
                m2  =  n  -  i ndx  +  1
c
c       express  data  dependenc i e s  for  c l ean  up  s t ep
c
                JOBTAG = k
                ICANGO = 0
                NCHEKS = 1
                MYCHKN( 1 )  =  k+1
c
                CALL   DEP ( j obt ag , i c ango , nc hek s , my c hkn )
                CALL   PUTQ( j obt ag , INPROD, m2 , a ( i ndx ) , b ( i ndx ) , t emp ( k ) )
c
                indx  =  i ndx  +  m1
c
                JOBTAG = k+1
                ICANGO = k
                NCHEKS = 0
c
        CALL   DEP ( j obt ag , i c ango , nc hek s , my c hkn )
        CALL   PUTQ( j obt ag , ADDUP, k , s i gma , t emp )
c
        r e t ur n
        e nd
```

The code that will execute in parallel has been derived from the serial code by replacing calls to **parprd, inprd, addup** with calls to SCHEDULE routines that invoke these routines. The modifications are signified by putting calls to SCHEDULE routines in capital letters. Let us now describe the purpose of each of these calls.

```
        CALL  SCHED( npr oc s , PARPRD, n , k , a , b , t emp , s i gma )
```

This replaces the call to **parprd** in the serial code. The effect is to devote *nprocs* virtual processors to the parallel subroutine **parprd**. The parameter list following the subroutine name consist of the calling sequence one would use to make a normal call to **parprd**. Each of these parameters must be called by reference and not by value. No constants or arithmetic expressions should be passed as parameters through a call to **sched**. This call to **sched** will activate *nprocs* copies of a virtual processor **work**. This virtual processor is a SCHEDULE procedure (written in C) that is internal to the package and not explicitly available to the user.

```
                JOBTAG = j
                ICANGO = 0
                NCHEKS = 1
                MYCHKN( 1 )  =  k+1
c
```

```
CALL   DEP(jogtag,icango,ncheks,mychkn)
CALL   PUTQ(jobtag,INPROD,m,a(indx),b(indx),temp(j))
```

This code fragment shows the *j–th* instance of the process **inprod** being placed on a queue. The information needed to schedule this process is contained in the data dependency graph. In this case, the *j–th* instance of a call to **inprod** is being placed on the queue, so *jobtag* is set to *j*. The value zero is placed in *icango*, indicating that this process does not depend on any others. If this process were dependent on *p*, other processes then *icango* would be set to *p*.

The mechanism just described allows static scheduling of parallel processes. In this program the partitioning and data dependencies are known in advance even though they are parameterized. It is possible to dynamically allocate processes; this capability will be explained later. It might be worthwhile at this point to discuss the mechanism that this package relies on.

5. The SCHEDULE Mechanism

The call to the SCHEDULE routines **dep** and **putq**, respectively, places process dependencies and process descriptors on a queue. A unique user supplied identifier *jobtag* is associated with each node of the dependency graph. This identifier is a positive integer. Internally it represents a pointer to a process. The items needed to specify a data dependency are non-negative integers *icango* and *ncheks* and an integer array *mychkn*. The *icango* specifies the number of processes that process *jobtag* depends on. The *ncheks* specifies the number of processes that depend on process *jobtag*. The *mychkn* is an integer array whose first *ncheks* entries contain the identifiers (i.e., *jobtag* s) of the processes that depend on process *jobtag*.

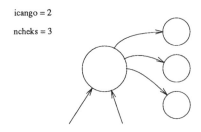

icango = 2

ncheks = 3

Figure 5.1.
A Node in a Dependency Graph

In Figure 5.1 a typical node of a data dependency graph is shown. This node has two incoming arcs and three outgoing arcs. As shown to the left of the node one would set *icango* = 2, *ncheks* = 3, and the first three entries of *mychkn* to the identifiers of the processes pointed to by the outgoing arcs.

The initial call to *sched* (nprocs,subname,<parms>) results in *nprocs* virtual processors called **work** to begin executing on *nprocs* separate physical processors. Typically *nprocs* should be set to a value that is less than or equal to the number of physical processors available on the given system. These **work** routines access a ready queue of *jobtag* s for schedulable processes. Recall that a schedulable process is one whose data dependencies have been satisfied. After a **work** routine has been successful in obtaining the *jobtag* of a schedulable process, it makes the subroutine call associated with that *jobtag* during the call to **putq**. When this subroutine executes a *return*, control is returned to **work**, and a SCHEDULE routine **chekin** is called which decrements the *icango* counter of each of the *ncheks* processes that depend on process *jobtag*. If any of these *icango* values has been decremented to zero, the identifier of that process is placed on the ready queue immediately.

We depict this mechanism in Figure 5.2 . The array labeled **parmq** holds a process descriptor for each *jobtag*. A process descriptor consists of data dependency information and a subroutine name together with a calling sequence for that subroutine. This information is placed on **parmq** through the two calls

```
CALL  DEP(jobtag,icango,ncheks,mychkn)
CALL  PUTQ(jobtag,subname,<parms>).
```

When making these two calls the user has assured that a call to *subname* with the argument list *<parms>* is valid in a data dependency sense whenever the counter *icango* has been decremented to the value zero. When a **work** routine has finished a call to **chekin** , it gets the *jobtag* of the next available schedulable process off the *readyq* and then assumes the identity of the appropriate subroutine by making a call to *subname* with the argument list *<parms>*.

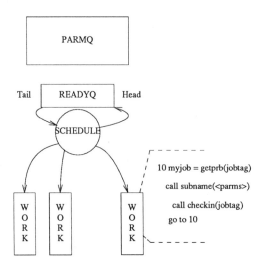

Figure 5.2.
The SCHEDULE Mechanism

6. Low-Level Synchronization

Ideally, the mechanism we have just described will relieve the user of explicitly invoking any synchronization primitives. Unfortunately, some powerful parallel constructs are not so easily described by this mechanism. It may be desirable to have two processes executing simultaneously that are not truly data independent of each other. A typical example is in pipelining a computation, that is, when several parallel processes are writing on the same data in a specified order which is coordinated through explicit synchronization. To provide this capability, two low-level synchronization primitives have been made available within SCHEDULE. They are **lockon** and **unlock**. Each takes an integer argument. An example of usage is

```
call lockon(ilock)
      ilocal = indx
      indx = indx + 5
call unlock(ilock)
```

In this example a critical section has been placed around the act of getting a local copy of the shared variable *indx* and updating the value of *indx*. If several concurrent processes are executing this code, then only one of them will be able to occupy this critical section at any given time. The variable *ilock* must be a globally shared variable and it must be initialized before any usage as an argument to **lockon** or **unlock** by calling the routine **lckasn**. In the above example the statement

```
call lckasn(ilock)
```

must execute exactly once and before any of the calls to **lockon** are made. After execution of this statement the lock variable *ilock* has been declared as a lock variable and has been initiated to be *off*. If there are low-level data dependencies among any of the processes that will be scheduled, then it will be necessary to enforce those data dependencies using locks. It is preferable to avoid using locks if possible. However, in certain cases such as pipelining, locks will be required.

7. Dynamic Allocation of Processes

The scheme presented above might be considered static allocation of processes. By this we mean that the number of processes and their data dependencies were known in advance. Therefore the entire data structure (internal to SCHEDULE) representing the computational graph could be recorded in advance of the computation and is fixed throughout the computation. In many situations, however, we will not know the computational graph in advance, and we will need the ability for one process to start or spawn another depending on a computation that has taken place up to a given point in the spawning process. This dynamic allocation of processes is accomplished through the use of the SCHEDULE subroutine **spawn**. The method of specifying a process is similar to the use of **putq** described above.

We shall use the same example to illustrate this mechanism.

Processes:

subroutine inprod .... same as above

```
      subroutine addup(n,k,a,b,sigma,temp)
      integer myid,n,k
      real a(*),b(*),sigma,temp(*)
c
c     declare local variables
c
      integer j,jdummy,m1,m2
c
      EXTERNAL INPROD
c
```

```
        m1 = n/k
        indx = 1
c
        CALL PRTSPN(myid)
        do 200 j = 1,k-1
c
c           replace the call to inprod with a call to spawn
c
            CALL SPAWN(myid,jdummy,INPROD,m1,a(indx),b(indx),temp(j))
            indx = indx + m1
  200   continue
        m2 = n - indx + 1
c
c       clean up step
c
        call inprod(m2,a(indx),b(indx),temp(k))
c
c       keep one of the inner product assignments and then call wait
c       to be sure all others have completed.
c
        CALL WAIT(myid)
c
c
c       All have checked in, now addup the results.
c
        sigma = 0.0
        do 100 j = 1,k
           sigma = sigma + temp(j)
  100   continue
        return
        end
```

The subroutine **parprd** must change somewhat.

```
        subroutine parprd(n,k,a,b,temp,sigma)
c
c       declare shared variables
c
        integer n,k
        real a(*),b(*),temp(*),sigma
c
c       declare local variables
c
        integer mychkn(1),icango,ncheks,jobtag
        EXTERNAL ADDUP
c
        JOBTAG = 1
        ICANGO = 0
        NCHEKS = 0
c
        CALL DEP(jobtag,icango,ncheks,mychkn)
        CALL PUTQ(jobtag,ADDUP,n,k,a,b,sigma,temp)
c
        return
        end
```

In this example the call to *spawn* invokes $k-1$ instances of the subroutine *inprod*. Note that the user does not specify data dependencies. Instead, the jobtags for these processes are assigned internally and they are required to report to the spawning process. More details of this mechanism are given in the next section.

8. Syntax Summary

In this section we summarize the SCHEDULE operations and their syntax. We begin with a list of these operations and follow up with a general description of their usage. With this we hope to supplement the examples given above.

Table 1

User Calls

```
CALL  SCHED  ( nprocs, subname, <parms> )
CALL  DEP    ( jobtag, icango, ncheks, mychkn )
CALL  PUTQ   ( jobtag, subname, <parms> )
CALL  PRTSPN( myid )
CALL  SPAWN  ( myid, jdummy, subname, <parms> )
CALL  WAIT   ( myid )
CALL  LCKASN( ilock )
CALL  LOCKON( ilock )
CALL  UNLOCK( ilock )
```

nprocs	- Number of virtual processors to be used on subname.
subname	- Subroutine to be executed.
<parms>	- list of parameters for subname.
jobtag	- A unique user supplied identifier.
icango	- The number of processes that must complete before this process starts.
ncheks	- Specifies the number of processes that depend on this process.
mychkn	- An integer array whose first *ncheks* entries contain the identifiers of the processes that depend on this process.
myid	- An alias process id assigned by SCHEDULE to record completion of the spawned processes.
jdummy	- A dummy integer parameter.
ilock	- A globally shared variable used as a lock.

There are three basic operations in SCHEDULE other than the lowest level synchronization provided through locks. They are **call sched**, **call putq**, and **call spawn**. Various additional calls must be made in association with these three basic operations to enforce data dependencies, and orchestrate the correct execution of the parallel program.

The statement

$$call \quad sched(nprocs,paralg,<parms>)$$

is used to initiate the execution of a parallel program *paralg* and will devote *nprocs* physical processors to the computation. The symbol *<parms>* stands for a set of parameters that would normally be supplied to the Fortran subroutine *paralg* through a Fortran call in a sequential program. One should think of the memory associated with the parameter list represented by <parms> as shared memory throughout the execution of *paralg*. Another way to share memory is through the use of *named common*.

The pair of statements

$$call \quad dep(jobtag,icango,ncheks,mychkn)$$

$$call \quad putq(jobtag,subname,<parms>)$$

are used to place a process in the large grained data dependency graph with *dep* used to record the data dependencies and *putq* to record the unit of computation. In this case the unit of computation will be a call to the subroutine *subname* with the calling sequence *<parms>*. When the data dependencies for this process have been satisfied the *icango* counter will have been decremented to zero and this process will be placed upon the ready queue. At some point in time a physical processor executing a *work* program will be free and will pick this process off of the ready queue and execute a *call subname*(*<parms>*). It is the user's responsibility to assure that the addresses passed in *<parms>* are addresses in shared memory. That is to say that these addresses must reference shared memory that has been declared in the program that made the call to *sched*. Hence these addresses must come from named common declared in that program and referenced by the subroutine *subname* or which have been passed as parameters in the call to *sched*.

The sequence of statements

$$call \quad prtspn(myid)$$

$$\cdots$$

$$call \quad spawn(myid,jdummy,subname,<parms>)$$

$$\cdots$$

$$call \quad wait(myid)$$

accomplish dynamic spawning of processes. Usually calls to spawn are made in a loop but they need not be. The call to *prtspn* secures an alias process id *myid* to record completion of the spawned processes. This id must be used in the subsequent calls to *spawn* and to *wait*. There is an implied barrier at the call to *wait* and execution does not continue past this point in the spawning process until all spawned processes have been completed. The syntax of the call to *spawn* is very similar to that of *putq*. The parameter *jdummy* is a temporary location that must be supplied to *spawn* and the result is a *call subname*(<*parms*>) as soon as there is an available instantiation of *work* available to execute this call. A key difference between a call to *spawn* and a call to *putq* is the lack of user specification of jobtags and data dependencies. The only data dependency indicated here is that a spawned process must report to the parent process and no other. This is taken care of automatically with use of the above mechanism. If communication is desired at a lower level among spawned processes this must be accomplished through the use of shared data and/or locks. Local variables in the spawning routine may be passed as arguments to the spawned routines with the understanding that their scope extends to all members of the subtree rooted at the spawning process. These variables must be treated as shared data within this subtree.

The implications of the mechanism used for spawning should be discussed further. We rely upon the stack associated with the calling tree rooted at the spawning process. If one wishes to do recursion to gain parallelism then there is potential for stack overflow. This is an unfortunate consequence of an attempt to avoid the burden of placing and keeping track of entry points and saving the state of an executing process which spawns other processes. The key problem is to retain the ability to execute the parallel program on a single physical processor. The call to *wait* does not result in a busy wait as one might suspect. Instead, a call is made to a routine that is quite similar to the *work* routines discussed earlier. This routine accesses the readyq makes a call and then "checks in" and returns to *wait*. Within the routine *wait* completion of spawned processes is monitored. If any have failed to complete another call to the special work routine is made. Obviously, this may result in stack overflow when spawned routines spawn other routines. An alternative was used in this package previously [11] which avoided stack overflow but was deemed overly complicated for the user.

9. An Environment for the Development of Explicitly Parallel Programs

As we mentioned at the outset, SCHEDULE is a programming aid that is intended to serve as a backbone to an environment for the development of explicitly parallel programs. Our goal is to provide a uniform interface to the parallel capabilities provided by existing and impending parallel systems. Included in this are tools for debugging and analyzing the performance of parallel programs. In this section we shall describe some of our goals in this area. Preliminary attempts at achieving these goals have been implemented or are in the design stage.

We regard SCHEDULE as a tool primarily designed as an aid to constructing an implementation of a "new" parallel algorithm. We do not think it is particularly well suited to converting an existing serial code to a parallel version, although this would be possible in some cases. The reason for this statement is that to program with SCHEDULE it is imperative that the large grain data dependencies are well understood by the programmer. This is of course required in any successful implementation of a parallel program. However, with SCHEDULE the data partitioning and construction of the data dependency graph are an explicit part of the programming effort. We expect that the designer of a parallel algorithm will have this information naturally at hand. At least this has been our experience in the design of our own parallel algorithms[10]. Enforcing this programming style goes a long ways towards avoiding bugs traditionally associated with parallel programming.

We have found it very useful to retain the capability of executing a parallel program in serial mode. A SCHEDULE program does not depend upon the number of physical processors available in a given system. It will execute with one processor active. Others have found this to be a useful property of such a programming tool [14]. However, Gentleman [12] does not feel that this is feature is essential to the dubugging of parallel programs. We find at least two reasons to provide it. First it allows one to develop and test code on a serial machine such as a VAX 11/780. This is quite important when the parallel machine is at a remote site or if the software development environment existing on the parallel machine is inadequate or difficult to use. Second, when developing a new algorithm one would like to be assured that the numerical properties of the algorithm are correct in serial mode before entering the parallel testing regime. Operating in this sequence tends to separate ordinary programming bugs from those associated with parallel programming. Moreover, due to the SCHEDULE mechanism one can be fairly confident that a parallel bug is due either to incorrect specification of the data dependency graph or to incorrect partitioning of data. Explicit synchronization is generally not a part of a SCHEDULE program and thus difficult synchronization bugs do not generally arise.

To aid in ascertaining a correct specification of the data dependency graph we envision a tool that will graphically represent the units of computation and the data dependency graph. A preliminary version of this tool has been implemented and experience with this tool is reported below. Similar ideas have been proposed in [2]. In our view the level of detail in those efforts is too fine. We expect the level of detail in such a representation of a parallel program to roughly correspond to the abstract level at which the programmer has partitioned his parallel algorithm. Let us illustrate this with a simple example. A favorite example of our's is the solution of a triangular linear system partitioned by blocks.

10. Triangular Solve Example

We can consider solving a triangular system of equations $Tx=b$ in parallel by partitioning the matrix T and vectors x and b as shown in Figure 10.1.

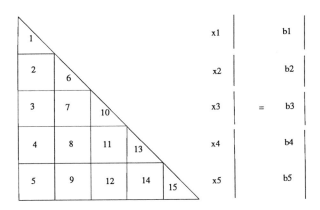

Figure 10.1

Partitioning for the Triangular Solve

The first step is to solve the system $T_1 x_1 = b_1$, this will determine the solution for that part of the vector labeled x_1. After x_1 has been computed it can be used to update the right hand side with the computations

$$b_2 = b_2 - T_2 x_1$$

$$b_3 = b_3 - T_3 x_1$$

$$b_4 = b_4 - T_4 x_1$$

$$b_5 = b_5 - T_5 x_1 .$$

Notice that these matrix-vector multiplications can occur in parallel, as there are no dependences. However, there may be several processes attempting to update the value of a vector b_j (for example 4,8,11 will update b_4) and this will have to be synchronized through the use of locks or the use of temporary arrays for each process. As soon as b_2 has been updated, the computation of x_2 can proceed as $x_2 = T_6^{-1} b_2$. Notice that this computation is independent of the other matrix-vector operations involving b_3, b_4, and b_5. After x_2 has been computed, it can be used to update the right hand side as follows:

$$b_3 = b_3 - T_7 x_2$$

$$b_4 = b_4 - T_8 x_2$$

$$b_5 = b_5 - T_9 x_2$$

The process is continued until the full solution is determined. The data dependency graph for this can be represented in Figure 10.2.

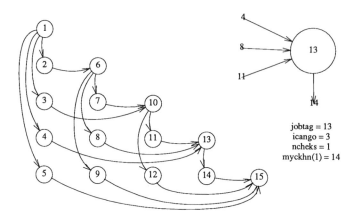

jobtag = 13
icango = 3
ncheks = 1
myckhn(1) = 14

Figure 10.2
Data Dependency Graph for Triangular Example

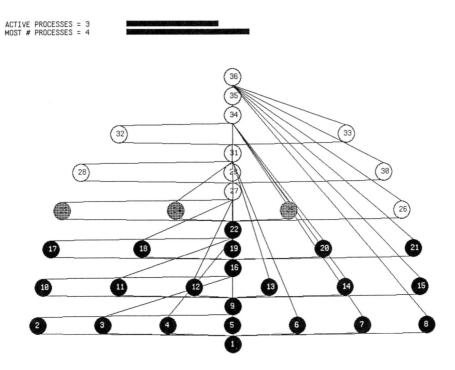

ACTIVE PROCESSES = 3
MOST # PROCESSES = 4

Figure 10.3
Output from SCHEDULE for Triangular Example

In Figure 10.3 we show the output from an executing SCHEDULE implementation of this example. The program was run on an Alliant FX/8. An output file was produced as the program executed which recorded the units of computation as they were defined and executed. The file was then shipped to a SUN workstation where a graphics program interpreted this output, constructed the graph and played back the execution sequence that was run on the Alliant. In the graph shown above in Figure 10.3 the black nodes show processes which have completed execution, the hatched nodes show executing processes and the white nodes show processes waiting to execute. The program was able to produce this playback output automatically simply by linking to SCHEDULE with a request for the graph option. The user's program did not change in any way. At the moment this is the only capability we have. It would be very useful have the user construct that part of the Fortran program that corresponded to the dependence graph, i.e. calls to the SCHEDULER routines, from a "mouse driven" graphics input device.

In addition to discovering bugs in the specification of the graph, this representation is useful in exposing more subtle aspects of the executing program. For example

when the graph produced by SCHEDULE is contrasted with the abstract user specified graph there are noticeable differences. The graph produced by SCHEDULE
exposes inherent serial bottlenecks in the algorithm. During execution the two bar
graphs at the top track the existing and maximum number of concurrently executing
processes respectively. In the ideal situation, the bottom bar has a value equal to he
number of physical processors available and the top bar is almost always equal to the
bottom bar. In addition to discovering serial bottlenecks, one can learn about load
balancing problems within an executing program. It may even be useful to put in
artificial data dependencies to force certain processes to complete before others to
achieve a better load balance. Another interesting feature is that this execution
sequence can be accelerated if the program took a long time to execute and slowed
down if it took a short time while maintaining the ratios of execution times among the
parallel processes.

A final feature that is possible but not fully implemented is the possibility of
observing the executing program within a node. We envision the capability of open-
ing a workstation window corresponding to a selected node. Within this window we
would see a display of selected variables being updated as execution takes place. At
the same time the currently executing lines of Fortran code would be displayed. This
capability currently exists for a serial Fortran program[8]. The output of such a
display is shown in Figure 10.4

```
n       =        10 i       =        3 small  =  1.19e-07 ierr  =         0
f       =  1.06e+01 b       =  1.36e-06 l       =         4 j     =         0
h       =  4.24e-07 c       =  1.85e-12 █m      █         7 l1    =         4
s       =  1.48e-06 g       = -1.54e-06 p       =  1.06e+01 q     =  1.12e-06
pythag  =  4.72e+02 r       =  6.88e+00 mm1     =         7 ii    =         2
t       =  4.00e+00
d          8.23e+00  9.40e+00  1.06e+01 -1.24e+00 -1.04e+00  1.29e+00 -3.54e+00
d          3.12e+00 -5.46e+00  4.41e+00
e2         4.36e-08  9.97e-14 -2.29e-12  5.33e+00  1.79e-02  5.68e+00  2.39e-01
e2         1.53e+00  1.55e+00  0.  e+00
```

```
105     do 110 m = 1, n
           if (e2(m) .le. c) go to 120
           if (e2(m) .le. c) go to 120
           t = 4.0e0 + r
 20 pythag = p
        s = sqrt(e2(1))
        h = small * (abs(d(1)) + sqrt(e2(1)))
      f = 0.0e0
        e2(i) = float(i)
        d(i) = 10.0
```

Figure 10.4
Screen Output from Debugger

The main problem to overcome here is selected display of data. A real program executing on large arrays could produce an overwhelming display of output.

11. Experience with SCHEDULE

At present the experience with using SCHEDULE is limited but encouraging. Versions are running successfully on the VAX 11/780, Alliant FX/8, and CRAY-2 computers. That is, the same user code executes without modification on all three machines. Only the SCHEDULE internals are modified, and these modifications are usually minor, but can be difficult in some cases. They involve such things as naming and parameter-passing conventions for the C - Fortran interface. They also involve coding the low-level synchronization primitives and managing to "create" the **work** processes.

On the CRAY-2 process creation is accomplished using **tskstart**, and the low-level synchronization is very similar to the low-level synchronization routines provided by the CRAY multitasking library[1]. For the Alliant FX/8 we coded the low-level synchronization primitives using their test-and-set instruction. To "create" the work routines, we used the CVD$L CNCALL directive before a loop that performed *nprocs* calls to the subroutine **work**.

In addition to some toy programs used for debugging SCHEDULE, several codes have been written and executed using SCHEDULE. These codes include the algorithm TREEQL for the symmetric tridiagonal eigenvalue problem [11], a domain decomposition code for singularly perturbed convection-diffusion PDE [4], an adaptive quadrature code[5], and a block preconditioned conjugate gradient code for systems arising in reservoir simulation [7].

The graph associated with TREEQL is probably the most illustrative of the capabilities of SCHEDULE so we shall discuss it in some detail. This is a parallel algorithm for computing the eigensystem of a real symmetric tridiagonal matrix T. The problem is to compute an orthogonal matrix Q and a diagonal matrix D such that $T = QDQ^T$. The crux of the algorithm is to divide a given problem into two smaller subproblems. To do this, we consider the symmetric tridiagonal matrix

$$(11.1) \quad T = \begin{bmatrix} T_1 & \beta e_k e_1^T \\ \beta e_1 e_k^T & T_2 \end{bmatrix}$$

$$= \begin{bmatrix} \hat{T}_1 & 0 \\ 0 & \hat{T}_2 \end{bmatrix} + \beta \begin{bmatrix} e_k \\ e_1 \end{bmatrix} (e_k^T, e_1^T)$$

where $1 \le k \le n$ and e_j represents the $j-th$ unit vector of appropriate dimension. The $k-th$ diagonal element of T_1 has been modified to give $\hat{T}_1$ and the first diagonal element of T_2 has been modified to give $\hat{T}_2$.

Now we have two smaller tridiagonal eigenvalue problems to solve. According to equation (11.1) we compute the two eigensystems

$$\hat{T}_1 = Q_1 D_1 Q_1^T, \quad \hat{T}_2 = Q_2 D_2 Q_2^T .$$

This gives

$$(11.2) \quad T = \begin{bmatrix} Q_1 D_1 Q_1^T & 0 \\ 0 & Q_2 D_2 Q_2^T \end{bmatrix} + \beta \begin{bmatrix} e_k \\ e_1 \end{bmatrix} (e_k^T, e_1^T)$$

$$= \begin{bmatrix} Q_1 & 0 \\ 0 & Q_2 \end{bmatrix} \left[\begin{bmatrix} D_1 & 0 \\ 0 & D_2 \end{bmatrix} + \beta \begin{bmatrix} q_1 \\ q_2 \end{bmatrix} (q_1^T, q_2^T) \right] \begin{bmatrix} Q_1^T & 0 \\ 0 & Q_2^T \end{bmatrix}$$

where $q_1 = Q_1^T e_k$ and $q_2 = Q_2^T e_1$. The problem at hand now is to compute the eigensystem of the interior matrix in equation (11.2). A numerical method for solving this problem has been provided in [2] and we shall discuss this method in the next section.

It should be fairly obvious how to proceed from here to exploit parallelism. One simply repeats the tearing on each of the two halves recursively until the original problem has been divided into the desired number of subproblems and then the rank one modification routine may be applied from bottom up to glue the results together again.

This *gluing* process is called updating and the general updating problem we are required to solve for each partition is that of computing the eigensystem of a matrix of the form

(11.3) $$\hat{Q}\hat{D}\hat{Q}^T = D + \rho zz^T$$

where D is a real $n \times n$ diagonal matrix, ρ is a nonzero scalar, and z is a real vector of order n .

A formula for computing an eigen-pair of the matrix on the right hand side of (11.3) is obtained by noting that if λ is a root of the equation

(11.4) $$1 + \rho z^T (D - \lambda I)^{-1} z = 0$$

and

(11.5) $$q = \theta (D - \lambda I)^{-1} z$$

then q , λ is such an eigen-pair, i.e they satisfy the relation

$$(D + \rho zz^T)q = \lambda q.$$

In (11.5) the scalar θ may be chosen so that $||q|| = 1$ to obtain an orthonormal eigensystem.

If we write equation (11.4) in terms of the components ζ_i of z then λ must be a root of the equation

(11.6) $$f(\lambda) \equiv 1 + \rho \sum_{j=1}^{n} \frac{\zeta_j^2}{\delta_j - \lambda} = 0.$$

Under certain assumptions this equation has precisely n roots, one in each of the open intervals $(\delta_j , \delta_{j+1})$, $j = 1,2,...n-1$ and one to the right of δ_n if $\rho > 0$ or one to the left of δ_1 if $\rho < 0$. We construct the eigenvectors corresponding to each of these roots using formula (11.5).

Due to this structure, an excellent numerical method may be devised to find the roots of the Equation (11.6) and as a by-product to compute the eigenvectors to full accuracy. It must be stressed, however, that great care must be exercised in the numerical method used to solve the secular equation and to construct the eigenvectors from formula (11.5). Theoretical results and numerical details may be found in [11].

The static partitioning of this algorithm results in a divide and conquer graph with a node corresponding to each split of the original matrix. There is of course a problem with such a scheme since parallelism is lost as the computation proceeds. Fortunately, in this problem the computation executing in a given node may be parallelized by splitting the root finding problems into a number of parallel parts. The location of each root is bracketed by the old eigenvalues and each of them may be computed independently. It is a simple matter, therefore, to divide the roots into groups which will be computed in parallel. An important numerical aspect of this algorithm which will not be discussed in detail here is the reduction of the number of roots to be computed in a given node through a deflation technique. This deflation is of considerable importance and has great impact both on the efficiency and numerical accuracy of the algorithm. Details are reported in [11], but we wish to emphasize here that this aspect of the computation is *problem dependent*. The number of roots to be found at any given step can only be known at the time of execution. Therefore, the number of groups that the roots can be reasonably divided into must be done at run time and assigning a parallel process to one of the groups requires dynamic allocation of processes. These aspects are illustrated in the following graph produced by SCHEDULE.

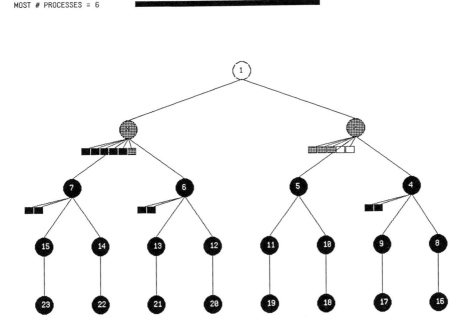

Figure 11.1
Output from SCHEDULE

Note that the graph produced by SCHEDULE corresponds very closely to the abstract graph a user might define for this algorithm. In Figure 11.1 we see a partially executed program. The square nodes show dynamically spawned processes, while the round nodes show the static partitioning. A particular point of interest is static node number 5 which had the potential for spawning processes just as the other nodes on that level but didn't because a computation done at run time indicated there was not enough work at this node to warrant the spawning of additional processes. This graph demonstrates the utility of such a capability because the maximum number of concurrently executing processes was met at nearly all levels of the computation. Without the potential of dynamically spawned processes the performance of the parallel algorithm would be very poor due to the loss of concurrency associated with the divide and conquer nature of the static partition. Without dynamic spawning the number of concurrent processes decreases by a factor of two at each level of the graph. This is even more disappointing when one realizes that the amount of work usually will be at a maximum in the nodes near the root of the tree. Moreover, with this particular algorithm dynamic spawning is essential because the amount of work at any given node is proportional to the number of roots to be found at that node, but this number will be problem dependent and determinable only at the time of execution.

Another interesting example is associated with an algorithm for solving singularly perturbed convection diffusion PDE. An algorithm based upon the ideas presented in [4] has been implemented using SCHEDULE on the Alliant FX/8. This algorithm solves

$$u_t + c(x,t)u_x = \varepsilon \Delta u \ , \ 0 \le t \le \infty, \ 0 \le x \le 1, \ 0 < \varepsilon < 1$$

given appropriate initial data and boundary conditions. The basis of the method is to use asymptotic analysis of the dominate scales of the equation in various subdomains as depicted in Figure 11.2. Such analysis is evident upon making a characteristics coordinate transformation $(x,t) \rightarrow (\xi,\tau)$ with

$$\frac{dx}{dt} = c(x,t) \ , \ x(0) = \xi \ , \ t \equiv \tau \ .$$

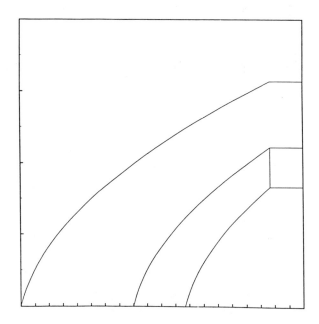

Figure 11.2

Subdomains of Convection Diffusion Equation

We refer the reader to [4] for details. The parallelism is obtained through solving for these characteristics in parallel block fashion with the number of characteristics associated with each block tuned to the number of vector registers on a physical processor.

For the Alliant computer this number was 32. The vector processor was used as an SIMD engine to compute these characteristics in lock step using the step size control associated with the ODE solver for a system of ODEs. The solver used in this case was the code of Shampine and Allen[15]. The domain decomposition consisted of two regions where the solution was taken to be constants on these characteristics, an internal layer where the full equation was solved, a boundary layer where two point boundary value problems were solved , and a region where the internal layer meets the boundary layer.

In the following graphics output, Figure 11.3, from SCHEDULE we show this program executing. Node 4 corresponds to gridding the internal layer with characteristics. Note that there is dynamic spawning of blocks of 32 characteristics made there. Nodes 2 and 3 correspond to convection only regions and blocks of 32 characteristics are spawned off there as well. These processes just continue to solve the boundary layer problem as soon as the characteristics meet the boundary at $x = 1$. Process 6 solves the full equation in the internal layer while process 5 grids the region where the internal and boundary layers meet. Both 5 and 6 must be completed before node 7 can execute to solve the equation where the internal and boundary layers meet. Node 1 is a dummy node used to collect all the results.

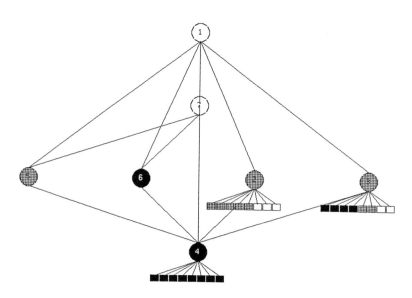

```
ACTIVE PROCESSES = 10
MOST # PROCESSES = 10
```

Figure 11.3
SCHEDULE Output from Domain Decomposition

12. Conclusions

The construction SCHEDULE has evolved naturally during the course of our experience with programming various parallel machines. We are primarily motivated by the lack of unformity and limited capabilities offered by vendors for explicit parallel programming. We are deeply indebted to our colleagues who are active in similar pursuits and we have used many of their ideas in constructing SCHEDULE. However, although we have been greatly influenced by their work, we feel that our experience with numerical software has guided us in subtly different directions. It is our view that the user interface at this point in time is paramount to the utility of such an effort. Our goal has been to provide a methodology that is within the grasp of a capable Fortran programmer and a to provide syntax which does not represent a radical departure in appearance from standard Fortran programming. We have already modified this interface several times based upon interaction with users. The current approach has evolved from a package written entirely in Fortran to one requiring a $C - Fortran$ interface. Our goal of transportability is made more difficult with this requirement due to the lack of standardization of such an interface. However, our goal of providing clean user interface with the parallel capabilities seems to require this mixed language approach.

Our experience with SCHEDULE has been encouraging for the most part. We do not view it as a "solution" to the software problem we face in parallel programming. However, we do think this will be useful in the short term and perhaps will have some influence on the development of a long term solution.

Acknowledgements

We would like to thank Terry Disz for introducing us to the graphic environment RT/1 and his assistance in making the graphics interface work.

References

1. *CRAY 2 Multitasking Users Guide,* Cray Research Inc, Minn, MN (1986).

2. R.G. Babb, "Parallel Processing with Large Grain Data Flow Techniques," *IEEE Computer* **17, 7**, pp. 55-61 (July 1984).

3. J.C. Browne, "Framework for Formulation and Analysis of Parallel Computation Structures," *Parallel Computing* **3**, pp. 1-9 (1986).

4. R. Chin, G. Hedstrom, F. Howes, and J. McGraw, "Parallel Computation of Multiple-Scale Problems," pp. 134-151 in *New Computing Environments: Parallel, Vector, and Systolic*, ed. Ed. A. Wouk, Siam Pub., Philadelphia (1986).

5. S. Comer, *Private Communication*, 1986.

6. Alliant Computer Systems Corp, *Alliant FX/Fortran Programmer's Handbook*, Acton Mass., 1985.

7. J.C. Diaz, "Calculating the Block Preconditioner on Parallel Multivector Processors," *Proceedings of the Workshop on Applied Computing in The Energy Field*, Stillwater, Oklahoma, (October 10, 1986).

8. J. Dongarra and D. Sorensen, "An Aid to Fortran Programming and Debugging," ANL MCS - TM in preparation (1986).

9. J.J. Dongarra and I.S. Duff, "Advanced Architecture Computers," Argonne

National Laboratory Report, ANL-MCS-TM-57 (October, 1985).

10.   J.J. Dongarra and D.C. Sorensen, "Linear Algebra on High-Performance Computers," pp. 3-32 in *Proceedings Parallel Computing 85*, ed. U. Schendel, North Holland (1986).

11.   J.J. Dongarra and D.C. Sorensen, "A Fully Parallel Algorithm for the Symmetric Eigenvalue Problem," *To appear SIAM SSISC* (March, 1987).

12.   M.W. Gentleman, *Private Communications*, 1985.

13.   H. Jordan, "HEP Architecture, Programming and Performance," in *Parallel MIMD Compuation: HEP Supercomputer and Its Applications*, ed. Ed. J. Kowalik, MIT Press (1985).

14.   E. Lusk and R. Overbeek, "Implementation of Monitors with Macros: A Programming Aid for the HEP and Other Parallel Processors," Argonne National Laboratory Report, ANL-83-97 (1983).

15.   L. Shampine and R. Allen, *Numerical Computing: An Introduction,* W. B. Saunders Company, Philadelpha (1973).

16.   John VanRosendale and Piyush Mehrotra, "The BLAZE Language: A Parallel Language for Scientific Programming," ICASE Report # 85-29 (1985).

17.   J.R. McGraw, et al., "SISAL: Streams and Iteration in a Single Assignment Language," Language Reference Manual, Ver. 1.2, Lawerence Livermore National Laboratory

The Force†

Harry F. Jordan
University of Colorado
Boulder, Colorado

Introduction

The Force is a parallel programming language and methodology which evolved in the course of implementing numerous algorithms, primarily of a numerical nature, on large scale shared memory multiprocessors. The number of processes supported was from 20 to 200. The Force was realized as a macro preprocessor extending the Fortran language and includes primitive operations supporting both fine and coarse grained parallelism. There is a unifying philosophy which determines the structure of algorithms implemented using the Force. In this programming paradigm, multiple processes execute a single program. The number of processes is arbitrary but fixed at run time and may be one. Work is not assigned to specific processes but distributed over the entire force of processes by parallel constructs. The variables on which work is performed are either uniformly shared among all of the processes or strictly private to a single process. Process synchronization is non–specific or generic. Generic synchronization primitives do not individually identify processes to be synchronized. The parallelism in a Force program is present from the beginning of the program; it is not encapsulated inside of some modules of the program hierarchy. The philosophy could thus be characterized as that of universal or global parallelism which suppresses explicit process management.

The Force is based on the shared memory multiprocessor model of computation. There are multiple instruction streams which could be thought to execute separate programs, but since all code resides in a common memory, a better formulation is that of a single program being executed by multiple processes, each of which has a separate program counter. The shared memory model implies that references to a variable need specify no associated process identity. The user does not explicitly write commands such as *send* and *receive* to communicate values between processes. Transfer of a datum between processes is based on coincidence in the shared name space

† This work was supported in part by ONR under Grant N00014-86-k-0204, by AFOSR under Grant AFOSR 85-1089 and by NASA Langley Research Center under Grants NAG-1-640 and NAS1-17070.

of the write and read references to a variable. The Force as a language is at a high enough level to suppress machine dependence but is not intended to be a "very high level" language. No attempt is made to suppress the differences between multiprocessors (MIMD) and vector (SIMD) computers, nor is the shared memory versus explicit communication distinction hidden. A language similar to the Force might be used as intermediate code output from a very high level language processor which did automatic dependency analysis, parallelization and/or scheduling.

The rationale behind the structure of the Force methodology is grounded in the desire to address large scale multiprocessing. With many processes, it is not feasible to write separate code for each process. Furthermore, independence from the number of processes is the only manageable structure for an algorithm to be executed by many of them. Any required specification of process identifiers significantly complicates the coding for many processes. If synchronization, variable naming, work partitioning or process control required process identification, the programmer might have to keep track of as many separate interactions as there are pairs of processes. As it is, there are no topological aspects of the parallel computer environment to be considered in algorithm design, and no processes are individually managed. It is interesting that this style of parallel programming seems to lead to programs in which a relatively local analysis is sufficient to determine correctness of synchronization. The inclusion of variables private to a process supports the need for temporaries and is particularly useful in coarse grained parallelism, where there are long sequences of serial code. The primitives which assign work can support load balancing by virtue of the fact that they are not process specific. Finally, the design of the Force as an extension to an existing sequential language provides all of the types, operations and control structures of that language and allows the user, or designer, to focus on the parallel aspects of programming.

Efficiency on existing multiprocessors corresponding to the abstract shared memory multiprocessor model of computation was also an important consideration in the design of the Force. It must be emphasized that the concept of efficiency used is based on minimum execution time for a single parallel program with a given number of processors. Throughput of a multiprogrammed multiprocessor system is simply not considered. Although multiprogramming throughput is an important topic in operating systems design, it can probably not be addressed reasonably at a programming language level. The purpose of parallel processing, after all, is to speed up the execution of individual programs. Multiprocessing throughput is probably optimized by running many multiprogrammed sequential machines simultaneously. To address a balance between one program completion speed and system utilization would require a specification of the relative importance of these aspects which is certainly not available at the language level. Efficiency is addressed in the Force both through its philosophy and through the selection, specification and design of individual parallel primitives. It was previously noted that the Force philosophy eliminates encapsulation of parallelism. Encapsulation would imply that all code above the level of encapsulation in the program hierarchy would be sequential, and it is well known that the efficiency penalty paid for a small amount of sequential code is high in a many processor system. As far as possible, the parallel primitives are kept conceptually small, each embodying only one concept. Complex parallelism is obtained by combining these elementary operations.

Implementations of the Force on various machines have certain features in common, many of which contribute to parallel performance. A parallel environment consisting of some shared and some private variables is maintained at run time. It contains implementation structures from which the user is insulated and its structure may help to specify and clarify the Force computational model. A shared variable always contained in the environment is the number of processes for this execution and a private variable always included is the process identifier. The user usually need not be concerned with these variables, but they are made available for algorithm performance determination and debugging. Other implementation details used, for example, in implementing barrier synchronization can be completely hidden from the user. As a result of the Force philosophy, even the implementation can avoid detailed process management; mutual exclusion and process counting are sufficient to implement most parallel primitives. The fact that processes are not created or destroyed in the course of a Force program implies that the implementation need not necessarily save and restore any process states.

The current implementations of the Force are constructed as macro preprocessors producing output consisting of Fortran augmented by whatever parallel extensions the manufacturer has supplied with the multiprocessor. The preprocessing requires a pattern matching macro processor. Since the macro processing has been done in a Unix environment which lacks a combined pattern matching macro processor, the pattern matching has been done with the stream editor (*sed*) and the macro expansion with the *m4* macro processor. As well as allowing a more syntactically consistent language, macro processing has some efficiency advantage over a subroutine library based implementation. Some macros can be expanded in line using information from the parallel environment, thus saving the performance cost of subroutine linkage. More or fewer macros can be handled this way depending on the suitability of the parallel extensions supported by the target multiprocessor. Disadvantages of the macro based implementation include the well known problem that compiler error messages refer to the macro expanded code, of which the user would rather not be aware, and the problem of name collision between user names and both macro keywords and internal names generated by the expansion.

To give the reader a flavor of programs written with the current implementation of the Force, Table 1 shows a complete Force program. This overly simple example generates and multiplies two real matrices. The

```
        Force MATMP of NP ident ME
        Shared REAL A(200,200), B(200,200), C(200,200)
        Shared INTEGER M, LDIM
        Private INTEGER I, J, IRES
        Private REAL SUM
        End declarations
        Barrier
C       The order of the matrices - M
        READ (5, 50) M
50      FORMAT(I4)
C       Echo the input with number of processes.
        WRITE (6, 75) M, NP
75      FORMAT(' Order',I4,' matrices using',I4,' processes.')
        End Barrier
C       Set up the matrices to be multiplied.
        Presched DO  10 I = 1, M
          DO  5 J = 1, M
            A(I, J) = 1./3.
            B(I, J) = 3.
5         CONTINUE
10      End Presched DO
        Barrier
        End Barrier
        Selfsched DO  300 I = 1, M
C         Produce all of row I of the C matrix.
          DO  200 J = 1, M
            SUM = 0.0
            DO  100 K = 1, M
              SUM = SUM + A(I,K)*B(K,J)
100         CONTINUE
            C(I,J) = SUM
200       CONTINUE
300     End Selfsched DO
        Barrier
C       Write the results.
        DO  20 I = 1, M
          WRITE (8, 500) (C(I, J1), J1=1,M)
20      CONTINUE
500     FORMAT(6E13.6)
        End Barrier
        Join
        END
```

Table 1: A Matrix Multiply Force Program.

constructs used will be described in detail later, but the program is presented early as a concrete example to which to tie the conceptual discussion.

The Force philosophy grew during work by the author [1] with the Denelcor HEP [2] computer starting in 1980. Work on the Force as a programming language extension [3] was begun while the author was on sabbatical leave at the Ballistics Research Laboratory of the U. S. Army during the 1983-84 academic year. It was used for fluid dynamics computations at BRL by Nisheeth Patel [4] [5] during the spring of 1984. The system was subsequently implemented on the Flexible Computer Corporation's Flex/32 multiprocessor at the NASA Langley Research Center with support from the Institute for Computer Applications in Science and Engineering (ICASE) [6] [7]. Support from Encore Computer Corporation enabled the completion and polishing of a Force implementation on the Encore Multimax at the University of Colorado, begun in the spring of 1986 by Muhammad Benten [8]. Mr. Benten is in the process of porting this implementation to the Sequent Balance 8000 multiprocessor at the time of writing.

Other macro based language extensions were produced for the Denelcor HEP and later ported to other multiprocessors. Lusk and Overbeek at Argonne National Laboratory produced a system [9] based on the monitor concept of C. A. R. Hoare [10], and Babb at the Oregon Graduate Center implemented a method known as Large-Grain Data Flow [11]. The idea of having a single body of code executed by many processes has also been employed by IBM in the RP3 project through the EPEX Fortran preprocessor [12] and by the BBN Corporation through the Uniform System [13] subroutine library for the Butterfly multiprocessor. The philosophy of the latter two systems is probably most closely related to that of the Force in that they tend to suppress process management while retaining the concept of instruction streams. Babb suppresses process management but also, at the top level, exchanges instruction streams for data flow scheduling of operations.

The Force Computational Model

It is important to characterize the model of parallel computation which underlies Force programs. It is this model which is loosely referred to as the Force philosophy. The computational model is based on, and strongly influenced by, the parallel machines used to execute Force programs, but for the purpose of machine independence, one should extract major architectural features of different machines to form more abstract models of parallel computation, each representing a number of machines. The partitioning of the class of all parallel machines into SIMD and MIMD, first proposed by Flynn [14], seems to have distinct impact on algorithm structure. The parallelism in SIMD or "vector" machines appears within single hardware instructions and is thus reflected at the lowest or leaf level of an hierarchically structured algorithm. MIMD parallelism is at the instruction stream (or process) level and could appear at any level of algorithm structure where code execution takes place. In practice, process management overhead favors algorithms in which parallelism appears at a high level.

The Force computational model retains the concept of instruction streams. The data flow [15] or functional programming models discard instruction streams entirely. The cost of eliminating the idea of instruction streams is that it assumes a completely automatic solution to the problem of scheduling operations. Programs written in a functional language contain no operation scheduling information, thus having the programmer specify only functional dependencies. This removes the issues of load balancing, variable access conflict, synchronization overhead and other performance related problems from the programmer's world entirely. Since these are not closed topics in the research community, there is probably room for a computational model which leaves the programmer some control of, and responsibility for, these performance issues. Furthermore, the program counter is a well understood and successful scheduling mechanism. The step from one to many instruction streams seems sufficient to introduce a number of interesting extensions without demanding a complete solution to scheduling.

Within the class of MIMD machines, those based on shared memory and those with explicit interprocessor communication form distinct subclasses. The shared memory designation is appropriate when hardware mechanisms such as local cache, high connectivity switching networks and memory request pipelining are used to make at least a portion of the memory address space equally accessible to all processors. Explicit interprocess communication is the correct designation when a subset of the total memory is associated with each instruction stream, either because it is only accessible to that stream or because the access time for that stream is very low compared to other streams referencing that part of the address space. Uniform accessibility can be supported at various levels: software library, microcode firmware or memory interface hardware. For the purpose of specifying a computational model, it is appropriate to base the distinction between shared memory and explicit communication on whether the programmer can use one process to directly reference a value produced by another process or whether he must write explicit send and receive operations to communicate values between processes.

The algorithmic implications of classifying computers on the basis of SIMD versus MIMD and shared memory versus explicit communications seem large enough to give meaning to a shared-memory, MIMD model of parallel computation. The discussion of algorithms and program structure to follow assumes this computational model and the performance trade-offs discussed would be significantly impacted by a change in the model. This model is of interest because many-process MIMD machines seem to offer the best speedup potential for unrestricted algorithms and because notable progress has been made in hardware mechanisms to support the shared memory concept. We will thus assume a class of machines consisting of multiprocessors with a substantial portion of their address space devoted to uniformly accessible, shared memory. Separate memory hardware, address mapping or software may also support memory private to each processor.

Implicit information on the shared memory, MIMD model of computation is inherent in the various programming language extensions supported by the manufacturers of different shared memory multiprocessors. In the Fortran for the Denelcor HEP [16], common variables were automatically shared while Fortran local variables became process private, thus confusing a parallelism attribute of variables with a textual name scope issue. New processes could be created, binding the process to a Fortran subroutine which it began executing in parallel with its creator. Return from that subroutine meant self termination of the created process. Processes could not terminate others. The only synchronization mechanism supported by the language was producer/consumer access to variables, directly implemented in hardware for every word of memory. The Flexible Computer Corporation's Flex/32 [17] multiprocessor has separate private and shared memory hardware, the distinction being made in extended Fortran at compile and link time. Thus either Fortran local or common variables may be of either parallelism class, private or shared. Processes can create and terminate other processes, and the code unit associated with a new process is again the Fortran subroutine. A variety of synchronization and communication primitives is implemented in the operating system, including process control, locks, events and messages. Initial support for parallel Fortran on the Encore Multimax [18] and Sequent Balance 8000 [19] multiprocessors consisted of standard Unix [20] fork and process control, dynamic sharing of memory with child processes, and spinlocks as the basic synchronization mechanism. A user library of parallel primitives built on this basic software support is available to Fortran programmers.

The Force computational model suppresses all explicit control of processes. An unspecified but fixed number of processes is assumed to be available at the start of execution and these processes execute the Force program to completion before terminating. The primary reason for this aspect of the computational model is that the explicit management of processes is not feasible in the case of a very large number of them, which is the case toward which the Force is directed. A secondary benefit is that many differences between the various machines' parallelism support lie in the process control area. No parent-child process relationship is required to support a Force implementation nor any control of one process by another. The Force makes few demands on the process model supported by a specific multiprocessor and has been successfully implemented on the four, rather different, systems mentioned above. The fact that no synchronization or communication primitives include an identifier for another process implies that there is no topological structure associated with process interaction.

In keeping with the lack of any topological structure to the set of processes executing a Force program, only two types of variables are distinguished: strictly private to one process and uniformly shared by all processes. This distinction is made on the parallel usage of variables by multiple processes and is independent of, and orthogonal to, any variable class specification inherited from the underlying sequential language, such as the Fortran local or common distinction. A private variable is a single name for a set of different variables, one per process. The name of a private variable appearing in a statement refers to the specific variable associated with the process executing the statement. Thus an assignment to a private variable which is executed by all processes represents as many separate assignments as there are processes. Private variables are normally used for process specific temporary storage by a programmer and will usually constitute a larger fraction of the variables in a Force program with large parallelism granularity. At the machine level, private variables are cacheable and might be implemented by cells in local memories associated with each processor if the machine had such storage in addition to its shared memory. A shared variable name refers to one storage item in the shared memory of the underlying machine. The variable is equally accessible to all processes of the Force and may be read or written by any of them. Most major problem data in a Force program is of this type. Of course, parallel writes to such variables must be synchronized and shared variables are only cacheable over program fragments in which use of them is read-only.

We have tried above to characterize the Force computational model from the bottom up, that is, to relate it to multiprocessor hardware. A top-down view of the principles of the Force comes from consideration of the structure of parallel algorithms and their representation as multiprocessor programs. The first issue in parallel algorithm structure is how the set of all operations to be performed is divided into sets of operations to be done in parallel. In a vector or SIMD machine there is only one way to decompose an algorithm into parallel parts. That is to decompose the problem data into items which can be processed in parallel by a single control structure which will become the single instruction stream. Decomposition of an algorithm on the basis of data is also possible in the MIMD environment, but in addition, one can partition the operations into classes or functions which can be applied in parallel. A simple division in the taxonomy of MIMD parallel algorithms may thus be made on the basis of data decomposition versus functional decomposition. Functionally parallel algorithms can perhaps be further classified by the way in which the parallel functions share data. For example, the macro-pipelining structure is well known from the use of coroutines [21] even on uniprocessors. A more important issue in the structure of functionally parallel algorithms, however, is the level at which the parallelism appears in an hierarchically structured computation.

The view of a computation as an hierarchically structured set of functions is well established and maps into the subroutine calling hierarchy in most programming languages. The level of the (usually tree structured) functional hierarchy at which parallelism enters the description of an algorithm is important. As noted, the leaf level is the only place where SIMD parallelism can be applied. We can denote this as fine grained parallelism. As MIMD parallelism is applied at higher levels, we speak of algorithms with coarser grained parallelism. The question of parallelism in programs is closely tied to programming language structure, but the current discussion is meant to be language independent. If parallelism is expressed at a high level in a program for an SIMD machine, and a clever compiler transforms it into fine grained parallelism to fit the execution environment, then the resulting program has fine grained parallelism, whether written by a person or produced by transformation from another form.

The usage of synchronization operations by a parallel program is related to its parallelism granularity. If a program has fine grained parallelism, the major issue in expressing the computation is to specify exactly what is to be done in parallel in each of the small grains. Very tight synchronization must be the rule for fine grained parallelism to make sense. In the SIMD case, this synchronization is part of the execution environment and does not appear in the code. In a program with coarse grained parallelism the amount of code devoted to expressing the parallelism may be very small and localized in a high level module. In exchange, the specification of synchronization becomes the major issue and may appear explicitly at any level of structure, all the way down to the leaf.

One possible way to fit MIMD parallelism into the calling hierarchy is to try to encapsulate parallelism below a certain level, or grain size. This has the advantage that the upper levels of the program can be written without knowing anything about parallel computation. Using the Fork/Join mechanism [22] to manage parallel processes, a single instruction stream would fork within some subroutine into multiple streams which would perform a parallel computation and then join into a single stream before returning from the subroutine. The drawbacks in this scheme lie in the area of performance. Efficiency is a sensitive function of the amount of sequential code in an otherwise parallel program on a highly parallel system. The encapsulation idea forces all code above a certain level of structure to be sequential, possibly reducing efficiency significantly if many processes are active. Furthermore, there is overhead associated with managing processes and execution environments in fork and join which is invoked whenever the program passes into or out of the parallel level of structure. Since encapsulation overheads tend to make larger grained parallelism more efficient regardless of the grain size, there is a good reason to locate MIMD parallelism at the highest level of program structure. Experience shows that it is feasible to write applications programs in which parallel processes are established at the beginning of execution and the process environment remains constant over the entire program hierarchy. We use the term "global" parallelism to refer to this method of managing processes. It is the most characteristic feature of the Force computational model.

In the global parallelism model one begins a program under the assumption that it may be executed by an arbitrary number P of processes. There is no explicit code for process management. The processes are managed by entry level, system dependent code which chooses the number of processes on the basis of hardware structure and available knowledge of algorithm needs. The code appearing explicitly to deal with parallelism is all related to process synchronization and distribution of work across processes. Work may be distributed on the basis of a data decomposition of the algorithm or on the basis of a functional decomposition. In algorithms with a strong Euclidean space connection, such as partial differential equation solution, data decomposition is often called domain decomposition. With many processes, some data decomposition of an algorithm is surely necessary since the number of independent functions is limited. Thus data decomposition is more fundamental to the structure of the Force than functional decomposition. Typically, a high level decomposition into a few parallel functions would involve a large amount of data based parallel decomposition within each function. Statements written in a Force program are implicitly executed by all processes in parallel, unless limited by an explicit parallel primitive. An assignment statement, for example, may combine the values of shared and private variables to produce a private or shared result. If the result is private, no assignment conflict is possible. If it is shared, then assignment conflict must be prevented, either by allocation of disjoint sections of a shared data structure to multiple processes or by synchronizing the assignment across processes, say by enclosing it in a critical section or by using producer/consumer synchronization on the variable assigned. Library or user subroutines which are either free of side effects or carefully synchronized can be invoked in parallel, one copy for each process.

The effect of removing process management from the user's area of responsibility by placing it above the top of the program hierarchy is in some cases quite similar to the effect of encapsulation. An example is the parallel loop, or DOALL [23] construct, in which data decomposition of work associated with a loop index is used to distribute parallel operations over processes. Using the encapsulation idea, processes would be forked on entry to the DOALL, perform the body computations for disjoint sets of values of the index and join at the end of the DOALL. With global parallelism, the DOALL construct may be expressed identically. The interpretation is that the processes, already established and with their own private environments, are to execute the body for disjoint sets of values of the index. The processes may either use synchronized access to a shared copy of the loop index to self schedule the work, or the process number, contained in a parallel environment managed implicitly by the Force, can be used to preschedule a fixed set of index values for the process with a given number.

Since parallel execution of statements is implicit in the Force, a construct is needed to satisfy data dependencies by forcing sequential completion of distinct parallel sections of code and to introduce strictly sequential operations, such as input or output on a sequential device. The "barrier" construct is used to supply both of these needs in the Force and has a strong influence on the structure of many Force programs. The barrier is a control oriented synchronization primitive and has an associated section of sequential code, which may be null. The semantics of this construct are that all processes pause when they reach the barrier. After all have arrived, one process executes the associated sequential code, after which all processes exit the barrier. Barrier synchronizations introduce process delay and some sequentialization, even if the body of the barrier is null. The semantics are inherently simpler, however, that those of a Join followed by a sequential code section followed by a Fork, which is another way to accomplish the sequentialization required. The difference is that the barrier does not require process environments to be allocated or released and private variables retain their values across a barrier.

Several advantages arise out of independence from the number of processes. It is not necessary to design algorithms with a detailed dependence on the, potentially very large, number of processes executing them. The choice of the optimal number of processes can be made at run time on the basis of system hardware configuration and load. Since complete independence from the number of processes implies correct execution with only one process, the issues of arithmetic correctness and multi-process synchronization can be separated in the testing of a program. Experience indicates that programs written to be independent of the number of processes executing them are not significantly less efficient than programs tailored to a specific number of processes, at least for the shared memory class of machines. In fact, this programming style evolved, to a large extent, from attempts to produce maximally efficient programs for this class.

The information required by the system to manage processes for the user, maintain independence from the number of them and to implement cooperative synchronization constructs is referred to as the Force's parallel environment. Part of the parallel environment is fixed, being the same for any Force program, and part is variable, depending on the specific Force primitives included in a program. The values of two items in the fixed portion of the environment are made available to the user through the Force program or subroutine header. These are the shared value of the number P of processes and the private value of a unique index between one and P assigned to each process. The process index is primarily useful for debugging, while the number of processes is useful in performance measurements and optimizations. The number of processes could, for example, be used to select one of two different algorithms to be applied to some subcomputation on the basis of the (possibly data dependent) size of the computation and the number of processes available to perform it. Other parts of the fixed portion of the environment are the shared variable required to initially assign process indices and the synchronization mechanism associated with implementing the barrier, which is the most complex and cooperative of the synchronizations provided. The variable part of the parallel environment is automatically generated by constructs appearing in a specific Force program. The items generated are implementation dependent but include such things as shared index variables for self scheduled DOALLs and full/empty locking variables for producer/consumer synchronized variables.

To summarize the characteristics of the Force computational model, it suppresses explicit process management by creating and terminating them at the top of the program hierarchy. Parallel computation is thus the normal mode of execution on entry to a Force program, and any sequential operation must be explicitly invoked using a control primitive. Programs are independent of the number of processes executing them, except of course, for performance, which should increase with the number of processes. Since the Force is devoted to optimizing single program completion time, the number of processes can remain fixed during execution, and operating system throughput issues are not addressed. All processes are identical in capability, and all execute the same program. Variables are either strictly private to one process or uniformly shared among all of them. The combination of process and variable structure implies that there is no topology associated with data access by processes of the Force. There is never any need for one process to identify another one explicitly. Because programs following the Force computational philosophy can be executed with one process, correctness of program execution can be tested independently of effects due to improper synchronization. Finally, the Force allows parallel constructs at any level of the program hierarchy and thus supports both coarse and fine grained parallelism.

Realization of the Concept

The programming language associated with the Force consists of some simple extensions to the Fortran language, which are currently implemented as macros expanded by a language independent preprocessor. The target Fortran system must, of course, include ways of creating multiple processes, sharing memory between processes, and of supporting synchronized access to shared variables. The macros interact through the variables of a parallel environment, which contains some general information such as the number of processes and some machine dependent items. A macro preprocessor has advantages over the use of a subroutine library to provide language extensions in that it allows a more readable syntax for the parallel operations and does not require subroutine linkage overhead in the case of simple, and potentially very efficient, macros. It is at a disadvantage, however, with respect to a compiler for the extended language. A key dividing point in capability is that the compiler builds a symbol table while the macro preprocessor does not. This not only makes it impossible for the preprocessor to check correct variable usage, but also requires the user to explicitly supply variable type information in some parallel constructs which would be available implicitly if the program were compiled.

The macros currently constituting the Force can be divided into several classes, as shown in Fig. 1. The first class deals with parallel program structure. The macros *Force* and *Forcesub* respectively begin parallel main programs and parallel subroutines. They make the parallel environment variables available to the macros within that program module as well as making the number of processes and a unique identifier for the current process available to the user at run time. An *End Declarations* macro marks the

Program Structure:
     Force <name> of <# of procs> ident <proc id>
          < declaration of variables >
          [Externf < Force module name >]
     End declarations
          <Force program>
     Join

     Forcesub <name>([parameters]) of <# of procs> ident <proc id>
          < declarations >
          [Externf < Force module name >]
     End declarations
          < subroutine body >
     RETURN

     Forcecall <name>([parameters])

Declaration of Variables:
     Private <FORTRAN type> <variable list>
     Private Common /<label>/ <FORTRAN type> <variable list>

     Shared <FORTRAN type> <variable list>
     Shared Common /<label>/ <FORTRAN type> <variable list>

     Async <FORTRAN type> <variable list>
     Async Common /<label>/ <FORTRAN type> <variable list>

Figure 1a: Parallel Constructs of the Force - Structure and Declaration

beginning of executable code and provides target locations for declarations and start up code which may be generated by the macros. A *Join* macro terminates the parallel main program. It is the last statement executed by all processes of the Force. The Fortran RETURN statement is sufficient to terminate a parallel subroutine, provided that all processes eventually execute it. The *Forcecall* macro is executable rather than declarative but must correspond to the semantics of the *Forcesub*. For correct operation of a *Forcesub* all processes must execute a *Forcecall* so that *Barrier*'s, for example, contained in the *Forcesub* will operate correctly. No implicit synchronization is forced on entry to, or exit from, a *Forcesub*. Synchronization constructs can be included explicitly, either inside or outside the subroutine, if it is required.

    Macros of the second class deal with variable declaration. The primary new variable characteristic which appears with the Force computational model is the shared versus private distinction. This is a specification of name scope across multiple processes, just as the Fortran local/common

Parallel Execution:

    Pcase on <variable>
    [Pcond (<Fortran IF condition>)]
        <code block>
    [Usect]
    [Pcond (<Fortran IF condition>)]
      . . .
    End Pcase

    Presched Do <n> <var> = $<i_1>,<i_2>[,<i_3>]$
        <loop body>
<n> End Presched Do

    Selfsched Do <n> <var> = $<i_1>,<i_2>[,<i_3>]$
        <loop body>
<n> End Selfsched Do

    Pre2do <n> <var1> = $<i_1>,<i_2>[,<i_3>]$; <var2> = $<j_1>,<j_2>[,<j_3>]$
        <doubly indexed loop body>
<n> End Presched Do

    Self2do <n> <var1> = $<i_1>,<i_2>[,<i_3>]$; <var2> = $<j_1>,<j_2>[,<j_3>]$
        <doubly indexed loop body>
<n> End Selfsched Do

Synchronization:

    Barrier
        < code block >
    End barrier

    Critical <lock-var>
        < code block >
    End critical

    Void <async variable>
    Produce <async variable> = <expression>
    Consume <async variable> into <variable>
    Copy <async variable> into <variable>
    ... Isfull(<async variable>) ...

Figure 1b: Parallel Constructs of the Force - Execution and Synchronization

distinction specifies name scope across program modules. Both Fortran local and common variables may be either private or shared across processes. The memory sharing mechanism in some machine implementations requires the size of shared variables to be known to the Force preprocessor, so the type information is required in the *Shared* declaration. Standard Fortran declarations and any implicitly declared variables will be taken as *Private* in order to maintain compatibility between sequential Fortran code blocks and Force programs which might incorporate them. Most implementations of the Force require that extra synchronization items be included with variables which are accessed using the data synchronizations, *Produce* and *Consume*, to be described shortly. Thus, a distinct variable type declaration, *Async*, for asynchronous variable, is introduced. Appropriate additional declarations are generated by *Async* to implement the full/empty variable state required by *Produce* and *Consume* operations.

Macros of another class distribute work across processes. The most familiar construct is the DOALL, which is employed when instances of a loop body for different index values are independent and can thus be executed in any order. Two versions are provided. The *Presched Do* divides index values among processes in a fixed manner which depends only on the index range and the number of processes. The *Selfsched Do* allows processes to schedule themselves over index values by obtaining the next available value of a shared index as they become free to do work. For situations in which it is desirable to parallelize over both indices of a doubly nested loop, both prescheduled, *Pre2do*, and self scheduled, *Self2do*, macros are available. Independence of the loop body instances over both indices is, of course, required for correct operation. Note that only one process will execute the body of a DOALL for any one specific value of an index, or index pair. Thus the body of a DOALL should be considered a sequential code block. A similar construct is the parallel case, *Pcase*, which distributes different single stream code blocks over the processes of the Force. Execution conditions can be associated with each block, and any number of conditions may be true simultaneously. No order of evaluation of the conditions is specified, and each is evaluated by an arbitrarily selected process, so conditions depending only on shared variables are most meaningful.

At the heart of the Force methodology are the synchronization macros. They characterize the approach to parallel programming and provide the means for controlling the Force so that coherent and deterministic computation can be performed. Two subclasses of synchronization are control flow oriented synchronizations and data oriented synchronizations. The key control oriented synchronization is the barrier since it provides control of the entire force. Its semantics are that all processes must execute a *Barrier* macro before one arbitrarily chosen process executes the code block between *Barrier* and *End Barrier*. When the code block is complete, the entire force begins execution at the statement following the *End Barrier*. Although all but one process are temporarily suspended by a barrier, no process termination or creation takes place, and private process states are preserved across the barrier. Operations which depend on the past computation, or determine the future progress, of the entire force are typically enclosed in a barrier. Another control synchronization is the critical section, familiar from the operating systems literature. Statements between *Critical <variable>* and *End Critical* may only be executed by one process of the Force at a time. This mutual exclusion extends to any other critical section with the same associated variable.

Data oriented synchronization is provided by the elementary producer-consumer mechanism, in which shared variables have a binary state, full or empty, as well as a value. Such variables are called asynchronous to indicate that the access mechanisms associated with them allow them to be updated asynchronously by different processes. Execution by some process of the macro, *Produce <async variable> = <expression>*, waits for the variable to be empty, sets its value to the expression and makes it full, all in a manner which is atomic with respect to the progress of any other process. The macro, *Consume <async variable> into <variable>*, sets the second variable, which should be private, to the value of the asynchronous variable when the latter becomes full and sets it empty. Variables in the wrong state may cause these macros to block the progress of a process. Auxiliary macros for asynchronous variables are *Void <async variable>*, which sets a variable empty regardless of its previous state, and *Copy*, which waits for the asynchronous variable to be full and copies its value to a private variable, but does not empty it. The primitive, *Isfull(<async variable>)*, is a logical function which can be used in conditions to test the state of an asynchronous variable without blocking on either full or empty. The test, however, will not be synchronized with any change in the state.

A weakness in the set of Force macros in Fig. 1, which is the set supported in current implementations, is that it does not smoothly support decomposition of a program into parallel components on the basis of functionality. The *Pcase* macro offers the rudiments of this, but only allows one process to execute each of the parallel functions. In discussing the computational model, we argued that data decomposition is more fundamental when the Force involves many processes, but that some functional decomposition was appropriate at the upper levels of the program hierarchy. What is desired is a macro, *Resolve*, which will resolve the force into components executing different parallel code sections. The section of code for each component would start with *Component <name> strength <number>*, which would name the component and specify the fraction of the force to be devoted to this component. The component strengths would be estimated by the programmer on the basis of any knowledge available about the computational complexity of each component. A macro, *Unify*, would reunite the components into a single force. The implementation of *Resolve* is complicated by the conflicting demands of generality and efficiency. It is quite simple if the number of processes of the Force is larger than the number of components of the *Resolve*; but complete independence from the number of processes implies that the number of components may be larger than the number of processes. Then inter-component synchronization may deadlock unless the components are co-scheduled over the available processes. An implementation which produces process rescheduling at every possible deadlock point, and is still efficient when the number of processes exceeds the number of components, is under development. Incorporation of a *Resolve* macro will make it useful to extend the barrier idea. A barrier should be able to specify whether only the processes in the current component are to be blocked or whether all processes in the parent force are to participate. In the case of recursively nested *Resolve* constructs, the barrier might specify a nesting level relative to the one in which it appears.

The *Resolve* idea promises a mechanism for functional decomposition of programs into parallel components, but there is one more capability of parallel programming environments with explicit process management which is not addressed by the Force. This is the ability to create new processes to handle new parallel tasks as they arise, which translates in the Force environment into the ability to give away work to "available" processes in a dynamic manner during execution. This ability is most called for by tree algorithms and dynamic divide-and-conquer methods. It would be desirable for the Force to have a mechanism for handling such algorithms without making the user responsible for process management or losing the benefits of independence of the number of processes. A mechanism related to resolve might be applied at each tree node but could lead to much process management overhead in cases where the correct thing to do is merely to traverse a subtree with the one remaining process.

A degree of structural consistency must be maintained in putting together Fortran code and parallel primitives from Fig. 1 to make a Force program. A Force program begins with a header which allows the user to declare variables to be private or shared. Fortran assignment and conditional statements are executed by all processes of the Force simultaneously unless limited by a process synchronization construct. Normally, parallel assignments are only made to private variables while both private and shared variables appear in expressions. If only shared variables appear in a conditional, the flow of control will be the same for all processes of the Force. Private variables appearing in a conditional can cause different processes to take different paths through the program. If processes are split among different execution paths, the paths must rejoin again before executing any primitive requiring the entire force. Such primitives include *Barrier*, *Pcase*, DOALL and *Forcecall*.

The semantics of the parallelism constructs in the Force imply that certain interrelationships among the primitives be observed when they are used together in a program. Several of the constructs restrict execution to a single stream within some code block. *Barrier* and *Pcase* limit execution of enclosed blocks to a single process while critical section code is eventually executed by all processes, but only one at a time. Thus constructs which depend on multiple, simultaneous execution, such as DOALL, *Pcase* or *Barrier* should not appear within such blocks. A critical section within a *Barrier* is meaningless, but critical sections have definite use within two or more code blocks of a *Pcase* construct. Nested critical sections have meaning when the associated locking variables are different. Data oriented synchronization primitives may occur within singly executed code without restriction, other than the natural possibility of deadlock. In fact, initialization of asynchronous variables is usually done within a singly executed block.

Parallel loops do not restrict the execution of their bodies to a single process, but they do limit execution of the body for each index value to one process. Thus constructs which depend on full parallel execution cannot appear within DOALLs. These include *Barrier, Pcase* and other DOALLs. The inconsistency in the parallelism requirements of nested DOALLs is the reason for supplying multiple index DOALLs for parallel execution of loop bodies which are independent over the Cartesian product of two or more index sets. Critical sections, *Produce* and *Consume* are quite useful within DOALLs, and often lead to programs in which the distributed nature of synchronization reduces its effect on program performance.

Subroutine invocation within a Force program can be done either with a *Forcecall* or an ordinary Fortran CALL. Only the *Forcecall* makes the parallel environment available to the subroutine called. Since a Force subroutine invoked by *Forcecall* assumes that all processes of the Force will enter it, a *Forcecall* must not appear within a code body in which parallel execution has been restricted. Thus, Forcecalls are not meaningful within *Barrier, Pcase, Critical* or DOALL constructs. An ordinary CALL implies execution of a subroutine in single stream on behalf of one or more processes. Since any Fortran based parallel system must support multiple independent execution of subroutines, such as those in the mathematical library, subroutines must have separate private variable states for all processes executing them. An ordinary Fortran subroutine or function call may thus appear within any code section of a Force program. The subroutines or functions so invoked contain no parallel constructs and access by them to any shared variables must be controlled externally if it is desired.

The *Resolve* construct is intended to produce a new parallel execution environment within each of its components, differing from the original only in the number of processes and their identifying indices. Thus all of the parallelism primitives have meaning within a Force component. The implementations of the primitives must, of course, refer to the parallel environment of the component rather than of the original force. The meaning of the barrier, as has been noted, can be extended to refer to higher levels of a nested component structure, but it retains its original meaning with respect to the immediate component with no modification of its semantics. *Barrier, Pcase* and the DOALLs have an action limited to the component in which they appear. Critical sections and data oriented synchronizations can synchronize operations within the current component with operations in any other components which share the corresponding variables.

It can thus be seen that a fairly small number of parallel primitive operations, superimposed on a sequential programming language, and used to produce programs which manage processes as a single entity is the essence of the current implementations of the Force computational model. The level of the parallel extensions is similar to the level of the Fortran language, making it intermediate across the range of parallel languages which have been considered in the literature. Implementation as a macro preprocessor to Fortran provides easy and efficient realizations on different machines, but has drawbacks in the areas of consistency checking, error messages and use of implicit information from the base language. Further work is under way to make the Force a coherent parallel programming language for shared memory multiprocessors, complete with compiler and run time environment.

Applications and Performance Issues

The Force system has been used to produce a parallel Gaussian elimination subroutine[3] identical in interface and operation to the SGEFA routine of LINPACK[23]. As well as being effective in this library subroutine type of application, it has been used to write large parallel fluid dynamics programs, including SOR algorithms for incompressible flow [4,5]. It has also been used to implement a new parallel pivoting algorithm for solving sparse systems of linear equations[24]. The structure of the Gaussian elimination and SOR algorithms is simple enough to serve to illustrate the use of global parallelism and the specific macros in writing applications code. The first example is an LU decomposition routine from the LINPACK package of subroutines for the solution of simultaneous linear equations. It has been tested extensively on a wide range of computer architectures [25]. The algorithm is Gaussian elimination with partial pivoting and leads to a Force program which uses control oriented mechanisms for most synchronizations. The second example is a fluid dynamics calculation relying on a Successive Over Relaxation (SOR) kernel to compute the stream functions in a driven cavity vortex. In contrast to the Gauss elimination, the SOR algorithm relies almost entirely on producer/consumer synchronization.

The SGEFA routine is implemented as a Force subroutine. As in a sequential Gauss elimination, the outer structure is a loop over pivots. This sequential DO loop is executed in parallel by all processes of the Force, each process having a private loop index variable. Within the loop body there are three computational phases, separated by barrier synchronizations. They involve finding the pivot element, swapping the pivot row with the top row of the sub-matrix, and reducing all non-pivot rows. In each phase a process does 1/P of the work, where P is the number of processes. The overall skeleton of the algorithm is shown in Fig. 2. It can be seen that all synchronization is done by the three barriers separating the computation phases along with one critical section in which each process updates the shared maximum pivot candidate on the basis of its own private maximum.

The first and second phases of the outer loop body contain order K DOALLs while the third phase has order K^2 computation, where K is the size of the square submatrix below the diagonal. None of the DOALLs has

DO loop over pivots.
   DOALL - search part of pivot column for private maximum.
   Critical section - update global maximum.
   Barrier - record pivot when all have finished.
   DOALL - swap part of pivot row into position.
   Barrier - take reciprocal of pivot.
   DOALL containing sequential loop over elements of a row:
     reduce part of non-pivot rows.
   Barrier - reset global maximum.
End of loop over pivots.

Figure 2. Structure of the Parallel Gauss Elimination

any explicit variability of execution time over processes, but the order K^2 phase may involve enough computation that asymmetric processor-memory interconnection could cause process speed to vary. Thus the first two DOALLs are implemented as pre-scheduled DO loops and the outermost one of the nested loops in phase three as a self scheduled DO. (Detailed measurements show that the synchronization required by self scheduling reduces performance for a large number of processes but improves it for fewer processes compared to that obtained by prescheduling the row reduction.)

The fluid dynamics application involves a complete user program of about 350 Fortran statements as opposed to a single subroutine and test driver. Since the Force main program is already parallel, overhead work such as array initialization and boundary condition calculation, which one might be tempted to do before forking parallel processes, is done naturally in parallel. The overall program structure is a sequential loop over simulated time with two sets of linear equations solved by SOR iteration at each time step. One set is for stream function and one for vorticity at the points of a rectangular grid. The method used for the parallel SOR is a simplified version of a more general algorithm [26] which is guaranteed to have the same convergence properties as a rowwise sequential sweep of the grid. The simplified algorithm involves processes sweeping different rows of grid points with interprocess synchronization used to guarantee that updated information (and old information) is used to compute a new value at a grid point exactly as in a single process rowwise sweep of the grid.

Consider only the Force subroutine to perform a single relaxation iteration for the stream function on the grid. It returns to the calling program the maximum change in stream function value over the grid. The overall structure of the parallel algorithm is shown in Fig. 3. One barrier separates the initialization of the boundary elements to full from the relaxation sweep. In addition, the code to zero the maximum change is contained in this barrier. Within the loop over rows, producer/consumer synchronization prevents a process from overtaking the one sweeping the previous row and

Entry.
Set the initial boundary row elements full.
Barrier - zero the maximum change.
Self scheduled DO loop over rows of the grid.
  Sweep row sequentially, waiting for corresponding elements of
    previous row to be full before updating and filling a value.
  Update private maximum change.
End of self scheduled DO over rows.
Return.

Figure 3: Force Subroutine for a Single SOR Sweep

thus ensures that the same values will be calculated as for a sequential row-wise sweep. Producer/consumer synchronization is also used to update the shared maximum, in contrast to the critical section used for a similar purpose in SGEFA.

Although fairly different in structure, both applications presented result in clear programs with parallelism constructs appearing in natural ways. The importance of DOALL type constructs opens the way to application of the research already done in parallelizing Fortran loops automatically [27]. The barrier makes possible the separation of a computation into sequential phases without invoking the process environment management overhead of Fork/Join. Producer/consumer synchronization and critical sections make it easy to deal with mutual exclusion type restrictions on access to shared variables. Each of the parallel constructs used seems closely related to the type of data dependencies considered by a programmer or an automatic parallelizer in producing parallel programs.

Various features of the Force methodology are related to the performance of a parallel computer system. An overall principle used in selection of primitive operations for inclusion in the Force was that the semantics of each primitive should be simple enough to admit of an efficient implementation across the range of shared memory multiprocessors. The simple process model, consisting of program counter, private variables and unique identifying index, also contributes to low overhead implementation on most shared memory machines. Process priorities and parent-child relationships, for example, can significantly complicate the implementation of a parallel programming system on some multiprocessors which do not directly support such features. The concept of efficient implementation of a parallel primitive through macro expansion and an implicit parallel environment is well illustrated by the *Presched DO* macro. Two elements always contained in the parallel environment are the shared variable containing the number of processes, here referred to as $<np>$, and the private variable, $<me>$, which contains the unique index, $1 \leq <me> \leq <np>$, for each process. In all implementations of the Force, the parallel construct,

Presched DO 10 I = I1, I2, I3

expands into the ordinary Fortran DO statement,

DO 10 I = I1+<me>-1, I2, I3*<np>.

The primitive operations of the Force define a virtual machine, and the generality of this machine yields independence from the details of the underlying hardware. This benefit of machine independence and portability need not, however, suppress all machine performance issues at the level of Force programming. Pratt [28] points out that a virtual machine for parallel execution should make "visible," as programming alternatives, distinctions which may reflect major hardware performance differences. The clearest example of such alternatives within the Force is the existence of both a prescheduled and a self scheduled DOALL.

At the level of the abstract machine, the process interactions implied by pre- and self scheduling are different. Prescheduling, since it allocates index values to processes in a fixed way as soon as the number of processes is determined, will split the workload evenly across processes only if processors run at similar speeds and the amount of computation specified by the DOALL body is independent of index value. On the other hand, no process interaction is required to allocate the index values; each process can determine its own portion of the work independently. In contrast, the self scheduling technique allows processes to load balance at execution time by obtaining further index values whenever they complete the work connected with previous values. This is done at the expense of a short critical section to obtain, increment and store a shared index variable, unless the host machine supports some cooperative synchronization mechanism such as fetch-and-add [29] by means of a combining network. In this case the price paid is the cost of using the combining network.

For a given underlying hardware, these distinctions at the abstract machine level can be translated into performance differences by using a few general characteristics of the hardware system. The most important parameters for the pre- versus self scheduling comparison are the size, in execution time, of a minimal critical section to access and update a shared index and the number of processes competing for this access. When combined with the program dependent parameters of the mean and standard deviation of the DOALL body size over the set of index values, they allow a determination of which type of scheduling will lead to better performance.

Implementations and Portability

As stated in the introduction, the Force was designed for implementation on the class of machines known as shared memory multiprocessors. Before giving examples of specific members of this class on which the Force has been implemented, it is useful to try to characterize a generic shared memory multiprocessor. As seen at the level of a user programming language, the system consists of two distinct layers: the hardware and the software. There are two distinct contributions to the software layer which are difficult, and usually unimportant, for the user to distinguish: the operating system calls and run time language support library functions built onto the system calls.

The hardware of any shared memory multiprocessor must support multiple instruction streams and will thus have multiple program counters. The different instruction streams control multiple execution units which yield parallelism in the processing of data. Multiple memory accesses must be able to progress simultaneously to provide instructions and data to the control and execution units. The feature described by the modifier "shared memory" means that at least a substantial portion of the data memory is accessible to all instruction streams. Program memory might be shared or private to a subset of the instruction streams, and some part of the data memory might be private. All of these memory variations have appeared in the machines described below on which the Force has been implemented.

In addition, the hardware of a shared memory multiprocessor must support synchronization between the instruction streams. Although synchronization is theoretically possible using only atomic read and write on a shared memory, no efficient multiprocessor program can be written without a hardware primitive which at least does a test and conditional change of value indivisibly on a shared memory item. Variations in support provided here lie in whether the indivisible test and update is performed by mutual exclusion or whether the effect of indivisibility is obtained through a combining network; in whether or not any instruction stream control features are included in the synchronization to support low overhead waiting; and in the details of the data test and update: test-and-set, compare-and-swap, fetch-and-add, etc.

The operating/run-time system software primitives available to the user of a generic multiprocessor must allow the user to do process management, sharing of data among instruction streams, and synchronization. Minimal process management support must allow the user to start parallel activity and terminate it when complete. Data is communicated among instruction streams using the shared memory, but the notion of an independent instruction stream implies that much data must be unique to that stream. Thus there must be user level support for the private versus shared data distinction. Software support for synchronization can either be very simple, supplying a single primitive on which the user can build more complex synchronizations, or very complex, supporting a wide range of expensive but powerful operations of the style familiar to the designers of operating systems.

We now describe the hardware of four computers on which the Force has been implemented; the HEP, the Flex/32, the Multimax and the Balance 8000. The HEP computer is a pipelined multiprocessor in which several processing units, called Process Execution Modules (PEMs), may be connected to a shared memory consisting of one or more memory modules as shown in

Fig. 4. Even within a single PEM, however, HEP is still a multiprocessor. Only the number of instructions actually executing simultaneously, about 12 per PEM, changes when more PEMs are added to a system. Separate memories store program and data with smaller memories devoted to registers and frequently used constants. Only data memory is shared between PEMs. We will concentrate on the architecture of a single PEM which implements multiprocessing by using the technique of pipelining.

There are several separate, interacting pipelines in a PEM but the overall effect is that, on the average, instructions from about 12 different instruction streams are being operated on simultaneously. Copies of process state, including program counter, are simultaneously available for all processes in the pipeline. A PEM is an MIMD processor in exactly the same sense in which a pipelined vector processor is an SIMD machine. In both, independent data items are processed simultaneously in different stages of the pipeline while in the HEP, independent instructions occupy pipeline stages along with their data.

It is important that data memory instructions occupy a separate, noninterfering pipeline consisting of the Storage Function Unit, pipelined switch and memory module. The relationship between the main execution pipeline and the SFU is shown in Fig. 5. An active process is represented in the hardware by a Process Tag (PT) which points to one of the 128 possible process states. When an SFU instruction (data memory access) is issued, the PT leaves the queues of the main scheduler and enters a second set of identical queues in the SFU. The PT then remains within the SFU-switch-

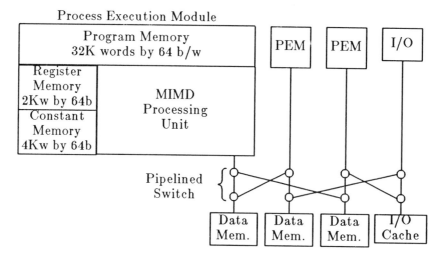

Figure 4: Architecture of the HEP Computer

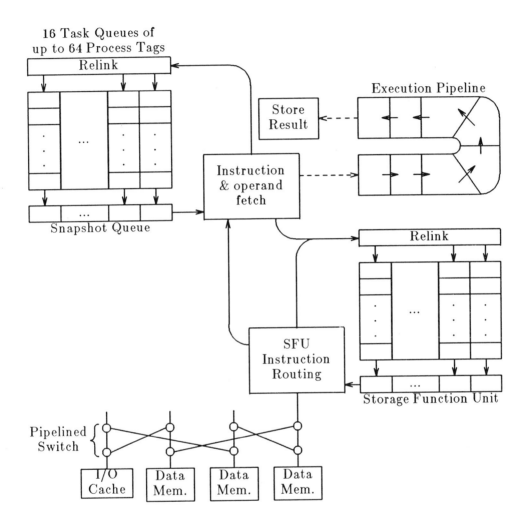

Figure 5: HEP Pipeline Architecture

memory pipeline until the memory reference is completed, and does not consume processor execution time.

HEP hardware support for process synchronization is based on
producer/consumer synchronization. Each cell in data memory has a
full/empty state and synchronization is performed by having an instruction
wait in the SFU-memory pipeline for a read cell to be full or a write cell
empty before proceeding. The synchronizing conditions are optionally
checked by the instruction issuing mechanism and, if not fulfilled, cause the
PT to be immediately relinked into its task queue with the program counter

of the PSW unaltered.

The architecture of the Flex/32 is conceptually simpler than that of the HEP, but the system support for parallelism is more complex. The machine consists of a set of single board microcomputers connected by several buses to each other and to some common memory and synchronization hardware. As shown in Fig. 6, there is a set of ten local buses, each of which can connect two boards. These are either single board computers, consisting of processor and memory, or mass memory boards. Two common buses connect the local buses together and to the common memory and synchronization hardware. The memory on the common bus is faster for a processor to access than that on the mass memory boards, but both are shared by all processors. The memory on a processor board is accessible only to that processor.

Hardware support for synchronization is supplied by an 8192 bit lock memory. This structure is meant to remove the requirement for repeated tests by a processor trying to obtain a lock. There is an interrupt system connected with each processor, which provides underlying hardware support for an event signaling mechanism between processors as well as for exception handling within a single processor. The processor/memory boards are

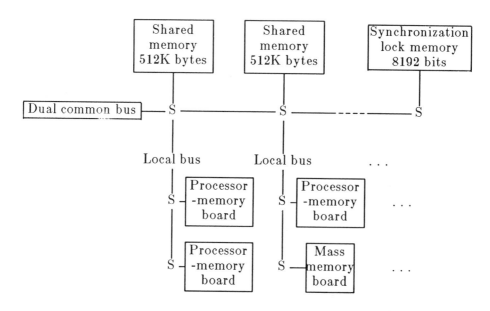

S -- bus switching

Figure 6: Flex/32 Architecture

based on the National Semiconductor 32032 microprocessor chip. There
may be one or four megabytes of memory on a board and a VME bus interface is provided to connect an individual processor to I/O devices. A self-
test system, connected to all processors, provides a mechanism for testing,
bootstrapping and initializing the multiprocessor.

The Encore Multimax and Sequent Balance 8000 are related to the
Flex/32 in that they all use the same microprocessor chip. The Multimax
and the Balance systems, however, are both organized around a single, high
speed bus instead of the local and common bus hierarchy exhibited by the
Flex/32. The high level diagram is simpler for the Multimax, as shown in
Fig. 7. The heart of the system is comprised of processor boards, each containing two complete processors and cache memory, and memory boards
containing a dual bank memory per board. These processor and memory
boards are connected with each other by means of a 96 bit pipelined bus
which is arbitrated by a Bus Arbiter and System Control board. Secondary
memory and peripherals are connected through Ethernet/Mass Storage
boards, which also communicate with the system through the single bus.

Hardware support for synchronization in the Multimax is supplied by
providing system support for the indivisible test-and-set instruction which
is present in the instruction set of the microprocessor chip used in the system. A busy wait can then be implemented on any addressable unit of

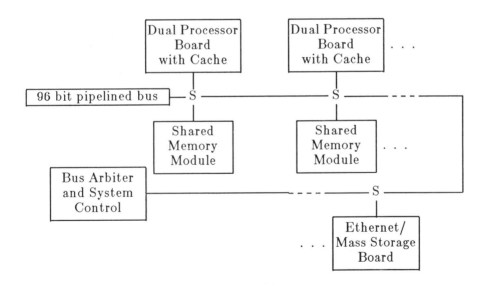

S - Switching

Figure 7: Encore Multimax Architecture

memory (byte) as the lowest level of synchronization. The busy wait consumes minimal resources since the byte being tested is moved to the cache of the waiting processor, and no bus traffic is generated until the byte is changed, giving the wait loop a chance to exit.

The central pipelined bus in the Sequent Balance 8000 is 32 bits wide and links processors, memories and I/O interfaces, as in the previous system. See Fig. 8. Processors are again packaged two to a circuit board, but there are two different kinds of memory boards. A memory controller board contains the controller along with 2-Mbytes of memory and is capable of controlling up to 8-Mbytes. Additional memory, up to 6-Mbytes more, is contained on a memory expansion board which is connected to a specific controller board. A second, single bit, bus, called the System Link and Interrupt Controller (SLIC) bus, is used for interprocessor interrupts and

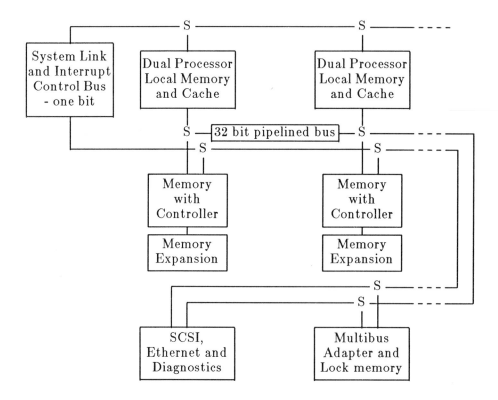

S - Switching   SCSI - Small Computer Systems Interface

Figure 8: Sequent Balance 8000 Architecture

other system control. Secondary memory and I/O is connected to the system through one of two different types of interfaces installed on the system buses. One is an adapter for the widely used MULTIBUS and the other a Small Computer System Interface.

There are two distinct hardware mechanisms used to support interprocessor synchronization in the system. The operating system kernel uses a set of 64 single bit "gates" contained in the SLIC subsystem to support mutual exclusion in access to shared kernel data structures. Since the 64 SLIC gates constitute a scarce resource, parallel user program synchronization is supported by a memory structure known as atomic lock memory. A 16K bit Atomic Lock Memory is contained on each MULTIBUS adapter board, and the indivisible test-and-set instruction of the processor is supported only on parts of the address space mapped into these memories.

Having seen examples of the hardware of four different shared memory multiprocessors on which the Force has been implemented, we now consider the range of parallel programming primitives offered to the user of these systems. Use of these parallel primitives from Fortran can be considered as a language extension. In two of the systems actual syntactic and semantic extensions of the Fortran language are implemented through preprocessing, compiler modification or a mixture of the two. In the other two, strictly semantic extensions are implemented as Fortran function or subroutine calls, perhaps supported by linker options. The degree to which the operating system is involved in carrying out a parallel primitive operation is important to performance. A system call usually requires about three orders of magnitude longer than an operation implemented directly by the hardware.

The HEP system was one of those which actually modified the Fortran language by adding parallel constructs, although this was retracted to some extent in a second software release. To allow for the fact that an independent process usually requires some local variables, the process concept is tied to the Fortran subroutine. Fortran local variables are automatically private to one process and all COMMON variables are shared by all processes. The Fortran extension is merely a second version of the CALL statment, CREATE. Control returns immediately from a CREATE statement, but the created subroutine, with a unique copy of its local variables, is also executing simultaneously. The RETURN in a created subroutine has the effect of terminating the process executing the subroutine. Parameters are passed by address in both CALL and CREATE.

The only other major conceptual modification to Fortran allows access to the synchronizing properties of the full/empty state of memory cells. Any Fortran variable may be declared to be an "asynchronous" variable. Asynchronous variables are distinguished by names beginning with a $ symbol and may have any Fortran type. They may appear in Fortran declarative statements and adhere to implicit typing rules based on the initial letter. If such a variable appears on the right side of an assignment, wait for full, read and set empty semantics apply. When one appears on the left of an assignment, the semantics are wait for empty, write and set full. To initialize the state (not the value) of asynchronous variables, a new statement, PURGE, sets the state of an asynchronous variable empty regardless of previous state. The wait for full, read and leave full semantics of the *Copy* operation in the Force are also supported by the HEP.

The HEP Fortran extensions of CREATE and asynchronous variables are the simplest way to incorporate the parallel features of the hardware into the Fortran language. Since process creation is directly supported by the HEP instruction set and any memory reference may test and set the full/empty state that is associated with each memory cell, the Fortran extensions are direct representations of hardware mechanisms requiring no operating system intervention. The parallel computation model supported by the Fortran compiler and run time system can thus be viewed as shown in Fig. 9. A process with its own program counter and registers may spawn others like it using CREATE, and the processes interact by way of full/empty shared memory cells.

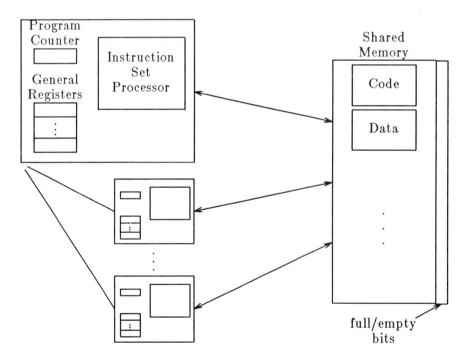

Figure 9: HEP Run Time System Model

The parallel programming primitive operations can be characterized as in Table 2. Note that all the parallel primitives are user level operations requiring no operating system intervention. Interrupts are not present in the HEP. Conditions which would normally lead to an interrupt, including supervisor calls, result in the creation of a supervisor process to handle the condition and may or may not suspend the process causing the condition.

The Flex/32 offers a fairly high level Fortran extension called Con-Current Fortran, which is implemented by a preprocessor. On this machine the Force only makes use of the preprocessor's facility for allocating variables in shared memory. Otherwise the Force constructs are implemented by Fortran calls which give the user access to operating system functions supporting process management and interaction.

The process model in the Flex/32 is somewhat different from that of the HEP and is shown pictorially in Fig. 10. Since not all of the address space is

Process
    Create   Quit and save state
Synchronization
    Produce   - Wait for empty, write and fill   Set location empty
    Consume   - Wait for full, read and empty
    Copy   - Wait for full, read

Table 2: HEP Parallel Primitives

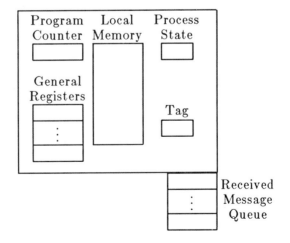

Figure 10: Flex/32 Run Time System - Process Model

shared, a process has a certain amount of strictly local memory. The system also manages a unique identifying tag for each process and maintains a process state which may be one of: running, non-existent, dormant, ready or suspended. There is also a received message queue for each process, which is managed by the system.

In addition to a slightly more complicated process model, the Flex/32 system supports more complex synchronization facilities. The total systems model is shown in Fig. 11. At the outset, processes are bound to individual processors. The processors may be multiprogrammed, so more than one process may be bound to a processor. The processes share communication and synchronization support supplied by the operating system. The Signaling Channels implement the Event mechanism and may be attached to a process as a receiver of the event, an originator, or both. Lock bits may also be connected to several processors for mutual exclusion enforcement. The message passing facility is represented by the received message queue in each process and is not shown separately.

The Flex/32 system provides numerous parallel processing primitives. They may be divided into classes dealing with four different parts of the

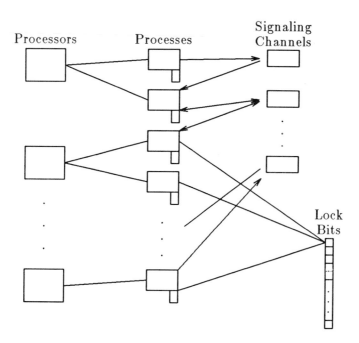

Figure 11: Flex/32 Run Time System - Overall Structure

system model: Processes, Messages, Events and Locks. The structures associated with each of these parts and the primitives which act on the structures are summarized in Table 3. The primitives are implemented through system calls since most interact with the multiprogramming of single processors. Only a small part of this extensive parallel programming model is needed to support the implementation of the Force constructs.

Both the Encore Multimax and the Sequent Balance 8000 build parallel programming support on the Unix [19] process model. Both process models are identical except that the Sequent machine has a portion of the address space mapped to the Atomic Lock Memory unit of that machine. Since the Encore machine performs the function of this unit using the standard shared memory, it need not appear separately in the hardware model. Otherwise the Sequent process model of Fig. 12 represents that for both machines. Unix process management primitives are accessed through Fortran compatible calls to the operating system. Memory is shared among Unix processes differently in the two systems. In the Multimax, a Fortran

Process

		Structure		
State:	•running	•dormant	Tag:	unique,
	•suspended	•nonexistent		system-wide
	•ready			identifier
		Primitives:		
-create	-startup	-kill		-wait for termination
		-get tag		-give up processor

Messages

Structure:	•type		•source id
	•length		•destination id
	•pointer		
Primitives:	-send		-receive/wait
			-receive/fail

Events

Structure:	list of sources and destinations		
Primitives:	-configure	-activate	-on event call
	-remove	-wait	-set timer
		-passive test	

Locks

Structure:	8192 single bits	
Operating mode:	polling or interrupt	
Primitives:	-allocate	-lock
		-unlock

Table 3: Flex/32 Parallel Primitives

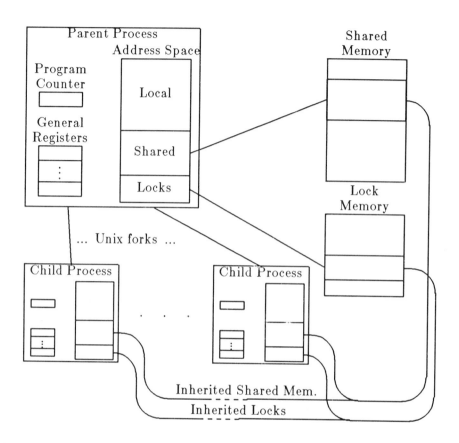

Figure 12: Sequent Process Model

function takes a common name and size as parameters and causes all subsequently forked child processes to inherit access to that same common area of memory. In the Balance 8000, a compiler/linker option accepts the common block names for the shared common blocks and generates the appropriate system calls to make them available to all processes. Synchronization in both machines is based on the so-called spinlock [30] which causes a process to busy wait until the lock is clear. Spinlocks can use any byte of memory in the Multimax, while in the Balance 8000 they must be done on a bit of atomic lock memory. The parallel programming primitives for the two machines are summarized in Table 4.

The Force implementations are built on top of the parallel process models supplied by the several machines. On each machine there is a parallel environment consisting of synchronization and process management

Process Management

•Unix fork   •Unix wait   •Unix exit

Memory Sharing

Multimax	Balance 8000
Function:	Compiler/linker option:
SHARE(<common name>, <size>)	-F /<common block>/

Synchronization

lock   -wait for zero and set   unlock   -set to zero, no wait

Table 4: Parallel Primitives Used by the Force on Unix Based Systems

variables which are suppressed and hidden from the user by the Force macros. The parallel environment is established by a driver program which also manages the processes by forking them at the beginning, and synchronizing their join at the end, of the parallel program. The driver is the "main" program from the point of view of the Fortran system, and the Force "main" program is executed as a subroutine by the processes created by the driver.

The main difference in the overall structure of the driver is between the Unix based and non-Unix based systems. In the Multimax and Balance 8000 machines, which use the Unix process model, a *fork* operation produces an identical replica of the executing process. The entire core image of the parent process is copied, with the exception of those portions of the address space which have been specified as shared prior to the *fork*. The latter are made to refer to a shared memory area by means of the address map of the child process. Copying the core image has the effect of treating any private variables used by the child process as call-by-value parameters and shared variables as call-by-reference parameters. The non-Unix systems, HEP and Flex/32, handle *fork* by creating a process executing a specified subroutine, rather than by replicating the entire core image. In this case one must use care in passing parameters to a created subroutine.

The usual implementation of Fortran supports call-by-reference, and this is the case for both the HEP and the Flex/32. In the HEP, all physical memory is shared so a reference to either a local or common variable of the Fortran program can be passed to a created subroutine. Potential problems arise from our tendency to think that call-by-reference and call-by-value are equivalent for parameters which are read-only for the created subroutine. This is not correct for a parallel program, since the parameters would have to be read-only, i.e. constant, for the future execution of both the creating program and the subroutine for the two parameter passing methods to be equivalent. On the Flex/32, physical memory is divided between shared memory and that private to a single processor, and only variables in shared memory can be used as call-by-reference parameters. Since the host Fortran system only allows Fortran common variables to be shared, the utility of passing parameters to a created subroutine is limited. The main use of parameter passing on subroutine creation by Force implementations is to supply a unique process index to each process forked by the driver. This can also be accomplished, however, by initialization code at the beginning of the Force main program which uses a shared numbering variable in the parallel environment to cooperatively compute unique indices.

A second difference in driver structure is imposed by the way the underlying Fortran supports the specification of variables which are shared. If shared variables are specified as part of the extended parallel Fortran language, as is the case on the HEP and Flex/32, or if shared common areas are specified as part of the linker command, as in the Balance 8000, the structure of the driver is not influenced. If, however, sharing of variables must be specified through an operating system call at execution time, as in the Multimax, the driver must invoke this operation for all of the shared variables of the Force program before forking the parallel processes.

When the process executing the driver has finished establishing other processes executing the parallel main program, it calls that subroutine itself and becomes one of the force of processes. Once all processes are executing the Force main program, the parallelism constructs of importance are executable primitives implemented by macros which expand into executable code of the underlying Fortran system, including the parallel extensions supplied by the manufacturer. The macros may introduce auxiliary implementation variables or use variables in the parallel environment. Apart from the private unique index and the shared number of processes, which are always present, the environment may contain shared variables associated with cooperative computation of the process index, the *join* at the end of parallel execution, with barriers and with other constructs, such as self scheduled DOALLs, which use shared variables for process cooperation.

A few selected examples of the implementation of executable macros will serve to illustrate some of the differences encountered using the various hardware and software primitives supplied by the different systems. A major hardware difference between the HEP and the rest of the systems influences even the lowest level synchronizations. This is that the basic synchronization mechanism on the HEP is the producer/consumer operation applied to any memory cell while the basic synchronization on the other three machines is an atomic test-and-set operation, which is used within a loop to implement the low level lock/unlock mechanism. Although the lock/unlock primitives are not supported directly by the Force they are closely related to the named critical section, which is. It is therefore interesting to look at the implementations of critical sections and asynchronous variables on the two different types of hardware. Table 5 compares the implementation of critical sections on the HEP and on the other machines. The table shows that the only real difference is that the full power of the HEP hardware primitive is not used, since the value of the asynchronous variable is ignored; only its state is used to perform locking. Of course, some of the machines implement the lock/unlock operations with spinlocks,

HEP	**Others**
System state and initialization:	System state and initialization:
Single full/empty variable - full	Single bit lock - clear
Critical section code:	Critical section code:
Consume critical section variable	Set critical section lock
Execute code body	Execute code body
Produce critical section variable	Clear critical section lock

Table 5: Implementation of Critical Sections

so that processes do a busy wait for the lock to clear, while others include a process suspension, possibly after a timeout. The effect of different waiting mechanisms can be seen in the performance of critical sections in tightly and loosely synchronized code sections but does not appear in the implementation. The implementation of *Produce, Consume* and *Copy* on the HEP are directly in terms of single user instructions, so only their implementation on other machines is of interest. Table 6 shows a method of implementing asynchronous variables using two locks to encode the full/empty state.

One of the most cooperative synchronizations supported by the Force is the *Barrier*, and its implementations illustrate several issues which result from the underlying process models and parallel primitives. One machine independent feature which must be taken into account in implementations of the barrier is that it must be possible to execute several barriers in a row which share implementation variables. This would be necessary even if

Shared variable - S   Private variable - P   Locks - L1, L2

L1	L2	Meaning
0	1	Empty
1	0	Full
1	1	Operation in progress
0	0	unused

Produce	**Consume**	**Copy**
lock L1	lock L2	lock L2
S = expression	P = S	P = S
clear L2	clear L1	clear L2

Table 6: Implementation of Asynchronous Variables with Locks.

separate barriers generated unique implementation variables because a single barrier might be enclosed in a sequential loop. Thus it is not enough to ensure that processes in the barrier do not conflict with those entering it, but also that processes entering the barrier for a second time do not conflict with those exiting from a prior execution of the barrier.

The HEP implementation is based on the use of asynchronous variables to implement the process delay at the barrier. If processes wait for an asynchronous variable to be full using *Copy* instead of *Consume*, then a single *Produce* can release many processes. This is the action needed by a barrier implementation, which always involves counting processes as they arrive and causing all but the last one to wait. The last arriving process then executes the code body of the barrier and releases the waiting processes. The lock/unlock synchronization is not as well suited to this because, if many processes are waiting to lock a single lock, only one at a time can succeed. Thus the straightforward implementation is for each process which successfully locks a lock unlock it for the next process. Another synchronization which embodies the idea of many waiting processes released by a single action is the event mechanism supported by the Flex/32. Here all but the last arriving process wait on an event which is activated (posted) by the last one after executing the body of the barrier. Proper implementation of the event mechanism by the underlying system ensures that each process "sees" a single activation of an event exactly once. Table 7 shows implementations of barriers on the machines discussed. Two implementations are shown for the HEP. The second illustrates what is done if resource utilization by processes doing a busy wait at the barrier is large enough to slow processes which have not yet reached it.

The only class of macros not yet mentioned in this implementation discussion is that labeled Parallel Execution in Fig. 1b. Both the *Pcase* ant the DOALLs have the function of allocating different code sections to different processors for parallel execution. In *Pcase*, the code sections are distinct bodies of text, while in the DOALLs the different sections are distinguished by the value of an index. Thus the only real implementation issue for this set of macros lies in the difference between pre- and self scheduling. The extreme simplicity of the prescheduled DOALL implementation has been shown above, so a few comments on the *Selfsched DO* are appropriate. In this construct, a shared index must be asynchronously initialized, updated and tested for exhaustion. As with the *Barrier*, it must be possible to execute a *Selfsched DO* repeatedly within an enclosing sequential loop. Thus no process may enter a *Selfsched DO* a second time until all processes have finished with the shared index mechanism on the previous execution. It is necessary that the index be initialized before any process attempts to access its value, but it is not strictly necessary that all of the processes enter the DOALL "together". A sufficient, but overly strong, synchronization mechanism is to enclose the loop index initialization in a barrier at the start

HEP - Active Waiting

System State		Initialization
Entry lock	-	full
Exit lock	-	empty
Counter	-	zero

Barrier Code
Copy entry lock
Count arriving process
If last process then
Execute code body
Consume entry lock
Produce exit lock
Copy exit lock
Count exiting process
If last process then
Consume exit lock
Produce entry lock

HEP - Process Suspending

System state		Initialization
Process state save area	-	empty
Counter	-	zero

Barrier Code
Count arriving process
If not last one then
Save state and quit
Else
Recreate other processes
Clear counter

Flex/32

System State	Initialization
Barrier event	- connected to all processes as source/destination
Counter	- zero

Barrier Code
Lock counter
Count arriving process
Clear counter if last
Unlock counter
If last process then
Execute code body
Activate barrier event
else
Wait for barrier event

Multimax & Balance 8000

System State	Initialization
Locks	- IN, OUT
Counter	- zero

Barrier Code
Lock IN
If not last in then
Count process in
Clear IN
Lock OUT
Else
Execute code body
Endif
If not last out then
Count process out
Clear OUT
Else
Clear IN
Endif

Table 7: Implementations of Barriers

of the DOALL. Weakening the synchronization so that processes can get started on the body before all have arrived is an interesting exercise, but it has little performance benefit in a tightly synchronized program, where processes make well coordinated progress through the code.

Conclusions

The Force is primarily a programming methodology for managing many processes on a shared memory multiprocessor. Secondarily, it is a parallel extension to Fortran which allows identical parallel programs to be run on several existing multiprocessors. Some key ideas embodied in the Force are independence of the number of processes executing a parallel program, high performance of a tightly coupled program on a dedicated machine, suppression of process management, and reliance on "generic" synchronizations. The Force is a true multiprocessor programming language in that it has no specifically vector constructs and does not avoid the idea of instruction streams, as is done in functional or dataflow languages. It is oriented toward shared memory multiprocessors, and there are basic problems in its structure which make implementation on a distributed memory machine not only difficult but somewhat inconsistent. It is difficult to reconcile send and receive operations in a distributed memory system with the need to suppress process identity. One might require all shared variables to be asynchronous, since *send* and *receive* are closely related to *Produce* and *Consume*, and variable identity might replace process identity. But the use of such small messages would entail severe performance penalties on currently available distributed memory systems.

The Force is an evolving system, as indicated by the discussion of the *Resolve* macro, which is not yet included in implementations. Pushing this minimal support for functional partitioning to accommodate support for divide and conquer type programs, without reverting to explicit process management is one direct line of evolution. Another is to address the requirement that the number of processes be constant during execution or that all processes need to be coscheduled on real processors for reasonable performance. Removing these requirements will be particularly important when the Force is used on a multiprogrammed multiprocessor. Of course there is some incompatibility between the use of parallelism in a single program, which is done to reduce that program's completion time, and multiprogramming, which is intended to maximize the throughput of a system running multiple jobs at some expense to the completion time of each job.

Implementation of the Force on systems involving three, rather different, process models has not been difficult, and portability between machines with similar system supported primitives is almost trivial. Given the fairly strong differences between the machines already hosting the Force, there should be no major difficulty in porting the system to any shared memory multiprocessor with which the author is familiar.

References

[1] H. F. Jordan, "Experience with pipelined multiple instruction streams," *Proc. IEEE,* Vol. 72, No. 1, pp. 113-123 (Jan. 1984).

[2]   M. C. Gilliland, B. J. Smith and W. Calvert, "HEP - A semaphore-synchronized multiprocessor with central control," *Proc. 1976 Summer Computer Simulation Conf.*, Wash., DC, pp 57-62 (July 1976).

[3]   H. F. Jordan, "Structuring parallel algorithms in an MIMD, shared memory environment," *Parallel Computing*, Vol. 3, No. 2, pp. 93-110 (May 1986).

[4]   Patel, N. R. and Jordan, H. F., "A parallelized point rowwise successive over-relaxation method on a multiprocessor," *Parallel Computing*, Vol. 1, No. 3&4 (Dec. 1984).

[5]   N. Patel, W. B. Sturek and H. F. Jordan, "A parallelized solution for incompressible flow on a multiprocessor," *Proc. AIAA 7th Computational Fluid Dynamics Conf.*, Cincinnati, OH, pp. 203-213 (July 1985).

[6]   H. F. Jordan, "Parallel computation with the Force," *ICASE Rept. No. 85-45*, NASA Langley Res. Ctr., Hampton, VA (Oct. 1985).

[7]   H. F. Jordan, "The Force on the Flex: global parallelism and portability," *ICASE Rept. No. 86-54*, NASA Langley Res. Ctr., Hampton, VA (Aug. 1986).

[8]   H. F. Jordan, M. S. Benten and N. S. Arenstorf, "Force User's Manual," *ECE Tech. Rept. 86-1-4*, Dept. of Electrical and Computer Engineering, University of Colorado, Boulder, CO (Oct. 1986).

[9]   E. L. Lusk and R. A. Overbeek, "Implementation of monitors with macros: A programming aid for the HEP and other parallel processors," *Tech. Rept. ANL-83-97*, Argonne National Lab., Argonne, IL (Dec. 1983).

[10] C. A. R. Hoare, "Monitors: an operating system structuring concept," *Comm. ACM*, pp. 549-557 (Oct. 1974).

[11] R. G. Babb II, "Parallel processing with large-grain data flow techniques," *Computer*, Vol. 17, No. 7, pp. 55-61 (July 1984).

[12] F. Darema-Rogers, D. A. George, V. A. Norton and G. F. Pfister, "VM/EPEX - A VM environment for parallel execution," *IBM Research Rept. RC11225(#49161)*, IBM T. J. Watson Res. Ctr. (Jan. 1985).

[13] "The Uniform System Approach to Programming the Butterfly Parallel Processor," BBN Laboratories Inc., draft (Nov. 1985).

[14] M. J. Flynn, "Some computer organizations and their effectiveness," *IEEE Trans. Comput.* Vol. C-21, No. 9, pp. 948-960 (1972).

[15] J. B. Dennis, "Data flow supercomputers," *IEEE Computer*, pp. 48-56 (Nov. 1980).

[16] J. S. Kowalik, Ed., *Parallel MIMD Computation: The HEP Supercomputer and its Applications*, MIT Press (1985).

[17] *The Flex/32® System Overview*, Flexible Computer Corp., Dallas, TX, (1986).

[18] *Multimax Technical Summary*, Encore Computer Corporation, Marlboro, MA, (1986).

[19] *Balance® 8000 System Technical Summary*, Sequent Computer Systems, Beaverton, OR (1984).

[20] D. M. Ritchie and K. Thompson, "The UNIX time-sharing system," *Commun. ACM*, Vol. 7, No. 7, pp. 365-375 (July 1974).

[21] M. E. Conway, "Design of a separable transition-diagram compiler," *Comm. ACM,* Vol. 6, No. 7, 396-408 (1963).

[22] J. B. Dennis and E. C. Van Horn, "Programming semantics for multiprogrammed computations," *Comm. ACM* Vol. 9, No. 3, pp. 143-155 (1966).

[23] J. J. Dongarra, J. R. Bunch, C. B. Moler and G. W. Stewart, *LINPACK Users Guide,* SIAM Publications, Phil., PA (1979).

[24] G. Alaghband and H. F. Jordan, "Multiprocessor sparse L/U decomposition with controlled fill-in," *ICASE Rept. No. 85-48,* NASA Langley Res. Ctr., Hampton, VA, 1985.

[25] J. J. Dongarra, "Performance of various computers using standard linear equations software in a Fortran environment," *ANL Tech. Memo.* (Argonne National Lab., 1983).

[26] L. M. Adams and H. F. Jordan, "Is SOR color-blind?," *SIAM J. Sci. Stat. Comput.,* Vol. 7, No. 2, pp. 490-506, April 1986.

[27] U. Banerjee, S. C. Chen, D. J. Kuck and R. A. Towle, "Time and parallel processor bounds for FORTRAN-like loops," *IEEE Trans. Comput.,* **C-28** (9) (1979) 660-670.

[28] T. W. Pratt, "Pisces: An environment for parallel scientific computation," *ICASE Rept. No. 85-12,* NASA Langley Research Center, Hampton, VA (February 1985).

[29] A. Gottlieb, R. Grishman, C. P. Kruskal, K. P. McAuliffe, L. Rudolph and M. Snir, "The NYU Ultracomputer - Designing an MIMD shared memory parallel computer," *IEEE Trans. Comput.,* Vol. C-32, No. 2 (Feb. 1983).

[30] W. A. Wulf, R. Levin and S. P. Harbison, *HYDRA/C.mmp: An Experimental Computer System,* McGraw-Hill, N. Y. (1981).

Index

The MIT Press, with Peter Denning, general consulting editor, and Brian Randell, European consulting editor, publishes computer science books in the following series:

ACM Doctoral Dissertation Award and Distinguished Dissertation Series

Artificial Intelligence, Patrick Winston and Michael Brady, editors

Charles Babbage Institute Reprint Series for the History of Computing, Martin Campbell-Kelly, editor

Computer Systems, Herb Schwetman, editor

Explorations in Logo, E. Paul Goldenberg, editor

Foundations of Computing, Michael Garey, editor

History of Computing, I. Bernard Cohen and William Aspray, editors

Information Systems, Michael Lesk, editor

Logic Programming, Ehud Shapiro, editor; Fernando Pereira, Koichi Furukawa, and D. H. D. Warren, associate editors

The MIT Electrical Engineering and Computer Science Series

Scientific Computation, Dennis Gannon, editor

RET